Topics in Current Physics 1

Topics in Current Physics Founded by Helmut K. V. Lotsch

Volume 1 **Beam-Foil Spectroscopy**
Editor: S. Bashkin

Volume 2 **Modern Three-Hadron Physics**
Editor: A. W. Thomas

Volume 3 **Dynamics of Solids and Liquids by Neutron Scattering**
Editor: S. W. Lovesey and T. Springer

Beam-Foil Spectroscopy

Edited by S. Bashkin

With Contributions by
S. Bashkin D. Burns L. J. Curtis L. J. Heroux
J. Macek R. Marrus I. Martinson I. A. Sellin
O. Sinanoğlu W. Whaling W. Wiese

With 91 Figures

Springer-Verlag Berlin Heidelberg New York 1976

Professor Stanley Bashkin

College of Liberal Arts, Department of Physics, University of Arizona,
Tucson, Arizona 85721, USA

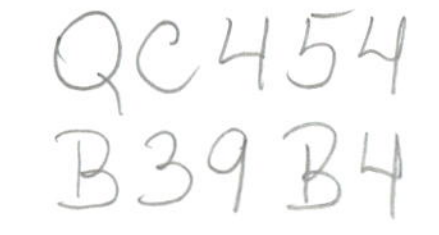

ISBN 3-540-07914-9 Springer-Verlag Berlin Heidelberg New York
ISBN 0-387-07914-9 Springer-Verlag New York Heidelberg Berlin

Offset printing: Beltz Offsetdruck, Hemsbach/Bergstraße

Bookbinding: Konrad Triltsch, Graphischer Betrieb, Würzburg

Introduction to the Series
Topics in Current Physics

Recent progress in pure research has produced vast quantities of results. The great need that this creates for authoritative reviews cannot be met by the scientific journals. Textbooks are, by their very nature, unsuitable for the discussion of specific topics in depth, and as a consequence the methods and results of newer research tend to receive rather cursory treatment. Advanced monographs usually appeal exclusively to experts in a particular field, and individuals find it expensive to purchase symposium proceedings or a collection of review papers for the sake of a particular article when the rest of the volume is of marginal interest.

"Topics in Current Physics" is a new series, published by Springer-Verlag and devoted to *critical reviews of subjects of current interest in fundamental physics*. This "monograph" series is intended to fill the gap described above and to supplement the well-established series "Topics in Applied Physics" in the field of basic research. Each volume deals with a particular topic, and contributions are invited by an editor who is a recognized authority in the field in question. The authors are scientists who are actively engaged in advancing the frontiers of research and thus write with the authority that comes from personal involvement.

This series is designed to provide the necessary background, theory, and working information on particular topics for physicists (and chemists and biologists as well), engineers, and advanced students. Furthermore, these critical reviews are definitive and intensive enough to be used by a specialist in some aspect of the topic wishing to update his knowledge on related areas. The publication periods are as short as possible to keep pace with the speed of scientific advance, and in this respect the new books are comparable with scientific journals.

Heidelberg, August 1976 Helmut K. V. Lotsch

Contents

Introduction. By S. Bashkin 1

1. Experimental Methods. By S. Bashkin 5
   - 1.1 Accelerators 5
   - 1.2 Ion Sources 6
   - 1.3 Beam Requirements and Limitations 8
   - 1.4 Mass Analyzers 9
   - 1.5 Target Chambers 12
   - 1.6 Targets 14
   - 1.7 Analytical Devices 17
   - 1.8 Detectors 19
   - 1.9 Detection Geometry and Line Width 22
   - 1.10 Beam Monitors 25
   - 1.11 External Fields 28
   - 1.12 Concluding Remarks 29

References 29

2. Studies of Atomic Spectra by the Beam-Foil Method. By I. Martinson 33
   - 2.1 Experimental Methods 33
   - 2.2 Results of Spectral Studies 38
     - 2.2.1 Previously Incompletely-Studied Systems 39
     - 2.2.2 Hydrogen-Like Levels 43
     - 2.2.3 Displaced Terms 48
     - 2.2.4 Multiply-Excited States 50

References 57

3. Lifetime Measurements. By L. J. Curtis 63
   - 3.1 Lifetime Studies as a Basic Area of Atomic Physics 64
     - 3.1.1 The Need for Lifetime Measurements 65
     - 3.1.2 Lifetime Measurements Prior to the Development of the Beam-Foil Technique 66
   - 3.2 Definitions of Basic Quantities 68
     - 3.2.1 Instantaneous Populations 68
     - 3.2.2 Transition Probabilities and Oscillator Strengths 69

3.3 Measurement of Beam-Foil-Excited Decay Curves 70
3.3.1 Strengths and Limitations of the Beam-Foil Technique 70
3.3.2 Details of Beam-Foil Apparatus and Measurement Procedures 72
3.3.3 Cascade Repopulation - A Tractable Problem 76
3.4 Time Dependence of the Measured Decay Curves 79
3.4.1 Solution of the Driven Coupled Linear Rate Equations 79
3.4.2 A Quantitative Indicator of Level Repopulation - The Replenishment Ratio 83
3.4.3 Intensity Relationships for an Aligned Source 83
3.4.4 Distortions Which Preserve the Mean-Life Content of a Decay Curve 85
3.5 Mean-Life Extraction by Exponential Fits to Individual Decay Curves 87
3.5.1 Maximum Likelihood Method 87
3.5.2 Non-Linear Least Squares Method 88
3.5.3 Differentiation and Integration of Decay Curves 90
3.5.4 Expansion About a Close-Lying Mean Life 91
3.5.5 Fourier-Transform Methods 92
3.5.6 Method of Moments 92
3.6 Mean-Life Extraction by Joint Analysis of Cascade-Related Decay Curves 93
3.6.1 Ambiguities in the Assignment of Fitted Mean Lives 93
3.6.2 Constrained Fits 96
3.6.3 Linearly-Fitted Normalizations of Cascade-Related Decay Curves 96
3.7 Cascade-Free Methods 100
3.7.1 Beam-Foil Coincidence Techniques 100
3.7.2 Use of Alignment to Discriminate Against Cascades 100
3.7.3 Laser Excitation 102
3.8 Concluding Remarks 104
References 104

4. Theoretical Oscillator Strengths of Neutral, Singly-Ionized, and Multiply-Ionized Atoms: The Theory, Comparisons with Experiment, and Critically-Evaluated Tables with New Results. By Oktay Sinanoğlu 111
4.1 The Non-Closed-Shell Many-Electron Theory 114
4.2 A Spectroscopic Interpretation of the Charge Wave Function 118
4.3 NCMET Calculations 120
4.3.1 The L^2, S^2 Symmetry of ψ_c 123
4.3.2 Dipole Length vs. Dipole Velocity 124
4.3.3 Semi-Internal Orbital Variations (Type A, Lowest-of-Symmetry, States) 125

4.4 States Not Lowest of Their Symmetry 126
4.4.1 Neutral and Singly-Ionized Atoms 127
4.4.2 Variational Collapse and Its Avoidance 129
4.5 New Oscillator Strengths for Intershell (KL → KL'[M]) Transitions to Pre-Rydberg Levels ($\bar{V}$ → pR) 134
4.6 Further Examination of Remaining Correlation Effects on Oscillator Strengths with NCMET 136
4.7 Conclusion 141
References 142

5. Regularities of Atomic Oscillator Strengths in Isoelectronic Sequences. By Wolfgang Wiese 147
5.1 Theoretical Basis 149
5.1.1 Definitions 149
5.1.2 Nuclear Charge-Dependence of the f-Value 150
5.1.3 Investigation of Lim 1/Z → 0 152
5.2 Discussion of Established Trends 153
5.2.1 Basic Trends 155
5.2.2 Curves With a Maximum 156
5.2.3 Curves With a Minimum 164
5.2.4 Anomalous Curves 166
5.3 Oscillator-Strength Distributions in a Spectral Series Along an Isoelectronic Sequence 167
5.4 Relativistic Effects and Corrections 169
5.5 Summary 174
References 175

6. Applications to Astrophysics: Absorption Spectra. By Ward Whaling 179
6.1 Branching Ratios 180
6.1.1 Light Sources 181
6.1.2 Spectrometers 182
6.1.3 Spectrometer Calibration 183
6.1.4 Selection of Branches to be Measured 184
6.2 Curve-of-Growth Analysis 185
6.2.1 Construction of a Curve-of-Growth 186
6.2.2 Internal-Consistency Test 187
6.2.3 Comparison of Transition Probabilities for Different Transitions 188
6.2.4 Solar-Abundance Determination 188
6.3 Beam-Foil-Spectroscopy Measurements Needed for Astrophysical Applications 189
References 190

7. Applications of Beam-Foil Spectroscopy to the Solar Ultraviolet Emission Spectrum. By Leon Heroux 193
7.1 Ionization Balance in the Chromosphere and Corona 196
7.2 Excitation Balance in the Chromosphere and Corona 196
7.3 Line-Ratio Measurements of Electron Temperature 198
7.4 Line-Ratio Measurements of Electron Density 203
7.5 The Determination of Chromospheric-Coronal Abundances 204
7.6 Beam-Foil Measurements Needed for Diagnostic Methods 206
References 207

8. Studies of Hydrogen-Like and Helium-Like Ions of High Z. By Richard Marrus 209
8.1 The Lamb Shift in the One-Electron System 209
8.1.1 Quenching Measurements on Fast Ion Beams of High Z 212
8.1.2 Lamb Shift in Hydrogen Using Separated Oscillating Fields 214
8.2 Lamb Shift in Two-Electron Systems 215
8.3 Radiative Decay of the $2S_{1/2}$ Metastable State of the One-Electron System 216
8.3.1 Theory 217
8.3.2 Experiments 220
8.4 Forbidden Radiative Decay in the n=2 State of the Two-Electron System 224
8.4.1 Radiative Decay from $2\,^1S_0$ 224
8.4.2 Radiative Decay from $2\,^3S_1$ 227
8.4.3 Radiative Decay from $2\,^3P_2$ 229
8.4.4 Radiative Decay from $2\,^3P_1$ 231
8.5 Study of Doubly-Excited Configurations in the Two-Electron System 232
References 233

9. Coherence, Alignment, and Orientation Phenomena in the Beam-Foil Light Source. By J. Macek and D. Burns 237
9.1 General Theoretical Considerations 239
9.1.1 The Emission Process 239
9.1.2 Symmetry Considerations 243
9.2 Alignment and Linear Polarization 246
9.2.1 Zero-Field Measurements 246
9.2.2 Electric Field 253
9.2.3 Magnetic Field 257
9.3 Orientation and Circular Polarization 260
9.3.1 Zero Field 260
9.3.2 Magnetic Field Measurements 261
9.3.3 The Quadratic Stark Effect 262
References 263

10. The Measurement of Autoionizing Ion Levels and Lifetimes by Fast Projectile Electron Spectroscopy. By Ivan A. Sellin 265

10.1 The Fast-Projectile Electron Spectroscopy (FPES) Method 269

10.1.1 Choice of an Analyzer 270

10.1.2 Properties of a Cylindrical-Mirror Analyzer Suitable for FPES 271

10.1.3 Kinematic Modification of Analyzer Optimization Criteria 274

10.1.4 Relativistic Corrections to Analyzer Performance 275

10.1.5 Broadening from Transverse Velocity Spread 276

10.1.6 Further Kinematic Considerations: Sample Estimates of Net Line Widths Observed in FPES 277

10.1.7 Summary of the Advantages of FPES 279

10.2 Examples of FPES 280

10.2.1 Spectra of Long-Lived States of the Li-Like, Be-Like, and B-Like Ions 280

10.2.2 Spectra of Long-Lived Core-Excited States of Sodium-Like Chlorine 282

10.2.3 Core-Excited States of the Neutral and Nearly-Neutral Alkali Metals 283

10.2.4 Electron Background in FPES with Foil Targets 286

10.2.5 Electron Background in FPES with Gas Targets 287

10.3 The Measurement of Auger Lifetimes by FPES 290

10.3.1 Auger Lifetimes of Metastable Lithium-Like Ions 290

10.3.2 Examples of Lifetimes from Optical Decay Channels of Auger-Emitting Levels 294

References 295

APPENDIX (Up-dated bibliography) 299

SUBJECT INDEX............ 311

List of Contributors

BASHKIN, STANLEY

Van de Graaff Laboratory, Department of Physics, University of Arizona, Tucson Arizona 85721, USA

BURNS, DONAL

Department of Physics, University of Nebraska, Lincoln, Nebraska 68508, USA

CURTIS, LARRY J.

Department of Physics, University of Toledo, Toledo, Ohio 43606, USA

HEROUX, LEON J.

Air Force Cambridge Research Laboratories, Office of Aerospace Research, Lawrence G. Hanscom Field, Bedford, Massachusetts 01731, USA

MACEK, JOSEPH

Department of Physics, University of Nebraska, Lincoln, Nebraska 68508, USA

MARRUS, RICHARD

Lawrence Berkeley Laboratory, University of California, Berkeley, California 94720, USA

MARTINSON, INDREK

Fysiska Institutionen, Lunds Universitet, Solvegatan 14, 223 62 Lund, Sweden

SELLIN, IVAN A.

Department of Physics, University of Tennessee, Knoxville, Tennessee 37916, USA

SINANOĞLU, OKTAY

Sterling Chemical Laboratory, Department of Chemistry, Yale University, New Haven, Connecticut 06520, USA

WHALING, WARD

Division of Physics and Astronomy, California Institute of Technology, Pasadena, California 91125, USA

WIESE, WOLFGANG

National Bureau of Standards, Washington, D. C. 20234, USA

Introduction

Stanley Bashkin

Beam-foil spectroscopy has enjoyed a rapid growth since the publication of KAY's first experiment [I.1] and my own first formal discussion of the possibilities inherent in a foil-excited particle beam [I.2]. In addition to fulfilling a number of the important promises, the beam-foil source has been found to hold substantial surprises, the unearthing of which has contributed to our knowledge of basic atomic physics. Since the early days, major extensions have been made in the range of wavelength and particle energies which have been used, but only the bare beginnings have been made in exploiting the potential of the beam-foil source.

Since there are many people who would like to turn their accelerator facilities to beam-foil problems or apply their theoretical techniques to calculations which bear on the beam-foil field, it seemed appropriate to assemble a discussion of the present status of beam-foil spectroscopy. The present volume attempts to summarize what has been learned and outlines a number of studies which remain to be made.

The first chapter deals with experimental details. In [I.2], it was cautioned that Doppler-induced line broadening was apt to cause serious difficulties to the spectroscopist, and that prediction has, unfortunately, been borne out by experience. However, some good work has been done to reduce the effect. Normalization of data on line intensities has been handled differently by different people. The advantages and limitations of various pieces of equipment from accelerators to spectrometers have been explored in many experiments. Chapter 1 puts these and related matters into one place for the convenience of all beam-foil workers.

Chapter 2 is concerned with what has been learned from the spectroscopy itself. Following a discussion of experimental details as related to the problems of spectroscopy, there is an extensive treatment of data for systems for which previous information was incomplete and for special kinds of states which are populated prolifically in the beam-foil source. The methods of classification of levels is described in detail. There is particular emphasis on hydrogenic and multiply-excited levels.

In [I.2], the hope was expressed of the possibility of using the beam-foil source for the measurement of the mean lives of excited electronic levels, and it is well-known that much success has been achieved. In Chapter 3, following a summary of the history of such measurements, a detailed analysis is given of how the mean lives are deduced from the data. Careful attention is paid to the effect of cascades, and methods are given for accounting for them.

At the same time that the experimental work has flourished, important developments have taken place in theory, especially in connection with the calculation of oscillator strengths. Chapter 4 describes recent advances in such calculations from a fundamental point of view, and presents new work not previously published. There are many suggestions for further experiments and calculations for the purpose of improving both kinds of attack on atomic properties.

Chapter 5 is based on years of work in which the systematic behavior of oscillator strengths with position in an isoelectronic sequence has been shown to be of great value. It is especially noteworthy that these trends make it practical to evaluate the errors in individual experiments on level lifetimes, and one can readily see those cases where further effort is required in order to remove the discrepancies between theory and measurement.

Some of the most interesting applications of the beam-foil technique have been to problems of astrophysics, especially in connection with understanding the nature of the solar corona and the relative abundance of the elements in the Galaxy. Chapters 6 and 7 are devoted to different features of the astrophysical consequences of beam-foil data. New experiments are proposed to answer outstanding questions.

Many beam-foil experiments are carried out at energies of a few hundred keV, and they are quite revealing of the electronic properties of the low stages of ionization of the accelerated elements. At the extreme upper end of the energy scale, that is, at 8 or 10 MeV per nucleon, entirely new phenomena come into view. At such energies, one creates one- and two-electron systems of high Z, and the transitions which their "optical" electron can undergo become sensitive tests of quantum electrodynamics. Chapter 8 surveys both the theory and experiments on the Lamb shift and various forbidden decays for one- and two-electron systems from $Z=14$ upwards.

An unexpected feature of the beam-foil source is the high degree of coherence and alignment which is exhibited by virtually every excited level, and of the orientation which accompanies a particular beam-foil geometry. Chapter 9 gives a thorough explanation of this feature, including a comparison of theory with experiment. There is a clear guide to further experimental work.

While the great bulk of beam-foil experiments have to do with optical spectroscopy, one cannot overlook the fact that electrons are also produced prolifically when heavy ions penetrate through thin foils. Indeed, it is noted that electron emission is intrinsically a far more common means of decay of excited systems than is optical emission. The study of the discrete energy spectra of such electrons has provided impressive information about metastable systems of high spin. Chapter 10 describes how the experiments are done and what is their basic significance as regards the electronic structures of highly-ionized atoms.

Even a cursory reading of this volume should make it clear that the beam-foil source is rich in information about monatomic systems, and that there is a great opportunity for both experimenters and theorists to obtain and codify it. It is the hope of the several authors and the editor that this book will stimulate new attacks on the still unknown characteristics of atoms.

The editor takes this opportunity to record his thanks to Mrs. Clarence Davis for her skill in typing the entire manuscript and for making many useful editorial suggestions. All readers of this volume stand in her debt.

References

I.1 L. Kay: Phys. Letters 5, 36 (1963)

I.2 S. Bashkin: Nucl. Instr. and Meth. 28, 88 (1964)

1. Experimental Methods

Stanley Bashkin

With 11 Figures

The intention of this chapter is to give practical information about the arrangements which are used for experiments in beam-foil spectroscopy (BFS). We begin with the production of the beam of energetic particles.

1.1 Accelerators

Any accelerator of positive ions can be used for BFS. Isotope separators, Cockcroft-Waltons, single-ended and tandem Van de Graaffs, and linear accelerators have been employed for such work, the Van de Graaff being the type used most frequently. While isotope separators tend to have relatively low voltages ($\leq$ 600 kV), their ion sources often produce multiply-ionized particles so that the ultimate particle energy may be $\sim$ 1.5 MeV. Such machines are used, for example, in Berlin, Denmark, and Sweden. The single-ended Van de Graaff, with an internal source of positive ions, is widely used for BFS, attractive features being the ease with which the particle energy can be varied and the available energy range.

A 2 MV Van de Graaff can easily operate at 150 kV. However, it is often desirable to use particle energies which are well below 150 keV. To do this, one may extract an appropriate molecular ion from the source. For example, a radio-frequency ion source may emit ions of H^+, H_2^+, and H_3^+ in roughly equal numbers. If these particles are all energized to eV, the energies of the component atoms are eV, 1/2 eV, and 1/3 eV, respectively. By using CH_4^+, the energy of one of the hydrogen atoms is reduced to 1/16 eV, etc. In this last example, of course, both C and H appear in the target chamber, but that is the price one has to pay for extending the particle downward, while operating the accelerator at a suitable voltage. Since molecules don't survive the collisions in the foil, only monatomic emitters can be seen, despite the complexity of the incident particle. The scattering may be somewhat worse for molecular constituents than for monatomic particles of the same velocity [1.1], but there are also some contrary results [1.2].

The tandem Van de Graaff is an excellent machine for making energetic ions of many different species [1.3-8]. Terminal voltages as high as 20 MV are in operation, and at least two 30 MV machines are under construction. At 30 MV, a two-stage tandem could generate ions with a final energy in excess of 400 MeV. As of this writing, the highest energy reported for BFS work with a tandem is 110 MeV [1.7,8].

The highest energies used to date have been achieved with linear accelerators, such as the machines at Orsay and Berkeley. At Orsay, particle energies of 1.15 MeV per nucleon have permitted the spectral examination of one-electron and several-electron ions of C, N, O, and Ne [1.9-13]. The high velocity of the particle beam also lends itself to the measurement of ultra-short mean lives, and data have been obtained on mean lives as short as 0.01 nanosec.

Some of the most imaginative work at very high energy has been done by MARRUS and his collaborators at Berkeley (see Chapter 8). With up to 10 MeV/nucleon available, hydrogenic and helium-like ions have been studied for a variety of elements from silicon to krypton. These experiments, which have concentrated on forbidden transitions, show that the high speed, useful for the measurement of short mean lives, is a bit of a disadvantage in measuring long mean lives, because the viewing length must be quite long. Thus, data have been taken at Berkeley over beam flight paths as long as two meters.

On the face of it, the heavy-ion cyclotron seems to be an attractive machine for BFS. Since the beams are pulsed, with widths as short as a few hundred picoseconds, one is tempted to think of time-delayed coincidence measurements of short mean lives. In fact, such experiments could be done with gaseous targets as well as foils. However, we do not know of a single cyclotron experiment of this nature, although a similar experiment at low energy has been described by DOTCHIN, CHUPP and PEGG [1.14]. It appears that this research area is wide open for exploitation.

1.2 Ion Sources

The rf ion source is the most common for the single-ended Van de Graaff. The standard High Voltage Engineering Company's source has been used for elements like B, S, and U by admitting $BC\ell_3$, SF_6, and UF_6, respectively, into the discharge bottle. The best performance has been obtained by running the source at high pressure ($\sim$ 50 microns) for 5 or 10 minutes and then turning off the molecular gas completely. The discharge is then operated with He or Ar; quite steady B or S beams of the order of 10 μa have been put on target, but the uranium beams have been only a nanoamp or so. The current falls slowly over a period of half a day, but can be restored by admitting some more of the mixture. Some metals have been accelerated by making the exit canal of the desired material, or merely by putting a small piece of the metal in the rf bottle, and running a beam of helium. Iron [1.15], aluminum [1.15], and beryllium [1.16] have been accelerated by the former method. The beams tend to be small and somewhat uncontrollable, but they're better than nothing.

It might be remarked that the Arizona group has run a wide variety of corrosive gases in its 2 MV Van de Graaff. In addition to $BC\ell_3$, SF_6, and UF_6, perhaps the most noxious substance tried was SiH_4, a gas which is spontaneously and explosively combustible on exposure to oxygen. These various gases have not yet had any deleterious effect on the accelerator tube, which has been used unchanged since the summer of 1964.

Another important ion source of positive ions is a furnace in which the substance of interest is volatilized. An electron beam ionizes the vapor, and positive ions are extracted. WHALING's work on Fe (see Chapter 6) has been carried out with such a source. In some sources one can inject $CC\ell_4$ into the hot furnace. This promotes the formation of metal chlorides, which have high vapor pressures at relatively low temperatures. However, this means that a lot of unwanted particles enter the accelerator tube; for a single-ended Van de Graaff, this introduces serious problems. One is that the electrons which are released by collisions between the heavy ions and residual gas in the accelerator tube or on slit edges prior to deflection of the beam, stream up to the high-voltage terminal, causing additional ionization on the way. When the electrons ultimately hit a stop at the end of the accelerator tube, a large flux of energetic X-rays is generated. Not only is this potentially hazardous, but the X-rays also promote voltage instability.

A second problem is that one usually pumps the ion source with a pump located at the exit end of the accelerator tube. For a 2 MV Van de Graaff, the pumping speed is tube-limited to ≤ 20 liter/s, which isn't much. Hence the admission of $CC\ell_4$ into the furnace could well drive the tube pressure up to the point of electrical breakdown.

In principle, a terminal deflector can be installed so that only the proper component of the output of the ion source can enter the accelerator tube. Such pre-selection is easy enough to arrange when the ion source is outside the accelerator proper, but, when the ion source is internal, space and power limitations make it hard to do, at least for atomic masses greater than perhaps 20.

Another kind of furnace source is useful for the alkali metals. It relies on the difference between the ionization energy of the alkalis and the work function of tungsten. Heating an appropriate salt, like lithium aluminum silicate [1.17], drives out the alkali constituent. When the alkali atoms touch the tungsten, they rebound as positive ions and can be directed into the accelerator tube.

The Arizona group has found it convenient to copy the old Wisconsin technique of admitting gas into the ion source from outside the accelerator. This is done by running a plastic tube alongside the accelerator tube and filling it with the desired gas at a pressure of $\sim$ 300 psig. For the HVEC 2 MV machine, this method, which allows one to use many gases without the need to install separate bottles and valves inside the high voltage terminal, works well for nearly all gases. However, if the gas pressure in the plastic tube gets too low, the tube may be disintegrated by an electrical discharge.

Not too long ago, MIDDLETON [1.18] developed a new type of negative ion source for a tandem. In this source, the element of interest is fabricated in the shape of a hollow cone. The cone is bombarded with cesium ions, and a large yield of negative ions of the cone material is produced. Some gaseous elements, like oxygen, can also be used by allowing a small stream of the gas to impinge on the cone.

Middleton's source has greatly extended the variety of elements which could be accelerated with a tandem. At the same time, the beam currents have been increased and the operation of the source simplified. About the most serious limitation still to be encountered is that the cesium reservoir is finite and one must replenish the cesium every 200-300 hours.

1.3 Beam Requirements and Limitations

The questions of how much beam is needed and how much can be used are not easily answered because of the great variety of experiments one might wish to do. However some kind of guide may be offered.

For experiments on spectra, say, where the lines are in the vacuum ultraviolet, it seems that currents of the order of a microamp are essential when the emitter has a net charge of under 4 or 5. However, when the net charge rises to, say, ten or more, the rate of radiation is so much higher than for the less-highly-charged ions that one can get by with far lower currents. In the work on iron [1.7,8], the current was only about 0.5 particle nanoamp, but it was still practical to obtain fairly good data. One can examine the transition probabilities listed by WIESE, SMITH, and GLENNON [1.19], and find that, for Ne IX, one may have A-values of 10^{12} or 10^{13}, whereas they may be around 10^9 or 10^{10} for Ne IV.

The same remark about current needs holds for lifetime experiments. Let one particle microamp, moving with 1% of the speed of light, be observed over a beam length of 0.5 cm. That viewing distance is reasonable for a one-meter, normal-incidence spectrometer, operated without lenses, with slit widths of 100 microns. The solid angle for the collecting system is about 10^{-5}. The detection efficiency is quite complicated, for it depends on the coefficient of reflectivity of the grating, the degree of polarization of the reflected light, the astigmatism of the system, and the response characteristics of the photomultiplier; all the foregoing depend, sometimes critically, on the wavelength of the light. As a very rough average over the foregoing factors, we take 2%. Next one must consider the rate of radiation from the excited particles. Let the Einstein A-value be $10^9 s^{-1}$, which corresponds to about the median mean life one might measure. Combining the above numbers, one finds about 2×10^6 photons per s reflected from the grating. Those photons are distributed among three or four stages of ionization with perhaps several hundred separate levels in each one. The final result is that a strong line gives rise to 10^3 or so counts per s. Of course, this number declines as the point of observation is translated downstream from the foil.

Other experiments can be more demanding. In attempts to detect rf-induced transitions between ℓ-states of high n in N V, beams of 5 microamp proved insufficient, given the available microwave power in the cavity through which the beam passed and the small throughput of the normal-incidence spectrometer which was used [1.20].

How much beam can one tolerate depends fairly critically on the kind of incident particle and its energy. A carbon foil with a thickness of 10 μgm/cm^2 can be bombarded by 10 microamp of 1-MeV protons for hours without failure, whereas 0.1 microamp of 5 MeV krypton destroys such a foil in a few minutes. SØRENSEN [1.21] has reported that foil life is extended considerably for heavy ions if some diffusion-pump oil is applied to the foil holder prior to picking the foil up from the water. He suggests that the foil essentially floats on the oily surface and moves under the beam so as to minimize any damage.

1.4 Mass Analyzers

When accelerators are applied to BFS experiments, they are apt to be limited by the available mass-analyzing power. We insist on the need for mass analysis, even if one is interested only in spectra, but especially for lifetime studies. In general, accelerators spew out at least several, and sometimes many, different particles at one and the same time, so that mass analysis is crucial to the acquisition of clean data. It is especially important to have good mass resolution when molecules are introduced into the ion source. Thus, the acceleration of boron, fluorine, chlorine, metal chlorides, or similar substances quickly lead to a complex beam, although the mass spectrum tends to simplify again if an inert gas or hydrogen is run for several days.

If one puts a (gas) stripper between the accelerator and the magnet, virtually all molecular ions can be dissociated into atomic ions. However, there is then a variety of ion energies. Thus, from CO_2 accelerated to energy E, one finds C^+ with energies of 0.04E (from $CO_2 \rightarrow CO + O$ followed by $CO \rightarrow C + O$) and 0.36E (from $CO_2 \rightarrow C + O_2$). In addition, there are oxygen ions with different energies. Add a touch of hydrogen to the beam, and the variety soars. Moreover, the incident beam is obviously fragmented so that the particle flux through the target is substantially reduced.

To illustrate what can happen, Fig.1.1 is a mass scan obtained [1.22] with the Arizona

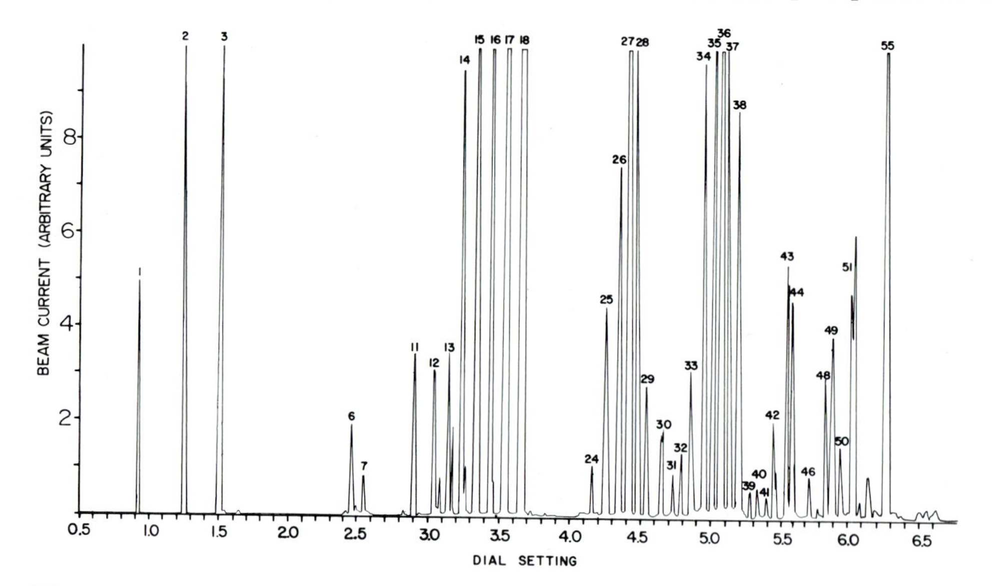

Fig.1.1 A mass scan (particle current vs. magnetic field) when a beam of hydrogen was energized to 1 MeV in the Arizona 2 MV Van de Graaff. The deflection angle was 45°. The abscissa is a dial setting, which is closely proportional to the magnetic field up to a setting of ∿ 5. The mass numbers are shown at the prominent peaks; probably "55" corresponds to iron, with an atomic mass of 56. [Courtesy of W. S. Bickel (unpublished)]

2 MV accelerator when a beam of protons was energized to 1 MeV. Figure 1.1 shows particle current as a function of magnetic field, the field being closely proportional to the dial setting (abscissa) up to about 5.0. The numbers on the peaks indicate atomic mass values. Mass number 18 is readily identified as due to H_2O^+, and 17 is probably due to OH^+. Mass number 34 might arise from $H_2O_2^+$ and 55 from iron, but it isn't easy to be sure of the identity of all the mass peaks without doing a lot of work. The complexity of Fig.1.1 illustrates well how difficult it can be to interpret the nature of the particle beam.

The two main factors which enter into the selection of an analyzing magnet are the mass resolving power and the mass-energy product. Mass resolving power of the order of 2000 is adequate to distinguish between a deuteron and H_2^+, or between CO and N_2. One way to describe the mass-analyzing capability in terms of the "mass-energy product", by which is meant ME/Z^2, where M is the atomic weight, E the kinetic energy in MeV, and Z the charge of the deflected particle. Thus, an analyzing magnet intended to be used for nuclear physics with a 6 MV Van de Graaff might have a mass-energy product of 42, which would allow one to use ions up to $^7Li^+$ at full energy. For BFS purposes, however, such a deflecting system is inadequate in comparison with the need. The Arizona 6 MV Van de Graaff, which is vertical, has a magnet which deflects particles with a mass-energy product of 700 into the horizontal plane. The magnet is mounted on a rotating base so that a number of different target arms can be used.

For a horizontal accelerator, one can use the design of Fig.1.2, which shows the double-focusing magnet used on the 2 MV Van de Graaff at Arizona. The mass-energy product is 80 at 60°. For other angles, $ME/Z^2 \sim \cot^2 \frac{\phi}{2}$, where ϕ is the angle of deflection.

Fig.1.2 A photograph of the 2 MV experimental area in Arizona's 2 MV Van de Graaff Laboratory. There are eight target arms on either side of the central position, the angular intervals being 7.5° out to a maximum deflection angle of 60°. [Courtesy W. S. Bickel (unpublished)]

Other kinds of mass-analyzers can be used. The velocity filter is favored by the Scandinavians. It can easily give a mass-resolving power of 1000 with reasonable slits. Furthermore it is essentially in-line with the accelerator. This may be a drawback if the laboratory is small or if the accelerator is vertical. Unless one is content to use a single experimental arrangement at a time, some type of switching device is needed also. The same can be said of electrostatic analyzers. These are nice in that the energy scale is linear in terms of analyzer voltage, but it's always a bother to have high D.C. voltages inside a vacuum system.

1.5 Target Chambers

Target chambers are all pretty much the same. They are cans with provisions for:

i) replacing a broken target with another without having to open the system. This is often done by mounting the targets on a wheel, as seen in Fig.1.3.

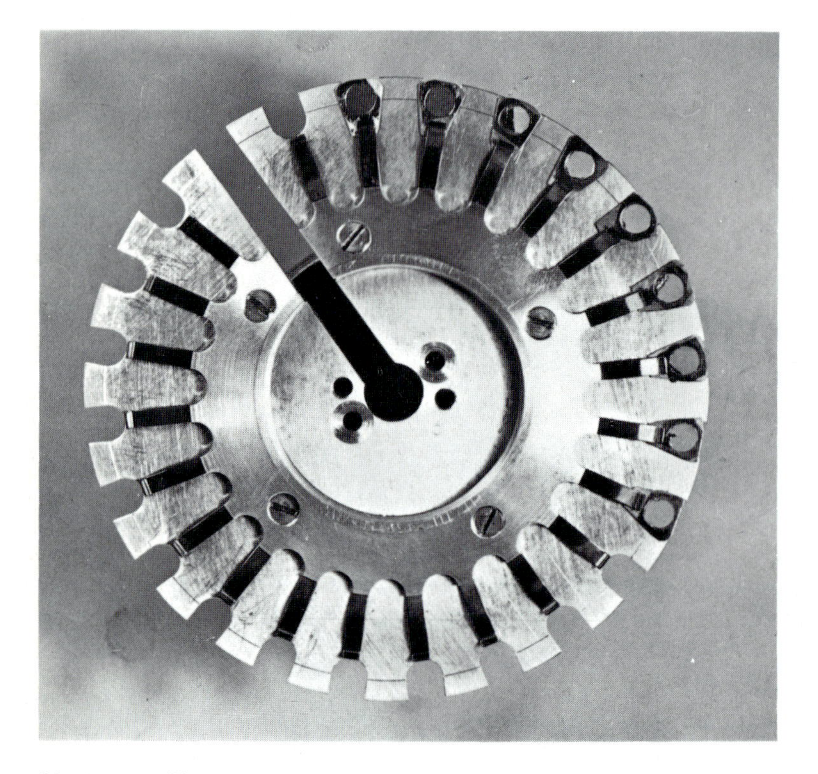

Fig.1.3 Photograph of a wheel on which targets are mounted. Four of our wheels hold 50 targets. [Courtesy W. S. Bickel (unpublished)]

ii) windows, which are necessary if the spectrometric device is an air instrument, and which may be convenient at other times as well. For example, a photomultiplier which is to be used as a monitor may look at the beam through a window. Many window materials are suitable, although we recommend against plastics because of their poor vacuum qualities. While LiF transmits down to 1000Å, air has a cut-off at 1800Å.

iii) pumps. The pressure in the target chamber should be as low as possible, and, preferably, below 5×10^{-6} torr. In some cases, this pressure can be reached only by mounting a diffusion pump and liquid-nitrogen trap as an integral part of the target chamber. One of the Arizona target chambers has a titanium pump as well as a diffusion pump connected to it. The titanium pump is useful for reducing the pump-down time; however, it must be turned off when taking data because scattered light from the filament adds to the background.

iv) translating the target parallel to the particle beam so that lifetime measurements can be made. This can be done by mounting the wheel on a track which moves on a good screw. See Fig.1.4. The screw may be driven manually or with a motor.

Fig.1.4 Photograph of a mount for linear translation of the wheel of Fig. 3 parallel to the particle beam. The translation may be accomplished manually or by a computer-controlled stepping motor. The screw is precision-made. [Courtesy W. S. Bickel (unpublished)]

v) a Faraday cup, so as to have a measure of the total charge collected during a run. Such measurements are convenient monitors for normalization purposes, a topic which is discussed later in this chapter.

vi) vacuum feedthroughs for gauge connections, the transmission of mechanical motion, or the introduction of electric current and high voltage. Such items generally present little problem, but the chamber should be large enough to accommodate them and permit their manipulation with ease. Removable flanges on which several items are mounted are frequently employed.

Target chambers are apt to exhibit reflections which can be troublesome when lifetimes are being measured. For example, CHUPP et al. [1.23] called attention to the problem which arose in the course of their measurements on the decay curve for Ly-alpha. Subsequently, BICKEL [1.24] showed how careful consideration of the chamber geometry makes it possible to reduce the intensity of the reflected signal to less than 10^{-4} of the direct signal.

vii) an occulter, which is sometimes placed in the target chamber so as to limit the length of beam which is viewed by the analyzing instrument. This may be more convenient than using the slits in the spectrometer or a mask over the grating.

viii) applying an external field to the radiating beam. The field may be established with electrodes which are inside the target chamber; this is obviously necessary for an electric field which is parallel to the particle velocity or when one wants to avoid Zeeman effects which would be present when generating a transverse electric field with a magnetic field. There are other situations, however, in which the entire target chamber is placed between Helmholtz coils. This introduces practical limitations on the size and complexity of the target chamber.

1.6 Targets

The targets are almost always thin foils of carbon. There are many reasons for choosing carbon. Carbon foils are simple to prepare; they are strong and easy to handle. Since carbon has a low nuclear charge, scattering of the incident particles is kept small. Foil life is generally reasonable. Occasionally, foils are mounted on grids in an effort to extend foil life. However, the results are variable. For Ar at 250 keV, foil life seems improved, but at 500 keV, there is no gain at all [1.25].

While foils can be purchased, they can be made by anyone who has a vacuum evaporator. The prescription has been given by DEARNALEY [1.26]. Briefly, one side of a microscope slide is coated with a thin layer of soap or detergent which, after drying, is polished with a lint-free cloth until no residue can be seen. The kind of soap or detergent is probably immaterial - when troubles arise, the reason is apt to be a poor vacuum in the evaporator and not the kind of coating.

Two rods of spectroscopically-pure carbon are pointed and held against each other by a light spring. A dozen or more of the microscope slides are placed on a tray, the coated sides facing the carbon rods. When the pressure has been reduced to $\leq 10^{-5}$

torr, a high-current arc is struck, and carbon evaporates freely. We normally throw a switch which promptly applies the full voltage (about 20 volts) to the rods, and allow the evaporation to proceed for less than a second. It may take several such "shots" before the desired carbon thickness is reached.

When the slides are removed from the evaporator, each is scratched into sizes and shapes as wished. We usually divide a 1" x 3" slide into about a dozen pieces. The slides are then dipped slowly into a deep pan of water, and the individual targets float away. A target holder may be made of a flat, metal plate, 1/32" thick with a 1/4" diameter hole in it. While Aℓ, Cu, brass, and stainless steel all work well as target holders, CURNUTTE [1.27] has mentioned that the carbon doesn't stick to titanium. With the exception of the latter metal, all one need do is slip the holder under one of the carbon pieces and lift it out of the water; the carbon adheres well to the holder. For titanium, CURNUTTE puts some grease on the holder and, instead of slipping the microscope slide into a dish of water, he puts a drop of water on the particular piece of carbon he wants to use. The water works its way under the carbon, detaching it from the glass. The greasy side of the titanium is lowered over that piece of carbon, and lifted away.

The foil thickness can be found, for example, from the energy loss or elastic scattering of alpha particles or other positive ions, but the Arizona laboratory makes use of an instrument built by J. O. Stoner, Jr. STONER [1.28] first burned the carbon in pure oxygen, and formed CO_2. He then measured the pressure of the CO_2 in a standard volume at known temperature; from this he determined the absolute amount of carbon in a piece of foil. Subsequently he measured the optical transmission of another piece of the same foil. After calibrating transmission against absolute carbon content over a range from $\sim$ 1-100 μgm/cm^2, he could simply use the transmission to establish the thickness of a foil to $\sim$ 5%. Stoner noted that there was an unidentified residue of non-carbon matter, amounting to $\sim$ 1 μgm/cm^2, independent of the carbon thickness.

There seem to be some problems in making good thickness measurements. For example, CURNUTTE [1.27] of Kansas State has reported that measurements made by Rutherford scattering at large angles indicate considerably greater foil thicknesses than the optical transmission method gives, at least for foils below 40 μgm/cm^2. Using proton scattering, CURNUTTE found that the foils contain a small amount of oxygen, the amount being roughly proportional to the carbon thickness, about as many hydrogen atoms as carbon atoms, and nothing heavier than oxygen. For foils having thicknesses between 40 and 120 μgm/cm^2, the optical transmission and Rutherford scattering techniques give the same values to within 1 μgm/cm^2. Some measurements by X-ray absorption [1.29,30] and energy loss [1.31] also agree with the optical transmission data for layers thicker than 40 μgm/cm^2. However, THOMAS [1.32] of Argonne National Laboratory reports that energy-loss data for thin foils (< 10 μgm/cm^2) suggest they are thicker than optical measurements give. The cause of the discrepancies for the thinner layers is currently under investigation. It is possible that a small amount of contamination by heavy elements could affect the energy-loss experiments more than the optical ones. Analysis of proton-induced [1.30] or argon-induced [1.33] X-rays suggests the presence of trace amounts of Na, P, S, and Cℓ, presumably from the parting agent.

Foil life under bombardment is a widely variable function of the energy and nuclear charge of the incident particle, as we have noted earlier. The exact cause of foil rupture is not known. Certainly there is some displacement of foil particles and, since there aren't very many of them to start with, the loss of a few may create internal stresses which tear the foil apart. It is possible that the beam heats the foil to the point of rupture. There are times when the foil looks bright red where the beam comes through, but it isn't clear that that color can be associated with a temperature or with absorbed energy. A third possibility is that microscopic electrostatic forces arise as the result of the loss (or gain) of electrons when a beam of positive ions goes through. YNTEMA [1.34-36], of Argonne National Laboratory, has markedly increased foil life by heating the foil to ∿ 525°C. For BFS, one might not be able to tolerate the light that accompanies such a temperature. YNTEMA also remarks that radiation damage and the temperature gradient seem to be major contributors to foil damage, with electrostatic effects being of lesser importance; all of these problems are mitigated (but others are introduced) by evaporating a thin layer of gold on the carbon [1.35,36]. We have already mentioned SØRENSEN's report [1.21] that it helps to apply a thin coat of diffusion-pump oil to the frame on which the foil is to be mounted.

While carbon is the substance used most frequently, foils have been made of other materials. We have used Be, B, Aℓ, Cu, and combinations of metals and carbon. Beryllium has several advantages over carbon - it has lower nuclear charge, so that scattering is reduced, and it appears to last longer than carbon. However, Be is harder to prepare and there is always some risk of beryllium poisoning. We take the precaution of venting the evaporator pumps into the air at such a height that the dilution to safe levels occurs well away from people, but there remains some concern over cleaning the evaporator and over foil fragments which are dropped on the floor and subsequently appear in the air as finely-divided powder. We therefore have given up the use of Be. Boron is hard to prepare, fragile, and exhibits little durability under the beam. For those interested in making foils of exotic elements, good prescriptions have been given by MAXMAN [1.37].

It would be a worthwhile project to investigate the relative yields of light from foils of different substances, but no one has yet undertaken such work. A comparison of conductors, like carbon, and non-conductors, like boron, looks to be worthwhile. Little work has been done on the change in light output for a foil as a function of bombardment time. Useful though such experiments would be, they don't have much sparkle.

1.7 Analytical Devices

Many kinds of instruments have been used to analyze the beam-foil light. In the early days, one had to use instruments which were originally intended for other kinds of light sources; recently it has been shown that certain modifications can be made so as to exploit the particular feature that the emitters of concern to us are moving very rapidly.

The simplest analytical device of all is a camera loaded with color film. It is quite instructive to photograph different beams at different energies. One sees directly that there are gross changes in color and in the length of the luminous part of the beam as the kind of particle is changed. Such pictures can be good guides to the design of quantitative experiments; in addition, the color photographs are quite beautiful. In taking such pictures, it is helpful to have some general, incandescent lighting so that the target chamber itself can be seen; this gives the viewer a good sense of scale. Of course, a double exposure can also be used.

Next in complication is an interference filter coupled to a photomultiplier tube. Such an arrangement is primarily useful as a monitor, but can also be employed for obtaining new data when well-isolated spectral lines are present. Of course, a simple filter puts a lower wavelength limit of $\sim$ 2300Å on the light which can be studied, although there is one [1.38] usable down to 1200Å, with 10% transmission. Measurements of mean lives in H have been made with a filter and photomultiplier. Other filters which have interesting applications are linear and circular polarizers.

For lifetime measurements, one can eliminate many mechanical complexities from the foil chamber by leaving the foil fixed in position and translating the filter-photomultiplier combination parallel to the beam but outside the chamber. Specifically, this simplifies the beam monitor; however, it also introduces its own difficulties. If the level lifetime is long (say 5×10^{-8} sec or more), the decay distance is large (for $v/c = 10^{-2}$, the e^{-1}-distance is 15 cm), which means that the foil chamber must be long and have a long window.

While interference filters are often useful, in particular permitting one to collect light through a large solid angle, most experiments require that the light be dispersed. An instrument which has proved useful for lines with wavelengths longer than

500Å is the normal-incidence grating spectrometer illustrated schematically in Fig.1.5.

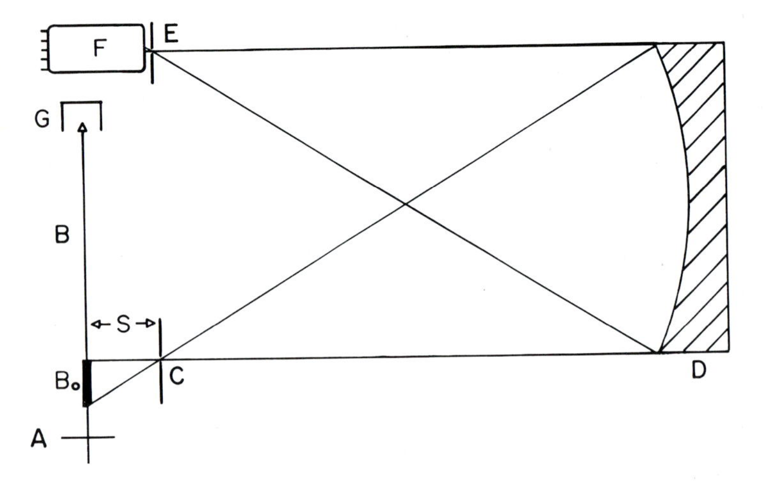

Fig.1.5 Representation of a normal-incidence grating spectrometer: A = Foil; B = particle beam; B_0 = observed segment of particle beam; C = entrance slit; D = grating; E = exit slit; F = photomultiplier; G = beam stop; S = distance from beam to entrance slit

There are certain important features one should note:

i) There are no optical elements in the system, other than the reflection grating. This means that one cannot focus the beam onto the entrance slit, so that the coupling efficiency from source to spectrometer depends primarily on S. The height and width of C and the grating size are other factors. It also means that the collection efficiency is not adversely affected by multiple reflections or by transmission elements.

ii) In the instrument of Fig.1.5, the slits are usually normal to the beam, and one examines only a limited segment, B_0, of the beam at any one time. The length of B_0 clearly depends on how close to the beam position one can mount the entrance slit. This limit is set by the diameter of the Faraday cup, for this essentially determines S. A practical value for S is 5 cm, which means that B_0 = 5 mm for a 1-meter grating 10 cm wide. BRIDWELL et al. [1.39] place the slits parallel to the particle beam and have S = 1 mm.

iii) Only one spectral line can be seen at a time. Therefore many scans in wavelength for many different pieces of the beam are needed in order to get all the information present in the beam.

iv) It is important to be able to move the slits in and out along radii of the grating. This is a highly desirable feature since it makes it possible to focus the spectrometer when, as in BFS, the source of light is moving at high speed. This point is further discussed below.

A second device, which lets one see lines down to ∿ 300Å, is the Seya-Namioka spectrometer. This spectrometer suffers from astigmatism, polarization, and poor line shape. Experience suggests that this kind of spectrometer is not really well suited to BFS.

For wavelengths shorter than 300Å, one must turn to a grazing-incidence spectrometer. Such a spectrometer may have a wavelength range which extends from perhaps 50Å up to 1000Å. Hence it accomplishes at least as much as the Seya-Namioka spectrometer and has a better line shape. Some results [1.40] appear in Fig.1.6.

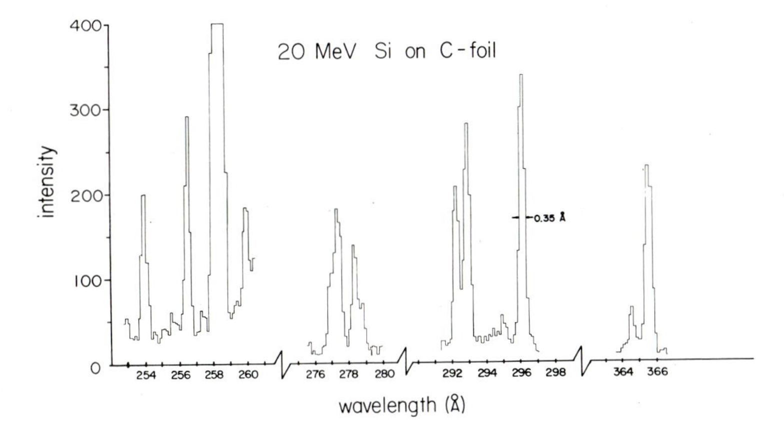

Fig.1.6 Spectral data from silicon [1.40] accelerated in the Oak Ridge tandem Van de Graaff to an energy of 20 MeV. The wavelength range covered is in the grazing-incidence region. Note the linewidth of 0.35Å

The foregoing spectrometers can be used well into the visible, given an appropriate grating. However, they are expensive and cumbersome. Often one can obtain quite useful information from much simpler instruments. For example, a small Czerny-Turner spectrometer - inexpensive and easy to manipulate - is a particularly useful tool.

1.8 Detectors

We consider the relative merits of photographic and photoelectric detection. A thorough study has been made by LOCHTE-HOLTGREVEN and RICHTER [1.41]. The photograph has much to recommend it, its main value being that it permits the capture of a wealth of two-dimensional data (distributions in space and in wavelength) in a single exposure. That is to say, one can record many spectral lines and the explicit spatial variation of intensity of each of those lines at the same time, by having the slits parallel to the beam. Also, by means of a decker, one can superimpose spatially-displaced calibration lines on the same plate that contains the data. For these reasons a spectrograph is an excellent survey instrument; sometimes it allows one to uncover curious events such as the unusual intensity patterns [1.42] which appear in the Balmer lines in Fig. 1.7.

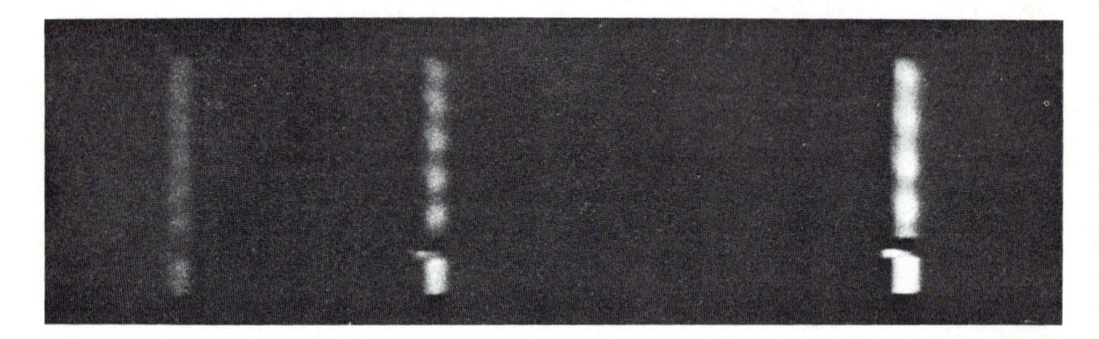

Fig.1.7 Quantum beats in some of the Balmer lines of hydrogen [1.42]. The beam was traveling upwards, parallel to the slit of a Meinel spectrograph. From left to right, the lines are H_ε, H_δ, H_γ. In this case the external field was a transverse magnetic field giving an equivalent electric field of 29 V/cm for 200 keV H_3^+ particles

Still and all, there are several serious drawbacks to a photographic record - it is not direct reading, the dynamic response has a narrow range, and the translation from plate darkening into line intensity is imprecise at best. Figure 1.8 illustrates these problems [1.43].

To get reasonable exposures for all of the lines of Fig.1.8 would be maddeningly costly in accelerator time; when one realizes that such work would have to be done for many different particle energies, it is seen that such an approach would not be fruitful. The foregoing difficulties are compounded by the insensitivity of film to radiations in the vacuum ultraviolet. The beam-foil light source is so faint that no successful photographs have been reported for $\lambda \leqslant 2300$Å.

Photoelectric detection is the answer to many of the above problems. It is direct reading, has a wide dynamic range over which it is linear, and, thanks to a good selection of low-noise, pulse-counting photomultipliers, either with or without windows, one can examine the wavelength range from 100Å up to the near infrared. A minor, but

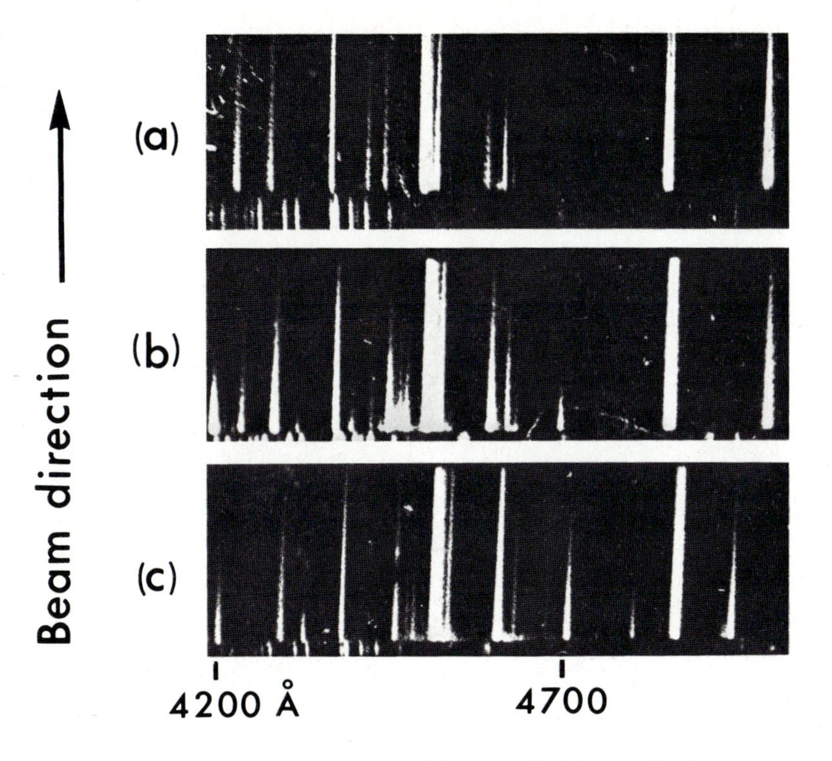

Fig.1.8 Photographic record of beam-foil lines of nitrogen at N_2^+ energies of 1.0, 2.0, and 4.0 MeV, respectively, from top to bottom. The beam was parallel to the slit of a Meinel spectrograph [1.43]

still useful, application of the direct-reading capability is that the signal drops instantaneously when the target foil is ruptured. In contrast, many an hour has been wasted exposing photographic plates while the light was out!

Direct reading simplifies the normalization of separate observations to a quantity like the total charge sent through the foil or the light output for some value(s) of λ. Unfortunately, even the photomultiplier has its faults. It is expensive in itself, some tubes costing $1800 or more, and it is generally used in conjunction with other costly electronic equipment, such as power supplies, amplifiers, multiscalers, and computers. There is not only the capital investment to worry about, but upkeep also takes a lot more money than a darkroom. Cooling the photomultiplier to dry-ice or liquid nitrogen temperature can be troublesome for some experimental arrangements, especially when frost might condense somewhere in the light path. As to use, calibration of the wavelength scale is a bit of a nuisance. Although a standard lamp can be observed simultaneously with the particle beam, the output is then the sum of the two, which is more complicated to interpret than the separate sets of lines on a plate. Generally the calibration is done independently of the data-taking, and one relies on the grating screw for linearity and reproducibility of the wavelength scale.

A major limitation of photomultiplier detection is that one gets only a point-by-point measure of the two-dimensional data array. Instead of seeing many lines at once, one looks at a single line at a time. While the light from a long length of beam can be detected, the spatial dependence of the intensity can be measured only point-wise. Therefore the labor involved in extracting information from the beam is increased. However, there is an enormous gain in the quantitative nature of the data when photomultiplier recording is employed. It's obviously advantageous to have digital data, rather than analogue; the former results from using photomultipliers, the latter, from photography. Our conclusion is that the primary application of a photographic instrument is for survey work; most numerically interesting results demand the use of photomultipliers.

A word might be said about image intensifier tubes, which are a kind of cross between a photomultiplier and a plate. Image intensifiers have been used for BFS by a group at the Carnegie Institution of Washington [1.44-47]. Non-linearity of the wavelength scale adds to the objections to photographic work and the gain in sensitivity does not seem to offset the other disadvantages of film. Also, the cost is rather high, $5,000 or more being the price of a commercial image intensifier. In addition, the final element is a photographic plate, the handicaps of which have already been mentioned.

1.9 Detection Geometry and Line Width

Simple detection geometries place the line of sight of the detector at 90° or 0° to the particle velocity, the former arrangement being the more common. There are other arrangements [1.48-57] in which a mirror or lens system collects light from some other angle and then directs that light into the detector; they have been used but rarely. The advantage of the 90° geometry is experimental simplicity, for the detecting equipment is then out of the way of the particle beam itself. Any geometry, however, introduces line-broadening which has the Doppler effect as its basic cause, and the Doppler broadening may be less for the 0° case than for any other.

It is well known that the Doppler contribution to line breadth which comes from a non-parallel particle beam can scarcely be controlled and may be substantial. The problem becomes particularly severe for particles of high atomic number and low energy, and thick foils. An analysis and review of this situation has been given by STONER and LEAVITT [1.58], who quote a line width of 1-2Å at 5000Å for 0.5 MeV particles of small Z passing through carbon foils with a thickness of 5 $\mu gm/cm^2$.

The line broadening which is due to the observation of a beam segment of finite length can be largely eliminated by the refocusing method [1.59].

Consider Fig.1.9. Light from a source at rest is shown (solid lines) to be focused

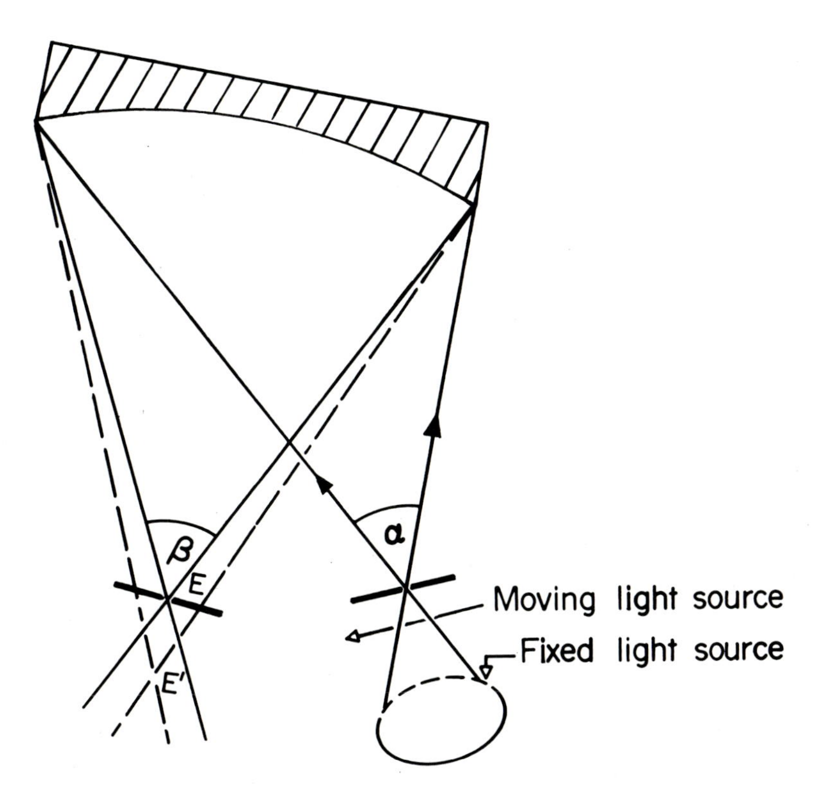

Fig.1.9 Schematic representation of focusing of light from a fixed source (solid lines) and a moving source (dashed lines). In the refocusing method [1.59], the exit slit is moved from E to E'

onto the exit slit, E. However, light from the moving particles is Doppler-shifted and the extreme rays follow the dotted paths towards the exit slit. These rays are clearly not focused at E. However, the extension of these Doppler-shifted rays shows a cross-over at the position E'. Hence, moving the exit slit to E' removes the Doppler effect, at the same time enhancing the ratio of signal to background. The distance, Δ, between E and E' is given by [1.59]

$$\Delta \approx \left(\frac{v}{c}\right) \left(\frac{\alpha}{\beta}\right) \left(\frac{\lambda}{k}\right) \quad . \tag{1.1}$$

where v is the particle speed, c is the speed of light, α and β are angles defined in Fig.1.9, λ is the wavelength of the spectral line, and k is the reciprocal linear dispersion of the grating. This technique of refocusing is quite useful, as is

illustrated in Fig.1.10 [1.59]

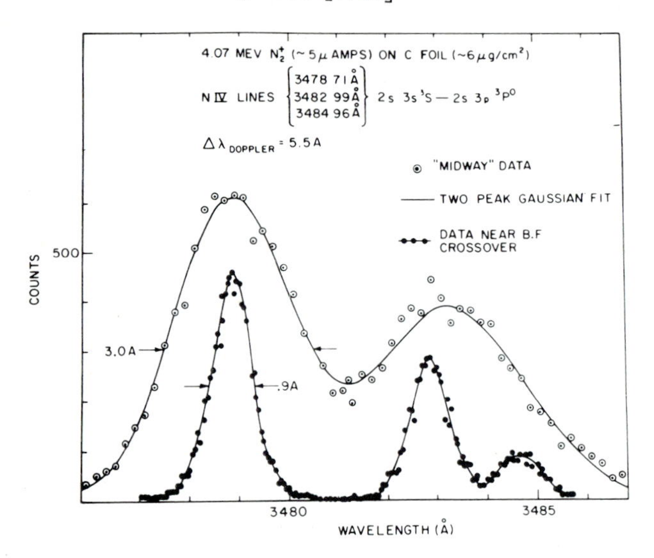

Fig.1.10 Comparison of spectral data taken with normal focus and refocused positions of the exit slit [1.59]

STONER and LEAVITT [1.60] have also shown that one can place a lens between the ion beam and the entrance slit of the spectrometer such that the angle of refraction from a point on the grating is independent of the angle of incidence at the grating. This means that the Doppler broadening associated with the finite width of the entrance slit may be eliminated, making possible the acceptance of light from a long length of the ion beam without any significant increase in line width. Of course, the angle of refraction varies from point to point on the grating, so that refocusing is needed. The focal length, f, of the lens is given by

$$f = (v/c)\, F \quad , \tag{1.2}$$

where F is the focal length of the grating (or collimator); f is usually but a few mm. Such lenses can be made conveniently from rods of glass or quartz. A general treatment of the foregoing methods has been given by BERGKVIST [1.61].

Another geometrical arrangement is to place a positive lens distant from the ion beam by the focal length of the lens, so that the spectrometer slit is illuminated with parallel light [1.56]. This has the advantage that only the 90° Doppler shift occurs. Moreover, one looks at a long length of the beam so that a lot of light is gathered. Disadvantages are that the arrangement can't be used if one wants to measure spatial effects and that, since the light doesn't have uniform intensity over the observed piece of the beam, there may be complicated vignetting in the spectrometer.

1.10 Beam Monitors

In order to deduce physically meaningful relationships from spectral data taken for different spectral ranges or particle energies, or from decay data obtained at different distances from the exciter foil to the point of observation, it is essential that one have a reliable method of normalizing the data to some common factor. One has a similar requirement if the investigation involves different foil thicknesses or materials. Since the signal being detected is obviously a linear function of the number of incident particles, a standard method of normalizing is to divide the numerical measure of the observed intensity by the number of particles which gave rise to the information. As we shall see, this technique is not always practical; a common alternative is to compare the signal to the light intensity as measured with a secondary detector. We consider these alternatives in turn.

To measure the number of incident particles, one uses the well-known Faraday cup and a current integrator. However, it is a matter of some surprise that all too often the Faraday cup is used improperly. Let us note at the outset that the difference between good and bad cups is perhaps only 15% so there may well be situations in which the complexity of a good cup is not justified by the experiment as a whole. Nonetheless, one should be aware of what distinguishes good from bad. Figure 1.11 shows a diagram

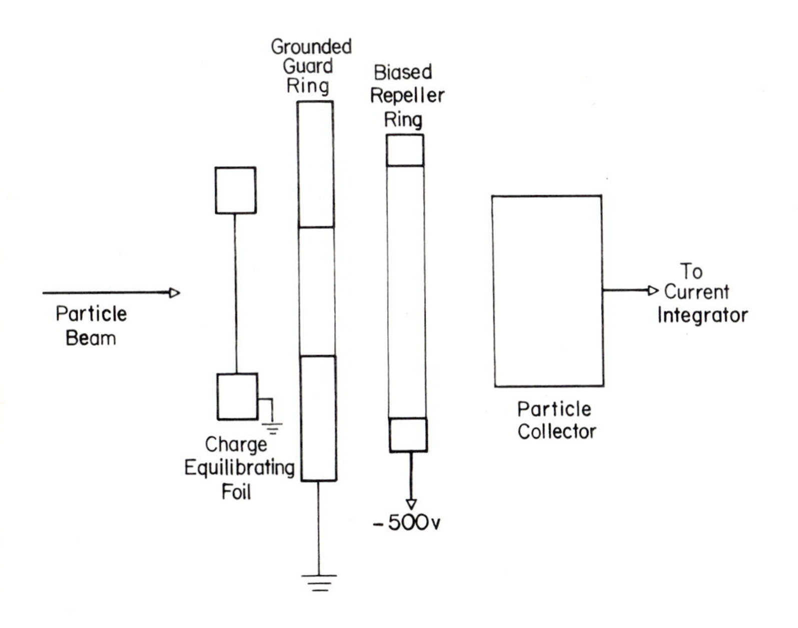

Fig.1.11 A shielded Faraday cup. Note, particularly, that the repeller ring must not be struck by the ion beam, and that a charge-equilibrating foil is in place

of a good cup. The requirements are:

i) A diameter large enough that few of the scattered particles are lost.

ii) A repeller ring, biased perhaps 500 volts negative with respect to the collector. This prevents secondary electrons, ejected from the cup by the impact of the incident ions, from leaving the cup and giving a spurious signal to the current integrator.

iii) A grounded shield which prevents the ions from striking the repeller ring. If ions do hit the repeller ring, secondary electrons liberated by those impacts are accelerated towards the collector.

iv) A good vacuum in the cup. If the pressure is 10^{-5}torr or worse, there may be ionization in the cup, with a drift of ions to the repeller ring and an incorrect collector signal. This problem is made increasingly severe as the energy of the incident particle is lowered and its charge raised.

Even when the foregoing are properly attended to, one must recognize that the cup signal is affected by the charge distribution of the foil-transmitted particles. Since that distribution is a complicated function of the energy and atomic number of the ion, as well as the nature and thickness of the foil, it is not so easy to normalize one's data for the usual variety of changing experimental parameters. In particular, the variation of charge distribution with particle energy makes it awkward to get good excitation functions simply because the cup response to the incident particles is not constant. A good way to circumvent this problem is to place a thin foil over the entrance to the Faraday cup, for then the particles reach charge equilibrium before entering the cup. Although windows have long been used over Faraday cups when gas targets were being bombarded, in so far as we know [1.62] gives the first description of a window for the purpose of producing charge equilibrium in the incident beam. The advantage is clear, since the cup current is now independent of the beam energy. Such a cup may also be used for the detection of neutral particles.

The Faraday cup of Fig.1.11 is not the only good design one could use. A magnetic field can also serve well to keep the electrons in or out of the particle collector. A deep cup may be quite satisfactory, with no fields at all, provided attention is paid to the loss or gain of electrons near the entrance to the cup.

One might point out that Faraday cup data suffer from a serious handicap - if the foil develops pinholes, the collected charge will generally decline (unless the cover foil is in place over the cup), but not necessarily in proportion to the decrease in the light under examination, so that the normalization of the data is in error. If the foil thickens with bombardment time, the charge distribution may be affected but little, whereas the beam light may vary substantially. An independent monitor of foil characteristics would be nice to have, but we don't know of any.

It is our attitude that one should always make good measurements of the particle flux. However, we recognize that one can't always do that. Consider, for example, the acquisition of data in the vacuum ultraviolet. One wants to examine a small segment of the particle beam, but lenses are out of the question. If an occulter is used over the beam, one may end up not filling the grating with light. The only recourse is to place the entrance slit of the spectrometer close to the particle beam. This means that there simply isn't enough space around the beam for one to have a good Faraday cup. In such a case, we turn to the alternative method of normalization - we use the beam light itself as a monitor.

Again, there are several possible procedures. The simplest is to view the entire radiating beam with a photo-diode or photomultiplier tube. A more elaborate system selects a particular spectral line, with either an optical filter or a monochromator, and compares the signal of interest with a numerical measure of the monitor light. In one case [1.63], a monitor photomultiplier tube collected light reflected from the grating itself. This signal was extremely sensitive to the position of the beam relative to the entrance slit of the grazing-incidence spectrometer then in use.

These methods are reasonably satisfactory for data taken at a single incident energy, but they are inadequate when the energy is changed during the course of the experiment. The reason for the latter difficulty is that the light, integrated over all wavelengths and over the beam length or examined over only one spectral line originating at a well-defined point in space, is far from proportional to the particle energy. Indeed, the behavior is quite complex. Some experiments have suggested [1.64] that the dependence of light intensity on particle energy follows the charge-state dependence for bound levels and the charge-dependence for the next higher ionization stage for autoionizing levels. Little good information is available, spanning a wide variety of atomic systems and particle energies, on which to base a reliable normalization procedure. This means that optical normalization techniques are not the answer to the problem.

One way out of the foregoing drawbacks is to rely on the steadiness of the beam over the time taken for the measurements. This approach, in conjunction with repeated runs to ensure reproducibility, is not so bad.

An especially troublesome situation arises when the flight path varies from short to long, as in the measurement of a long mean life over several decades. If the foil-to-monitor distance changes, as it usually does, geometrical corrections must be incorporated into the normalization factor. There are two separate problems. One, of course, is that the normalization detector subtends a varying solid angle at the foil. The other is that the spectrometer may not receive all the light it should. For example, scattering in the direction of the slit may prevent the spectrometer from viewing all the emitters. Also, scattering in the plane of the experiment may lead to a poor focus because of the finite depth of field for the spectrometer.

Clearly, the attempt to get good normalization data is increasingly hampered as the particle energy decreases and its atomic number goes up, and as the foil thickness increases. While we raise these problems, no good solution to them is known to us at this time.

1.11 External Fields

External fields, electric and magnetic, d.c. and time-dependent, have been used for many purposes. One recalls the determination of the stages of ionization appropriate to the emission of a given spectral line [1.44,65-70], of inducing resonance transitions between states of differing magnetic quantum numbers [1.71-73], of quenching hydrogenic levels [1.7,42,74-83], of generating quantum-beat effects [1.84-93], and of Zeeman-effect work [1.94]. For the most part, the experimental arrangements have been well described in the literature, and we make only a few comments.

If high, d.c. electric fields are to be applied to parallel-plate electrodes, there is the serious risk of electrical breakdown because of the large number of post-foil electrons which are apt to be present. If these electrons strike the positive electrode, numerous secondary electrons may be liberated. The negative electrode may be a source of photo-ejected electrons because of the presence of XUV radiation in the vicinity of the electrodes. Also, of course, the positive ions may release secondary electrons from the negative plate. Thus it is important to prevent any charged particles from hitting the plates. While this is sometimes done by mounting the electrodes at a large angle to each other, we feel that a better geometry is one in which the plates are slotted in line with the particle beam. In either case, the field necessarily deviates from uniformity, but that's a small price to pay for avoiding breakdown. Fields of the order of 75 kV/cm have been used successfully [1.70] in BFS. The power supplies need a current capacity of the order of several hundred microamp.

One might note that the normalization of the data is especially tricky when the beam is under the influence of an external field. In our experience, the reliance on the steadiness of the beam is as good a method of normalization as any.

1.12 Concluding Remarks

In reviewing the experimental aspects of BFS, it is somewhat surprising that only a small number of instruments and techniques have been developed with the specific properties of the beam-foil light source in mind. Refocusing and intensity enhancement are the most important of such developments; we expect that the next generation of papers in BFS will include novel methods which take advantage of the moving light source.

Support of NSF, ONR and NASA in the preparation of this chapter is gratefully acknowledged.

References

1.1 W. S. Bickel: Phys. Rev. A 12, 1801 (1975)

1.2 R. Laubert: Bull. Am. Phys. Soc. II, 21, 32 (1976)

1.3 R. Hallin, J. Lindskog, A. Marelius, J. Pihl, R. Sjödin: Physica Scripta 8, 209 (1973)

1.4 I. A. Sellin, B. Donnally, C. Y. Fan: Phys. Rev. Letters 21, 717 (1968)

1.5 I. A. Sellin, M. Brown, W. W. Smith, B. Donnally: Phys. Rev. A 2, 1189 (1970)

1.6 C. L. Cocke, B. Curnutte, J. R. MacDonald: Phys. Rev. Letters 28, 1233 (1972)

1.7 S. Bashkin: In Proceedings of 4th Intl. Conf. on Beam-Foil Spectroscopy (to be published)

1.8 S. Bashkin, P. W. Griffin, K. Jones, J. A. Leavitt, D. J. Pegg, D. Pisano, I. A. Sellin: Bull. Am. Phys. Soc. II, 20, 1452 (1975)

1.9 J. P. Buchet, M. C. Buchet-Poulizac, G. DoCao, J. Desesquelles: Nucl. Instr. and Meth. 110, 19 (1973)

1.10 J. P. Buchet, M. C. Buchet-Poulizac, A. Denis, J. Desesquelles, G. DoCao: Physical Scripta 9, 221 (1974)

1.11 M. Dufay, A. Denis, J. Desesquelles: Nucl. Instr. and Meth. 90, 85 (1970)

1.12 A. Denis, J. Desesquelles, M. Dufay: Compt. Rend. Acad. Sci. Paris 272B, 789 (1971)

1.13 J. P. Buchet, A. Denis, J. Desesquelles, M. Druetta, J. L. Subtil: In Pro ceedings of 4th Intl. Conf. on Beam-Foil Spectroscopy (to be published)

1.14 L. W. Dotchin, E. L. Chupp, D. J. Pegg: J. Chem. Phys. 59, 3960 (1973)

1.15 S. Bashkin, W. S. Bickel, H. D. Dieselman, J. B. Schroeder: J. Opt. Soc. Am. 57, 1395L (1967)

1.16 E. Norbeck, R. C. York: Nucl. Instr. and Meth. 118, 327 (1974)

1.17 S. K. Allison, J. Enevas, M. Garcia-Munoz: Phys. Rev. 120, 1266 (1960)

1.18 R. Middleton, C. T. Adams: Nucl. Instr. and Meth. 118, 329 (1974)

1.19 W. L. Wiese, M. W. Smith, B. M. Glennon: *Atomic Transition Probabilities*, NBS, NSRDS-NBS4 (Govt. Printing Offc, Washington, D.C. 1966)

1.20 D. Dietrich, J. A. Leavitt: private communication

1.21 G. Sørensen: In Proceedings of 4th Intl. Conf. on Beam-Foil Spectroscopy (to be published)

1.22 W. S. Bickel: private communication

1.23 E. L. Chupp, L. W. Dotchin, D. J. Pegg: Phys. Rev. 175, 44 (1968)

1.24 W. S. Bickel: Appl. Opt. 13, 1295 (1974)

1.25 J. O. Stoner, Jr.: private communication

1.26 G. Dearnaley: Rev. Sci. Instr. 31, 197 (1960)

1.27 B. Curnutte: private communication

1.28 J. O. Stoner, Jr.: J. Appl. Phys. 40, 707 (1969)

1.29 B. L. Henke: private communication

1.30 H. Oona: private communication

1.31 J. O. Stoner, Jr.: private communication

1.32 G. Thomas: private communication

1.33 E. M. Bernstein, L. C. McIntyre, E. Middlesworth: private communication

1.34 J. L. Yntema: Nucl. Instr. and Meth. 98, 379 (1972)

1.35 J. L. Yntema: IEEE Trans. Nucl. Sci. 19, 272 (1972)

1.36 J. L. Yntema: Nucl. Instr. and Meth. 113, 605 (1973)

1.37 S. H. Maxman: Nucl. Instr. and Meth. 50, 53 (1967)

1.38 E. T. Fairchild: Appl. Opt. 12, 2240 (1973)

1.39 L. Bridwell, L. M. Beyer, W. E. Maddox, R. C. Etherton: Nucl. Instr. and Meth. 90, 187 (1970)

1.40 P. W. Griffin, S. Bashkin, K. Jones, J. A. Leavitt, D. J. Pegg, D. Pisano, I. A. Sellin: In Proceedings of 4th Intl. Conf. on Beam-Foil Spectroscopy (to be published)

1.41 W. Lochte-Holtgreven, J. Richter: In *Plasma Diagnostics*, ed. by W. Lochte-Holtgreven (North-Holland Publ. Co., Amsterdam 1968)p.135

1.42 S. Bashkin, W. S. Bickel, D. Fink, R. K. Wangsness: Phys. Rev. Letters 15, 284 (1965)

1.43 S. Bashkin, D. Fink, P. R. Malmberg, A. B. Meinel, S. G. Tilford: J. Opt. Soc. Am. 56, 1064 (1966)

1.44 L. Brown, W. K. Ford, Jr., V. Rubin, W. Trächslin: In *Beam-Foil Spectroscopy*, ed. by S. Bashkin (Gordon and Breach, New York 1968)p.45

1.45 G. E. Assousa, L. Brown, W. K. Ford, Jr.: J. Opt. Soc. Am. 60, 1311 (1970)

1.46 G. E. Assousa, L. Brown, W. K. Ford, Jr.: Nucl. Instr. and Meth. 90, 51 (1970)

1.47 C. K. Kumar, G. E. Assousa, L. Brown, W. K. Ford, Jr.: Phys. Rev. A 7, 112 (1973)

1.48 J. A. Jordan, Jr.: In *Beam-Foil Spectroscopy*, ed. by S. Bashkin (Gordon and Breach, New York 1968)p.121

1.49 G. S. Bakken, A. C. Conrad, J. A. Jordan, Jr.: J. Phys. B 2, 1378 (1969)

1.50 A. Denis, J. Desesquelles, M. Dufay: J. Opt. Soc. Am. 59, 976 (1969)

1.51 M. Dufay: Nucl. Instr. and Meth. 90, 15 (1970)

1.52 E. H. Pinnington: Nucl. Instr. and Meth. 90, 93 (1970)

1.53 G. S. Bakken, J. A. Jordan, Jr.: Nucl. Instr. and Meth. 90, 181 (1970)

1.54 J. O. Stoner, Jr.: Appl. Opt. 9, 53 (1970)

1.55 A. Denis, O. Ceyzeriat, M. Dufay: J. Opt. Soc. Am. 60, 1186 (1970)

1.56 G. W. Carriveau, M. H. Doobov, H. J. Hay, C. J. Sofield: Nucl. Instr. and Meth. 99, 439 (1972)

1.57 L. Henke, H. J. Andrä: In Proceedings of 4th Intl. Conf. on Beam-Foil Spectroscopy (to be published)

1.58 J. O. Stoner, Jr., J. A. Leavitt: Optica Acta 20, 435 (1973)

1.59 J. O. Stoner, Jr., J. A. Leavitt: Appl. Phys. Letters 18, 477 (1971)

1.60 J. O. Stoner, Jr., J. A. Leavitt: Appl. Phys. Letters 18, 368 (1971)

1.61 K. E. Bergkvist: In Proceedings of 4th Intl. Conf. on Beam-Foil Spectroscopy (to be published)

1.62 W. S. Bickel, H. Oona, W. S. Smith, I. Martinson: Phys. Scripta 6, 71 (1972)

1.63 P. Griffin, D. Pegg, I. A. Sellin: private communication

1.64 N. Andersen, W. S. Bickel, R. Boleu, K. Jensen, E. Veje: Physica Scripta 3, 255 (1971)

1.65 P. R. Malmberg, S. Bashkin, S. G. Tilford: Phys. Rev. Letters 15, 98 (1965)

1.66 U. Fink: J. Opt. Soc. Am. 58, 937 (1968)

1.67 U. Fink: Appl. Opt. 7, 2373 (1968)

1.68 G. W. Carriveau, S. Bashkin: Nucl. Instr. and Meth. 90, 203 (1970)

1.69 I. Martinson, W. S. Bickel, A. Ölme: J. Opt. Soc. Am. 60, 1213 (1970)

1.70 S. Bashkin, G. W. Carriveau, H. J. Hay: J. Phys. B 4, L32 (1971)

1.71 T. Hadeishi, W. S. Bickel, J. D. Garcia, H. G. Berry: Phys. Rev. Letters 23, 65 (1969)

1.72 H. J. Andrä: Phys. Letters 32A, 345 (1970)

1.73 C. H. Liu, S. Bashkin, W. S. Bickel, T. Hadeishi: Phys. Rev. Letters 26, 222 (1971)

1.74 W. S. Bickel, S. Bashkin: Phys. Rev. 162, 12 (1967)

1.75 W. S. Bickel: J. Opt. Soc. Am. 58, 213 (1968)

1.76 I. A. Sellin, C. D. Moak, P. M. Griffin, J. A. Biggerstaff: Phys. Rev. 184, 56 (1969)

1.77 I. A. Sellin, C. D. Moak, P. M. Griffin, J. A. Biggerstaff: Phys. Rev. 186, 217 (1969)

1.78 S. Bashkin, G. W. Carriveau: Phys. Rev. A 1, 269 (1970)

1.79 H. J. Andrä: Nucl. Instr. and Meth. 90, 343 (1970)

1.80 H. J. Andrä: Phys. Rev. A 2, 2200 (1970)

1.81 M. J. Alguard, C. W. Drake: Nucl. Instr. and Meth. 110, 311 (1973)

1.82 E. H. Pinnington, H. G. Berry, J. Desesquelles, J. L. Subtil: Nucl. Instr. and Meth. 110, 315 (1973)

1.83 G. W. F. Drake, A. Van Wijngaarden: In Proceedings of 4th Intl. Conf. on Beam-Foil Spectroscopy (to be published)

1.84 C. H. Liu, D. A. Church: Phys. Letters 35A, 407 (1971)

1.85 D. A. Church, M. Druetta, C. H. Liu: Phys. Rev. Letters 27, 1763 (1971)

1.86 C. H. Liu, M. Druetta, D. A. Church: Phys. Letters 39A, 49 (1972)

1.87 D. A. Church, C. H. Liu: Phys. Rev. A 5, 1031 (1972)

1.88 C. H. Liu, D. A. Church: Phys. Rev. Letters 29, 1208 (1972)

1.89 D. A. Church, C. H. Liu: Physica 67, 90 (1973)

1.90 D. A. Church, C. H. Liu: Nucl. Instr. and Meth. 110, 147 (1973)

1.91 C. H. Liu, R. B. Gardiner, D. A. Church: Phys. Letters 43A, 165 (1973)

1.92 D. A. Church, C. H. Liu: Nucl. Instr. and Meth. 110, 267 (1973)

1.93 M. Gaillard, C. Carre, H. G. Berry, M. Lombardi: Nucl. Instr. and Meth. 110, 273 (1973)

1.94 J. O. Stoner, Jr., L. J. Radziemski, Jr.: Appl. Phys. Letters 21, 165 (1972)

2. Studies of Atomic Spectra by the Beam-Foil Method

Indrek Martinson

With 8 Figures

Experimental studies of atomic energy levels are usually performed by applying a suitable excitation mechanism to the atoms or ions and determining the energies and intensities of the de-excitation radiation. Besides the inherent value of obtaining atomic data, the results are often also of great theoretical interest, serving as tests of various approximations used in quantum mechanical calculations of atomic structure. The atomic data are also needed in astrophysics for identifying chemical elements and determining their abundances in the sun, stars, nebulae, comets, and interstellar matter [2.1]. Knowledge about the structure of highly-ionized atoms is also important in plasma physics, e.g., in determining impurities in thermonuclear plasmas [2.2].

The first beam-foil publications [2.3-5] emphasized the applicability of this method for studies of wavelengths and ionization states of atomic spectra. It was shown that the beam-foil technique possesses a number of attractive features, such as high purity, time resolution, possibilities of obtaining high ionization states, and the absence of perturbing interionic fields. These advantages are partly neutralized by Doppler effects, associated with the high-velocity ion beam, which make spectral studies at high resolution quite difficult. It is worth noting, however, that the excitation mechanisms in ion-foil collisions strongly populate certain types of atomic levels, e.g., inner-shell excited states and orbits with high angular momenta, which are excited to a much lesser extent in most other light sources, or, if so excited, decay by collisional rather than radiative processes. In this chapter we will first briefly discuss the characteristic features of the beam-foil light source [2.6-9]. The detailed spectroscopic results are then presented in a systematic way, and comparisons are made to theory as well as to data obtained using other light sources.

2.1 Experimental Methods

More than 60 elements have already been accelerated for beam-foil or beam-gas experiments. New information about energy levels has been obtained for many light elements ($Z<30$) whereas beam-foil studies of heavier elements have usually concentrated on lifetime determinations for known levels.

The experimental techniques have been described by several authors [2.6-10]. The accelerators used include isotope separators, Van de Graaff generators, cyclotrons and heavy-ion linear accelerators and the energy range has been approximately 10 keV - 400 MeV. Typical ion currents are 0.1 - 10 μA.

The acceleration and subsequent momentum analysis, usually in a magnetic field, should in principle guarantee a chemically and isotopically pure beam through the exciter. Contaminations may occur in several cases, however. It is well known that, e.g., a mass 28 beam may contain CO^+, Si^+ and N_2^+, a mass 32 beam O_2^+ and S^+, and a mass 51 beam Ti^+ and $C\ell O^+$. The contaminations can be studied by optical spectroscopy, either by recording the beam-foil spectra [2.11,12] or by directing the beam into a gaseous target [2.13]. By carefully adjusting the discharge conditions in the ion source it is often possible to reduce these problems. Even if the beam entering the foil contains ions of only one element, the beam-foil spectra may display impurity lines. These are due to transitions in excited atoms or ions ejected from the foil. If carbon foils are used, the spectra usually show strong C I lines at 1561, 1657, and 1930Å and C II lines at 1335 and 4267Å, as well as a number of weaker C lines [2.14-16]. Recently HALLIN et al. [2.17] found that the ejected C ions may also be in very high ionization states. Using 20 MeV F beams from a tandem Van de Graaff the authors thus observed transitions in C V and C VI in their spectra. When the vacuum in the target chamber is but moderately good, the spectra also may show impurity lines from Si I and the CH molecule [2.18]. When a gaseous target is used, the transitions in the target atoms also appear in the spectra. They can often be recognized from their narrow Doppler widths. A simple way of studying the target lines consists of using different exciter gases.

By varying the beam energy, it is possible to reach a large number of ionization states after the foil. If the ion velocity is low, neutralization in the foil is highly probable. With an incoming beam of 100 keV Li^* ions, the charge distribution after the foil is approximately 71% Li, 28% Li^+ and 1% Li^{**}. High ionization states are generally reached with the use of high energies. With 20 MeV O^+, the charge distribution after the foil is 6% O^{5+}, 40% O^{6+}, 42% O^{7+} and 12% O^{8+} [2.19]. Such charge distributions have been reviewed by BETZ [2.20,21]. It is a well-known fact that the use of a foil exciter gives much higher charge states than can be obtained with the gaseous exciter of equivalent thickness. This is due to the so-called density effect [2.22].

The foil-excited beam is usually viewed with an optical spectrometer, equipped with a photomultiplier tube at the exit slit. Some authors also use the photographic method, sometimes combined with image intensifiers [2.23,24]. Various geometrical arrangements are possible, the most frequently used being the side-on configuration in which the spectrometer views the beam at 90° to the particle velocity. Some authors have also used the end-on configuration [2.24,25] in which the light observed is emitted in the forward direction, parallel to the beam.

Examples of beam-foil spectra are shown in Figs.2.1 and 2.2.

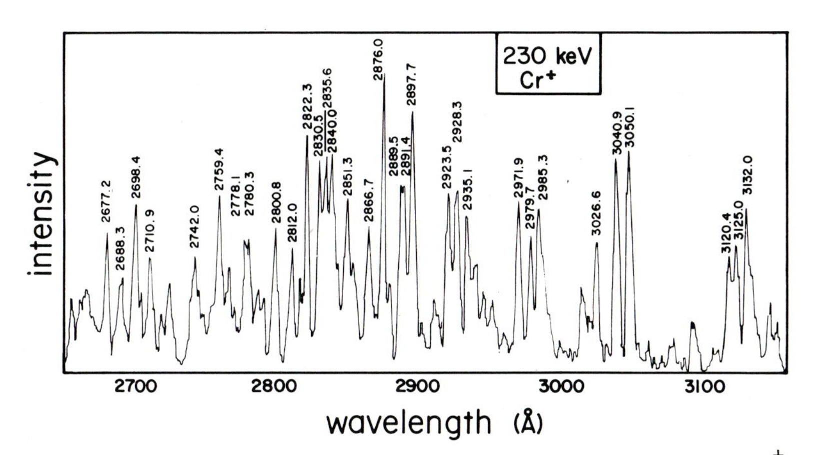

Fig.2.1 Beam-foil spectrum of Cr, measured with a 230 keV Cr^+ beam through a 5 $\mu g/cm^2$ foil. The light was dispersed with a Heath 35 cm grating monochromator. All strong lines in this region are from transitions in singly-ionized Cr, Cr II [2.160]

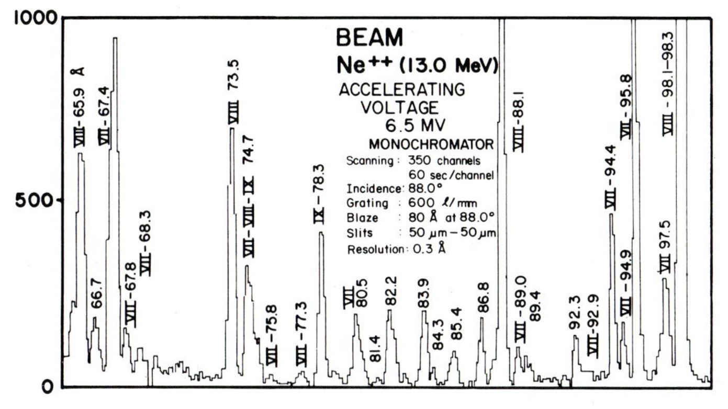

Fig.2.2 Beam-foil spectrum in the extreme ultraviolet. A 13.0 MeV Ne^{++} beam was directed through the foil and spectra were taken with a 2.2 m McPherson grazing-incidence monochromator [2.161]

The linewidths (above 2000Å) are typically a few Å, i.e., considerably larger than those obtained using, e.g., hollow-cathode light sources. Part of these linewidths is of instrumental origin. The beam-foil light source is a comparatively weak one; typical particle densities are 10^4 to $10^5/cm^3$ and thus orders of magnitude lower than in most other light sources. Relatively fast optical systems with corresponding sacrifice of spectral resolution are therefore needed. However, the major part of the line broadening is usually of a different origin. The light source consists of particles moving such that v/c is of the order of 10^{-2} and this leads to appreciable Doppler effects. When the foil-excited beam is viewed at an angle of 90°, the first-order Doppler shift vanishes, while the second-order shift, proportional to v^2/c^2, has to be taken into account in wavelength determinations. However, the Doppler broadening due to the finite acceptance angle of the spectrometer is more serious. With a typical acceptance angle of 0.05 rad and v/c of 1%, a line at 6000Å exhibits a broadening of 3Å, which makes high-resolution work very difficult.

Several methods have been used to reduce this Doppler broadening. KAY and coworkers [2.26,27] use an anamorphotic condensing system, which gives different magnifications in the vertical and horizontal planes. The foil-excited beam is viewed at 90° using two cylindrical lenses. STONER and LEAVITT [2.28-30] developed the principle of refocusing the spectrometer for a movable light source. For given values of wavelength λ and of v/c, there is an optimum setting of the exit slit position which minimizes the linewidth. An example is shown in Fig.2.3.

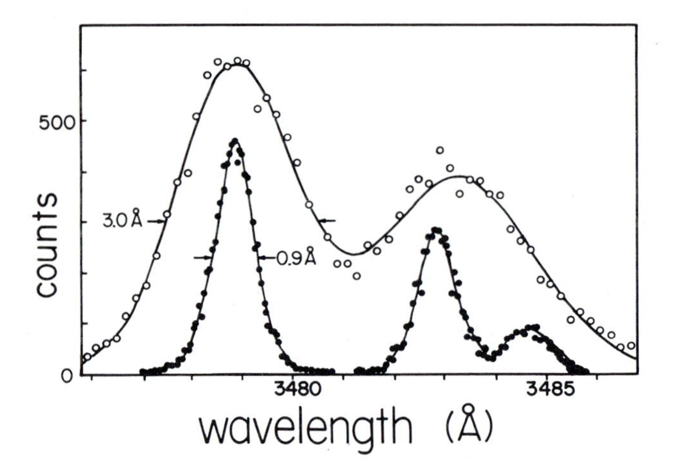

Fig.2.3 The effects of refocusing the spectrometer [2.28-30]. By optimizing the position of the exit slit for a movable light source the line widths for the N IV triplet 3s ^{3}S - 3p ^{3}P (3478.71, 3482.99 and 3484.96Å) are reduced from 3 to 0.9Å

This refocusing can also be accomplished by the use of quartz plates that change the effective position of the exit slit [2.31]. By simple optical devices it is also possible to admit only parallel light into the spectrometer [2.32]. It is also interesting to note that the line-broadenings can be substantially reduced by using the above-mentioned end-on configuration, as demonstrated by BAKKEN, CONRAD and JORDAN [2.24]. The first-order Doppler shifts affecting this technique can easily be taken into account because the beam velocity is known. Additional methods suggested for line narrowing include the use of an off-axis Fabry-Perot spectrometer which compensates for Doppler effects by a small tilt angle [2.33] and a modified Lamb-dip technique [2.34].

These techniques reduce the linewidths to below 1Å in a typical case. The remaining broadenings are usually largely due to scattering in the foil, which leads to a diverging foil-excited beam. STONER et al. [2.28-30,35] and KAY et al. [2.26,27] have made detailed studies of the linewidths due to multiple scattering processes in the foil and they have generally found good agreement with the theoretical treatments of MEYER [2.36]. By using the beam-gas technique this line broadening can be substantially reduced, as shown, e.g., by BEYER and coworkers [2.37-39]. STONER and RADZIEMSKI [2.40] also report linewidths of 0.1 - 0.2Å with this method, which even makes Zeeman studies possible. Although the linewidths have been significantly reduced, the determination of absolute wavelengths is not trivial. For example, the refocusing of the spectrometer for a movable light source will make reference lines from a stationary source appear very broad.

The wavelength analysis of a beam-foil or beam-gas spectrum may frequently show the presence of transitions which have not been reported earlier. It is a challenge to classify such lines. After a careful wavelength determination it is now necessary to determine the ionization states of the observed new transitions. Several techniques are available. Perhaps the most direct method, developed by BASHKIN and coworkers [2.41-43], consists of applying a transverse electrostatic field to the foil-excited beam. The optic axis of the spectrometer is parallel to the field direction. Depending on the field direction one observes a blue or red Doppler shift in the wavelength; the shift is proportional to the charge state. With a 4 MeV N_2^+ beam BASHKIN and LEAVITT [2.44], observed a red shift of 2.5Å for a N IV line at 1718Å when the field strength was 60 kV/cm. Another method, introduced by KAY [2.45], utilizes the variations of spectral line intensities with the beam energy. Both methods have their merits and drawbacks, as, e.g., discussed by DUFAY [2.25] and BASHKIN [2.7,9]. The electrostatic method is a rather time-consuming one and it cannot easily be applied to transitions from very short-lived levels. The energy-intensity method suffers from the fact that the excitation functions for levels within a given spectrum may not be identical.

The subsequent classification is usually accomplished by various semi-empirical techniques, used in atomic spectroscopy. These include studies of isoelectronic sequences, intensity ratios for multiplets, use of the polarization concept for nonpenetrating orbits, Ritz combination principle, etc. [2.46]. In many cases, quantum-mechanical calculations are of vital importance. For example, the classification of transitions from multiply-excited levels, often observed in beam-foil spectra, has benefited greatly from the existence of accurate theoretical calculations of energies and intensities. As will be noted later, lifetime measurements also often facilitate spectral identifications.

2.2 Results of Spectral Studies

Experiments based on beam-foil or beam-gas excitation have provided new information as to wavelengths and energy levels in a large number of elements, among them, He, Li, Be, B, C, N, O, F, Ne, Na, Mg, Aℓ, Si, P, S, Cℓ, Ar, K, Ca, Sc, Fe, Ni, Kr and Xe. Much heavier elements have also been accelerated, e.g., Pb and Bi, but the emphasis for those elements has usually been placed on lifetime measurements [2.12].

The new spectroscopic material which originates from beam-foil or beam-gas experiments may schematically be divided into two groups. The first category includes new information for atomic systems which have been incompletely studied by other techniques. Analyses of atomic spectra are complicated and time-consuming, and there are many examples of insufficiently investigated spectra, as summarized in a recent review by EDLÉN [2.47].

The second category includes a few types of energy levels which are abundantly populated in the beam-foil light source; their decay may thus be studied from strong radiative transitions. The transitions may be markedly weaker or absent in spectra observed with other light sources. Examples of such cases are states with high n and ℓ quantum numbers in multiply-ionized atoms and multiply-excited levels. The former, often called hydrogen-like levels, are susceptible to the first-order Stark effect whenever electrostatic fields are present. In spark light sources, where the particle density is rather high and the inter-ionic electric fields are appreciable, the high n,ℓ states with their comparatively long lifetime may either de-excite by collisions or decay by forbidden mechanisms. The low particle densities in the beam-foil source facilitate observation of the radiative decay of these hydrogenic levels. To produce an appreciable population of multiply-excited levels, it is usually necessary to have multiple collisions within a time shorter than the relaxation times. This is also provided in the ion-foil interaction where the fast ion, during a time interval of less than 10^{-14}s, traverses several hundred atomic layers.

In discussing the results obtained by beam-foil studies of spectra, we will group the material according to new data for (a) previously incompletely-studied systems, (b) states with relatively high n and ℓ quantum numbers, (c) displaced levels, (d) inner-shell excited levels. Such a systematization is admittedly crude. The categories

(a) to (d) are not always completely separable and the separations may therefore appear as slightly artificial. Whenever needed, modifications will be brought forth in the following discussion.

2.2.1 Previously Incompletely-Studied Systems

Early studies of atomic spectra by the beam-foil technique frequently showed transitions that were not tabulated in the literature. A good example is the work of BASHKIN et al. [2.48] who studied N and O spectra (2700-6600Å), using 0.5 - 2 MeV beams. Many lines in the spectra belonged to well-known N and O multiplets, but the authors also observed many new N and O transitions. Some of these could be identified from other transitions, since both the upper and lower terms were already known. Even today the spectra of O II and O III are incompletely known and several beam-foil lines could thus be transitions in these ions. The N work was later continued by FINK et al. [2,49], who were able to determine the ionization states for several of the new lines. Both N and O have also been studied by LEWIS et al. [2.50,51], whose results were similar to those of BASHKIN et al. [2.48]. Using a computer program which calculated all allowed combinations between known terms in ionized N and O, LEWIS et al. were able to explain several of their new lines as transitions connecting known terms in N II, N III, N V, O IV and O V.

Beam-foil studies of oxygen in the vacuum ultraviolet also brought forth new spectroscopic material. MARTINSON et al. [2.52] thus found several new transitions in the region 450-2200Å. The emphasis was placed on lifetime measurements for already known terms and therefore too little attention was paid to classifications of new spectral lines. In a beam-gas experiment, BEYER et al. [2.37-39] excited 0.4 - 2 MeV O ions in Ar gas. The spectra showed more than 50 new lines (500-1100Å) that were ascribed to transitions in O II - O V. The ionization states were determined from spectral-line intensity ratios at 0.5, 1.0 and 2.0 MeV. It is particularly valuable that the wavelengths in [2.37-39] could be determined as accurately as ±0.1Å. There clearly exists valuable material for future analyses of the level structure in oxygen, particularly O II and O III.

Neon is another element for which beam-foil studies have made interesting contributions. In an introductory study, DENIS et al. [2.53] investigated the region 2000-6000Å, using 0.5 - 3.7 MeV beams. A large fraction of the observed transitions remained unclassified, although the authors were able to suggest ionization states. In a subsequent paper [2.54], where the wavelength region was extended to the vacuum ultraviolet, several classifications were proposed. It was then possible to compare the data to independent high-resolution studies of LINDEBERG [2.55], who used a theta-pinch light source. New spectroscopic data have also been obtained for highly ionized C, N, O and Ne. Using the Orsay linear accelerator (1.15 MeV/nucleon energy), BUCHET et al. [2.56,57] have made an extensive study of the beam-foil spectra of those elements. More than 60 new transitions in these elements were observed. The classifications were often accomplished by isoelectronic comparisons, made possible by the rich experimental material. In analyzing the observed transitions which mostly belonged to the H I, He I and Li I isoelectronic sequences, advantage could also be taken of theoretical calculations for these few-electron systems.

Identification problems are generally more intricate for transitions in heavier atoms, but even here, valuable analyses based on beam-foil data have appeared. A representative example is found in a beam-foil experiment by DUFAY et al. [2.58], who were able to explain several unknown lines in the Na beam-foil spectra of BROWN et al. [2.23] as transitions in Na II. However, DUFAY et al. also reported a number of new Na transitions which were ascribed to Na III - Na V. The authors observed, e.g., two strong lines, at 2387Å and 2395Å, which they ascribed to the 3s' ^2D - 3p' ^2F multiplet of Na III, an identification that cast doubt on the previously accepted Na III level structure. However, this revision was later supported by MINNHAGEN and NIETSCHE [2.59], who made a thorough investigation of the Na III level structure, using a sliding spark light source. It is generally valuable to obtain confirmation of the beam-foil identifications from measurements with other light sources which allow work at higher wavelength resolution.

The improved experimental conditions have in recent years enabled one to make comparatively accurate wavelength determinations in beam-foil and beam-gas experiments. Detailed classifications and level structure analyses are now possible in a large number of cases, as already noted in the previous examples. From the theoretical point of view, the He I isoelectronic sequence is particularly interesting, because here we have a simple case of a quantum-mechanical problem that cannot be exactly solved. Quite elaborate procedures to solve the Schrödinger equation approximately have instead been applied. It is an interesting situation, since the theoretical accuracy forms a challenge to the experimentalists.

The energy levels of He I of the type 1s $n\ell$ ^1L and 1s $n\ell$ ^3L have been accurately known for a very long time and beam-foil work has not been able to make new contributions here, in contrast to the situation for the He I doubly-excited levels (see Section 2.2.6). Already for Li II BERRY et al. [2.60] found new transitions from ns, np, nd and nf levels with n=6-12. The theoretical accuracy should exceed the experimental one for these levels, however. The beam-foil studies of Be III have mostly concentrated on level lifetimes. The situation, from the level-structure standpoint, is more interesting for B IV. In this spectrum, experimental information has for a long time been limited to the $1s^2$ ^1S - 1s np ^1P resonance series and the 1s2s ^3S - 1s2p ^3P multiplet. In an introductory beam-foil experiment, MARTINSON et al. [2.61] obtained additional material (n=3-4 and n=3-5 transitions), whereas BUCHET and BUCHET-POULIZAC [2.62] identified a number of n=2-4 transitions in the far UV. More recently, BERRY and SUBTIL [2.63] observed many n=4-5 transitions in this spectrum. The wavelengths for S-P transitions were in good agreement with the elaborate calculations of ACCAD et al. [2.64,65], who included relativistic effects.

The latest beam-foil study (200-450Å) of this interesting spectrum, by TO et al.[2.66], was carried out simultaneously with EIDELSBERG's [2.67] investigations, in which the light source was of the plasma-focus type. It is interesting to compare the results of these two independent experiments. The beam-foil wavelength uncertainties were ±0.3Å, whereas the plasma light source yielded 5-15 times higher accuracies in the common wavelength region. The two data sets always agree to within ±0.1Å, which indicates that the error limits of TO et al. may be unnecessarily large. The latter authors further observed 15 B IV transitions that were not listed by EIDELSBERG. Several spectral series could be investigated in detail in the beam-foil experiments, among them 2p ^1P - nd ^1D (n=3-8), 2s ^3S - np ^3P (n=3-8) and 2p ^3P - nd ^3D (n=3-9), and, for the first time, term values for the highest levels were derived.

It is also interesting to study the singlet-triplet energy separations in the two-electron spectra. EDLÉN [2.46] has shown that for the S and P terms, the singlet-triplet interval (e.g., 3p ^1P - 3p ^3P) is approximately proportional to $(n^*)^{-3}$ (n^* is the effective quantum number) while it varies as $(n^*)^{-2}$ for the D terms. For B IV, this was experimentally verified by both investigations. In a recent theoretical analysis, CHANG [2.68] discusses the singlet-triplet separation and suggests that the n^*-dependence is actually more complex than the simple power laws proposed by Edlén. Additional beam-foil studies to verify these calculations are certainly called for.

The He I sequence was further investigated by PEGG et al. [2.69], who observed transitions in O VII and F VIII, using 5-22 MeV O and F beams from a tandem Van de Graaff. In the region 80-140Å, the wavelengths were determined to ±0.05Å. The majority of the lines studied have also appeared in experiments with other light sources, e.g., the theta-pinch or laser-induced plasma sources. All data are in very good agreement with calculated energies, which include relativistic corrections. In a few cases, PEGG et al. [2.69] claim better accord with theory than found by previous investigators and they assume this circumstance to reflect the cleaner experimental conditions in the beam-foil light source, i.e., absence of strong interionic fields.

Accurate wavelength measurements in the He I isoelectronic sequence also provide information about the radiative corrections to energy eigenvalues in two-electron systems. The radiative corrections to the $1s^2$ ground state have been calculated by KABIR and SALPETER [2.70] and later by ERMOLAEV [2.71,72]. These Lamb shifts are around 30 cm^{-1} for Be III and 140 cm^{-1} for C V, and they are experimentally determined by comparing the measured ionization energy to the calculated one [2.73]. However, the Lamb shifts for n=2 levels are also appreciable in the He I sequence.

Using a combination of traditional spectroscopic methods [2.74] and beam-foil work [2.75], BERRY and coworkers have been able to study these radiative corrections for the 2S and 2P levels in a few members of the He I isoelectronic sequence, namely, He I, Li II and N VI. While the Lamb shifts in one-electron atoms can be measured by resonance techniques or by mixing levels of different parity in an electrostatic field [2.76], such methods are not applicable to two-electron systems, where the ℓ-states are not degenerate. The energy separation between, for example, 2S and 2P levels, is largely due to the Coulomb interaction, to which fine and hyperfine effects as well as the Lamb shifts can be considered as small corrections. BERRY et al. obtained information about the fine and hyperfine effects from, e.g., beam-foil measurements using the zero-field quantum-beat technique [2.77]. The experimental energy-differences, corrected for hyperfine structure effects, were then compared to calculated theoretical values. The difference was interpreted as the Lamb shift, which had not been included in the theoretical description.

The foregoing procedure is illustrated by the following numerical example: For the Li II 1s2s $^3S_1$ - 1s2p $^3P_1$ transition (5485Å) BERRY and BACIS [2.74] made a very accurate wavelength study by means of a high-resolution scanning spectrometer. Using as corrections the hyperfine coupling constants A(2S) = 0.264 cm^{-1} and A(2P) = 0.240 cm^{-1} as found in the quantum-beat experiment [2.77], the authors obtained a 2s $^3S_1$ - 2p $^3P_1$ energy difference of 18226.093 ± 0.001 cm^{-1}, which was compared to the theoretical value of 18227.367 ± 0.001 cm^{-1}. The latter was given by ACCAD et al. [2.64,65], who used very elaborate wave functions with up to 1500 terms. The difference of the energies, 1.274 cm^{-1}, was interpreted by BERRY and BACIS to account for Lamb-shift corrections. It was further suggested that the radiative corrections raise 2s $^3S_1$ by 0.99 cm^{-1} and lower 2p $^3P_1$ by 0.28 cm^{-1}. This interpretation is in agreement with more recent QED calculations of ERMOLAEV [2.71,72].

New information about the level structure has also been obtained from beam-foil studies of the Na I, Mg I, and Aℓ I isoelectronic sequences. A few examples for the Mg I sequence will be discussed here. In this interesting sequence, which has two valence electrons, the levels of Mg I, Aℓ II and Si III have been extensively investigated by means of hollow-cathode and sliding-spark light sources [2.47]. The highly-ionized members of this sequence, e.g., Ar VII - Ti XI, have also been studied by means of laser-induced plasma or sliding-spark light sources [2.47,78-80]. In the intermediate region, P IV - Cℓ VI, it has been possible for beam-foil and beam-gas investigators to provide complementary data. The identifications of the spectral lines have often been based on isoelectronic comparisons. An example is shown in Fig.2.4. Here the reduced energies of three terms 3s3d ^1D, $3p^2$ ^1D and $3p^2$ ^1S (ξ is the net charge of the core, i.e., ξ=1 for the neutral atom, ξ=2 for the +1 ion, etc.) have been displayed. On the basis of such regularities, it has, e.g., been possible to identify the 3s3p ^1P - $3p^2$ ^1S transition in S V with a S-line at 924.08Å, observed

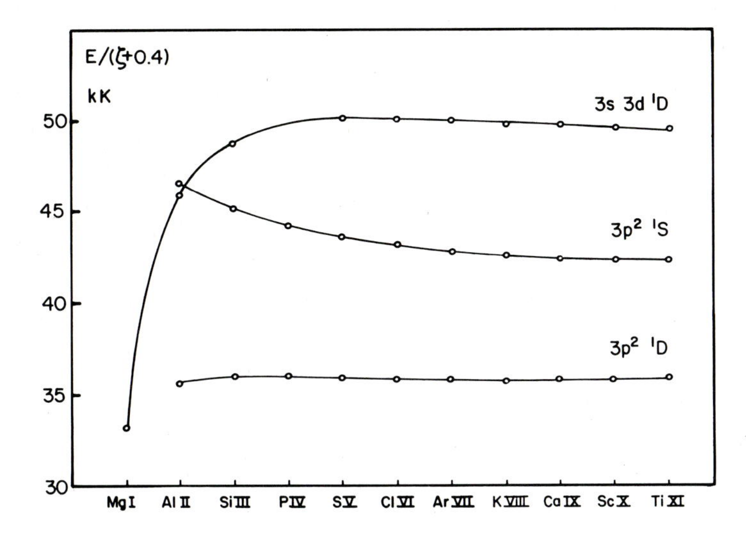

Fig.2.4 Isoelectronic comparisons for three levels, 3s3d ^1D, $3p^2\ ^1S$ and $3p^2\ ^1D$ in the Mg I isoelectronic sequence. The excitation energies are divided by $\xi + 0.4$, where ξ is the net charge of the core and 0.4 an arbitrary parameter. The values for P IV, S V and Cℓ VI originate mostly from beam-foil and beam-gas experiments [2.37-39,81-83,94,95]

by BEYER et al [2.37-39] in a beam-gas experiment. Also some of the points for P IV and Cℓ VI originate from beam-foil experiments [2.81-83].

Such classifications, particularly when they are combined with lifetime measurements, provide valuable data as to configuration-mixing effects. In a single-configuration picture the $3p^2\ ^1D$ and $3p^2\ ^1S$ term act as perturbers of the 3s nd ^1D and 3s ns ^1S Rydberg series, respectively. Good examples of new identifications in heavier systems are found in the beam-foil studies of K by POULIZAC and BUCHET [2.84] and of Ar by BUCHET et al. [2.85]. The former authors classified 15 new transitions (4s-4p and 4p-3d combinations) in K III, while BUCHET et al. [2.85] identified several n=2-2 and n=3-3 combinations in Ar XI - Ar XIV. Interesting comparisons to spectra from theta-pinch or laser-induced light sources could also be made.

2.2.2 Hydrogen-Like Levels

The term values T of atomic levels can generally be written as

$$T = R\,\xi^2/(n - \delta)^2 \quad , \tag{2.1}$$

where R is the Rydberg constant, ξ the net charge of the core, n the principal quantum number and δ the quantum defect. The quantum defects decrease with increasing angular momenta ℓ. For sufficiently high ℓ ($\ell \geq 3$ in the Li I isoelectronic sequence and $\ell \geq 4$ in the Na I isoelectronic sequence) the orbit for the excited electron does not penetrate the core. The term values can then be expressed as

$$T = T_H + \Delta_p \quad . \tag{2.2}$$

In (2.2) T_H is the relativistically-corrected hydrogenic term value

$$T_H = \frac{R\,\xi^2}{n^2}\left[1 + \frac{\alpha^2\,\xi^2}{n^2}\left(\frac{n}{\ell + 1/2} - \frac{3}{4}\right)\right] \quad , \tag{2.3}$$

and Δ_p the polarization energy. The latter can be written as

$$\Delta_p = \alpha_d\, R\langle r^{-4}\rangle + \alpha_q\, R\langle r^{-6}\rangle \quad , \tag{2.4}$$

where α_d and α_q, respectively, are the dipole and quadrupole polarizabilities of the core and $\langle r^{-4}\rangle$ and $\langle r^{-6}\rangle$ are the expectation values of $1/r^4$ and $1/r^6$ respectively, calculated with hydrogenic wave functions. These expectation values have been tabulated by EDLÉN [2.46]. Recently BOCKASTEN [2.86] has derived explicit expressions for $\langle r^s\rangle$ with $-8 \leq s \leq 5$. There is great theoretical interest in dipole and quadrupole core polarizabilities and their connections to shielding factors, which are important in studies of hyperfine structure. The calculations were reviewed by DALGARNO [2.87]. Recently VOGEL [2.88] has theoretically studied the dipole polarizabilities for the nf series in the Ne I, Na I and Ar I sequences, using also a penetration shift for the term values.

From (2.2) and (2.4) we can conclude that α_d and α_q can be determined experimentally from measured wavelengths for hydrogen-like transitions. Only two transitions are needed in principle, but the usual procedure is to determine the wavelengths of as many such transitions as possible and then perform a least-squares fit. The polarization formula can most conveniently be used for systems with one electron outside a closed core, e.g., in the Li I and Na I isoelectronic sequences. For the case of two valence electrons, perturbations may occur from other terms of the same parity and angular momentum. In the spectrum of O V, e.g., BOCKASTEN and JOHANSSON [2.89] found strong configuration interaction between the 2s6g ^{1,3}G and 2p4f ^{1,3}G levels.

A study of early spectral data by the beam-foil technique shows that the published wavelength lists often include hydrogen-like transitions. Often these lines were left unclassified. In their extensive investigation of beam-foil spectra of N and O, BASHKIN et al. [2.48] thus observed nitrogen lines at 3433 and 5289Å and oxygen lines at 3137, 3155, 3433, and 4480Å. These are hydrogen-like transitions in ions with +4 and +5 charge (N VI, O V and O VI). Several similar transitions in multiply-ionized N and

O also were seen by LEWIS et al. [2.50]. DRUETTA et al. [2.90] observed O VI transitions at 2072 (n=5-6), 3433 (n=6-7) and 5291Å (n=7-8). Note that when the core polarization effects are small the wavelengths for the hydrogen-like transitions are similar, independent of the element; they depend only on the ionization state and principal quantum numbers. For Na, BROWN et al. [2.23] list a line at 5292Å, which is the n=7-8 transition in Na VI, while a line at 4649Å, observed by BROWN et al. and DUFAY et al. [2.58], has been classified by LENNARD et al. [2.91] as the Na IV n = 5-6 transition. Using end-on observations, BAKKEN et al. [2.24] measured accurate wavelengths (to ± 0.2Å) for several lines between 4100-5200Å in the beam-foil spectrum of carbon. EDLÉN and LÖFSTRAND [2.92] subsequently showed that some of those lines are hydrogen-like transitions in C V. Using a tandem Van de Graaff accelerator, HALLIN et al. [2.93] saw hydrogen-like transition in O V - O VIII. All such transitions with Δn=1 and Δn=2 that were expected in their wavelength region (2000-5700Å) appeared in their beam-foil spectra, along with some Δn=3 combinations. Transitions from levels as high as n=14 were observed. The wavelength accuracy was unfortunately not sufficient to permit determinations of α_d and α_q.

Hydrogen-like transitions also appear in heavier systems. Figure 2.5 is a suggested

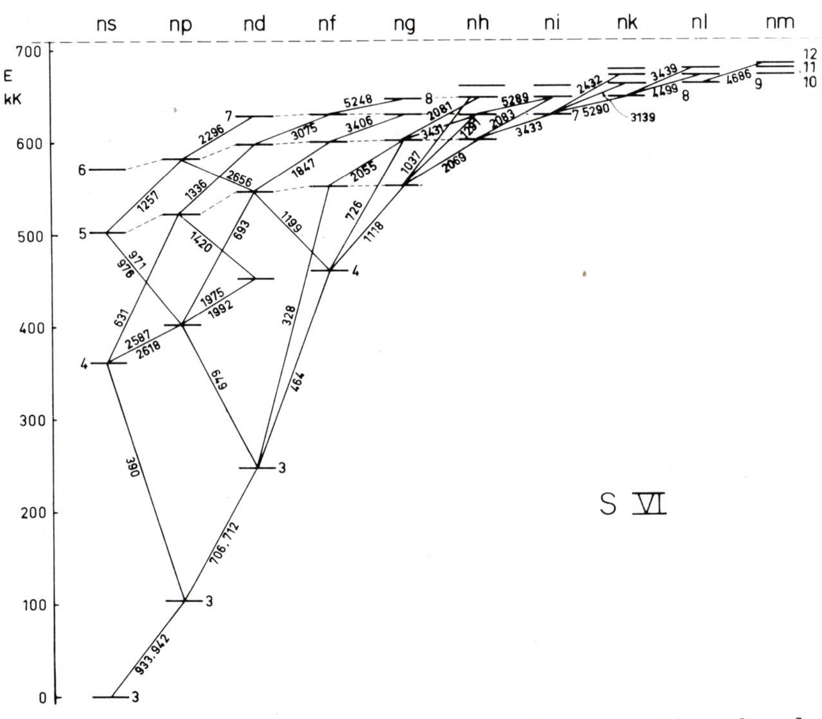

Fig.2.5 Energy level diagram for S VI. Most transitions from levels with $\ell > 3$ were identified from beam-foil experiments [2.94-96]. Transition wavelengths are in Angstrom units

energy-level diagram of S VI, based on the work of DYNEFORS and MARTINSON [2.94], who studied the beam-foil spectra of sulfur at 200-700 keV beam energies. That work followed earlier investigations of S-spectra [2.38,39,95] which had shown a strong transition at 3433Å (note the similarity to the above-mentioned examples). It is interesting that the S VI transitions between G and higher terms have only been observed by beam-foil methods. The lines were classified by means of the polarization formula. The agreement between measured and calculated wavelengths is excellent.

In beam-foil studies of Cℓ spectra, using 1-5 MeV $Cℓ^+$ beams, BASHKIN et al.[2.83,97] observed a large number of hydrogen-like transitions in Cℓ V - Cℓ X. The classifications are particularly easy for Cℓ VII, which belongs to the Na I isoelectronic sequence. Here α_d = 0.0434 and α_q = 0.0478 (in units of a_0^3 and a_0^5, respectively, where a_0 is the Bohr radius) can be deduced from beam-foil data, in excellent agreement with analyses by EKBERG [2.98].

In comparison to Cℓ VII, where the core is the closed n=2 shell, the situation is more complicated for Cℓ VI, because of the additional 3s electron, which causes a large dipole polarizability. The wavelengths of the observed transitions therefore generally display marked shifts from the purely hydrogen-like values. For example, the 6-7 transition that for negligible quantum defects occurs at 3435Å has in Cℓ VI been shifted to 3408Å, as found by BASHKIN et al.[2.83] and BHARDWAJ et al. [2.99]. A number of Cℓ VI transitions were classified [2.97,99]. The experiments were consistent with the values α_d = 1.62 a_0^3 and α_q = 9 a_0^5. However, it is clear that the two-parameter polarization formula does not describe the situation very well, because of configuration interaction. BHARDWAJ et al. thus find that the 3s nℓ terms with n=5 and 6-10 will be substantially perturbed bv 3p 4ℓ and 3p 5ℓ terms of the same parity and angular momentum.

Several hydrogen-like transitions in Cℓ VIII were also given in [2.83,97,99]. The n=7-8 fine structure was resolved into the following components (in Angstroms) 2952.3 (7f-8g), 2970.3 (7g-8h), 2973.7 (7h-8i) and 2975.9 (7i-8k) [2.99]. These and other measurements yield α_d = 0.085 a_0^3, whereas the data were not sufficiently accurate to allow determinations of α_q. There is unfortunately very little calculated material available to compare with the experimental determinations in such highly-ionized systems.

Parallel to these studies with medium-energy beams, the Cℓ-spectra were also studied by HALLIN et al. [2.93], at tandem energies (6-42 MeV). Hydrogen-like transitions in Cℓ VII - Cℓ XIV were observed: indeed, all expected transitions in the wavelength region (2000-6000Å) were present in their beam-foil spectra.

A similar study for argon was reported by BUCHET et al.[2.85] who studied the region 150-5500Å, using 46 MeV Ar beams. A large number of hydrogen-like transitions with

$\Delta n=1$ were observed in Ar X - Ar XV. In several cases, deviations from purely hydrogen-like wavelengths were reported, which may reflect appreciable dipole or quadrupole polarizabilities or series perturbations. Unfortunately no detailed analyses appear in [2.85].

Using 1-4 MeV Ni and Fe beams LENNARD et al. [2.91] observed hydrogen-like transitions belonging to Fe IV - Fe VIII and Ni IV - Ni VIII. Substantial deviations from purely hydrogen-like values were noted in a few cases which permitted estimates of the dipole polarizabilities.

These examples illustrate that the beam-foil light source is well suited for studies of hydrogen-like transitions and the subsequent deduction of dipole (and sometimes quadrupole) core polarizabilities. Several authors have discussed the excitation conditions which favor the creation of these levels by BFS. The high n,ℓ states in multiply-ionized atoms have radii of 20Å or more and such states would not survive inside the foil, where the atomic spacings are orders of magnitude smaller. Instead these states are believed by some [2.100] to be formed by electron capture at the exit side of the foil. LENNARD and COCKE [2.100] have studied this process in detail by measuring cross-sections for electron capture into high n,ℓ states of ionized Fe. The experimental data were compared to theoretical cross-sections, which were based on a surface capture model with the cross-section proportional to Z^2/n^3. The authors noted relatively good agreement with this model. However, it is interesting to note that an ℓ-dependence stronger than $2\ell+1$ could be inferred from the experiments. In comparing BFS with other light sources, BROMANDER [2.101] emphasizes that the high n,ℓ states are underpopulated in most other light sources because there the excitation is often caused by electron impact with ions in their ground states with subsequent low cross-sections. In plasma light sources, the ion densities, of the order of 10^{14} cm^{-3}, are perhaps 10^9 times higher than in a typical beam-foil experiment. As already noted, the high n,ℓ states are therefore affected by inter-ionic fields, leading to first-order Stark effects which broaden the lines and cause forbidden transitions. Collisional de-excitation may also play an important role in depopulating the lines. Already for Be III in the He I sequence, one can note that the beam-foil accuracy in wavelength determinations for hydrogen-like transitions is comparable to that obtained with a sliding-spark light source. For the Be III 4d ^3D - 5f ^3F and 4f $^{1,3}F$ - 5g $^{1,3}G$ transitions, LÖFSTRAND [2.102] obtained the wavelengths 4487.30 ± 0.1 and 4497.8 ± 0.3Å, respectively, in good agreement with the beam-foil values [2.103], 4486.8 ± 0.3 and 4497.8 ± 0.2Å.

The foregoing identifications of the hydrogen-like transitions have been based solely on spectroscopic analysis of the observed wavelengths. However, it has been pointed out by BASHKIN [2.104] that the hydrogen-like terms exhibit the same ℓ-degeneracy that is characteristic of levels in hydrogen. The mean lives of such levels are generally quite long, and are rapidly increasing functions of ℓ. On the other hand, the application of an electric field gives rise to Stark mixing of the ℓ-states; an immediate consequence is that the effective lifetime is sharply reduced. Thus one can make an independent determination of whether a spectral line indeed comes from the decay of a hydrogen-like term by observing the line intensity at a convenient point downstream from the foil, and then turning on an electric field. If the line intensity is reduced, there is excellent evidence that the mean life has been reduced, meaning that the Stark mixing of the degenerate ℓ-states has occurred, and that the parent level is hydrogen-like in character.

Tests of this nature have been carried out by CARDON and LEAVITT and their collaborators [2.105]. In that work, it was found that a number of lines exhibited strongly field-dependent intensities while other lines did not. It had previously been established, on the basis of spectroscopy alone, that the field-dependent intensities were associated with transitions from hydrogen-like states.

In experiments of the above character, it is important to take account of the possible change in optical collecting power, because the field, if transverse to the ion beam, may cause that beam to be deflected with respect to the spectrometer. This factor can be eliminated from consideration by making the field parallel to the ion beam. Here it is convenient to let the beam enter the field through a small hole in a plate perpendicular to the beam, and to cover that hole with the exciter foil. This makes the field quite uniform. Field strengths up to 50 kV/cm should be used.

2.2.3 Displaced Terms

Beginning with Be I, the atomic energy levels often contain so-called "displaced terms" which converge to a series limit higher than the first limit. In Be I, where the ground state is $2s^2\ {}^1S$ and the "normal" excited terms are $2s\ n\ell\ {}^{1,3}L$, there also occur excited terms of the type $2p\ n\ell\ {}^{1,3}L$, which converge towards the first excited term in Be II $2p\ {}^2P$. These displaced terms, which are well known in atomic spectroscopy, (see e.g., KUHN [2.106]), can often also be conveniently studied by beam-foil spectroscopy. Indeed, there are cases which indicate that this technique is very favorable in populating such levels, which are a special case of multiply-excited terms.

Very often, the displaced terms lie energetically above the first ionization limit and they are therefore likely to autoionize via the Coulomb interaction process. The operator for this electrostatic interaction is e^2/r_{ij} and the selection rules are $\Delta J=0$ (and $\Delta L = \Delta S = 0$ within the LS coupling approximation) and no parity change. Whenever these rules cannot be obeyed, radiative decays become competitive, due to the absence of allowed final states having a free electron.

As a first example, we consider displaced terms in Be I. A strong Be I line at 3455Å has until recently been labeled as the 2s2p ^1P - $2p^2$ ^1S transition, on the basis of isoelectronic comparisons. However, decay measurements for this line by the beam-foil technique [2.107,108] yielded extremely poor agreement with the theoretically-expected f-value for the Be I 2s2p ^1P - $2p^2$ ^1S transition. This fact motivated a theoretical analysis by WEISS [2.109] of the Be I spectrum. WEISS suggested that the 3455Å line instead arises from the 1s3p ^1P - 2p3p ^1P combination. This suggestion was experimentally confirmed by BERRY et al. [2.110] and HONTZEAS et al. [2.111], who also observed transitions downward from the 2p 3p ^1P term to 2s 4p ^1P and 2s 2p ^1P in their beam-foil spectra. (The two-electron jump was made possible by configuration interaction). HONTZEAS et al. [2.110] further established identical decay times for the 2s 4p ^1P - 2p 3p ^1P and 2s 3p ^1P - 2p 3p ^1P transitions. Such a lifetime measurement is often an asset in beam-foil work, being able to support spectroscopic assignments. HONTZEAS et al. also observed a large number of transitions from previously-unknown 2p np $^{1,3}P$ and 2p nd $^{1,3}D$ terms in Be I. These 2s $n\ell$ - 2p $n\ell$ transitions with $\Delta n=0$ appear as satellites to the Be II 2s-2p resonance line, and converge towards it from different sides. These beam-foil observations have later been confirmed by JOHANSSON [2.112], who made a renewed study of Be I, using a hollow-cathode light source.

Similar results have also been found in beam-foil studies of B II, isoelectronic to Be I. In this way the 2p 3ℓ levels were localized. Although these levels do not lie above the first ionization limit in this spectrum, their population by beam-foil seems to be favored when compared to other light sources, such as the sliding spark, as pointed out in the high-resolution study by ÖLME [2.113]. A particularly interesting case is a line at 2125Å which is very strong in beam-foil spectra [2.61] but did not appear with the sliding spark. BERRY and SUBTIL [2.114] have identified this line as the 2s3d ^3D - 2p3d ^3F transition. These authors also identified several transitions from the 2p 4ℓ levels in B II which were not previously known. These levels lie energetically above the 2s ^2S limit of B III, but a closer analysis shows that those 2p $n\ell$ $^{1,3}L$ levels with ℓ=L cannot autoionize with the Coulomb interaction due to the parity selection rule. BERRY [2.115] has quoted additional examples, besides Be I and B II, which show how easily the displaced levels are excited in the beam-foil source. In the B I sequence, such conclusions can be drawn for O IV, where MARTINSON et al. [2.52] were able to study several 2p 4ℓ levels in this way. Another interesting case is found for Ne VI. Here KERNAHAN et al. [2.116] identified the previously unknown $2p^3$ ^2D level by combining lifetime and wavelength measurements. The authors observed two lines, at 554 and 914Å, which they tentatively ascribed to the 2s $2p^2$ ^2D - $2p^3$ ^2D and 2s $2p^2$ ^2P - $2p^3$ ^2D combinations, respectively. The assignments were confirmed by lifetime studies, which yielded $\tau(554)$ = 0.35 ± 0.04 ns and $\tau(914)$ = 0.33 ± 0.03 ns. The $2p^3$ ^2D lifetime is further consistent with that expected from f-value systematics.

A number of examples demonstrate that displaced terms are also abundantly populated at beam-foil or beam-gas experiments with heavier projectiles. In the spectrum of S IV, BERRY et al [2.82,96] found the $3p^3\ ^2D$ and 3p 4p ^4D terms whereas BASHKIN et al. [2.83,95] have identified a number of displaced terms, e.g., 3p 3d ^3F, $3p^2\ ^1S$ and $3p^2\ ^1D$ in S V. For some of the corresponding transitions very accurate wavelengths have been available from beam-gas studies [2.37-39].

2.2.4 Multiply-Excited States

Under this category we include such levels which are formed by exciting two (or more) electrons in an atom to states with higher principal quantum number than the ground configuration. With the exception of negative ions, e.g., H^-, the simplest example of such a case is He I, where both electrons can be excited from the $1s^2\ ^1S$ ground state. The lowest doubly-excited level, $2s^2\ ^1S$, is situated 57.9 eV above the ground state and 33.4 eV above the He I ionization limit (He II ground state). Additional doubly-excited He I configurations of the type 2s $n\ell$ and 2p $n\ell$ lie in the interval between 57.9 and 65.4 eV above the $1s^2$ ground state. Early spectroscopic studies of He revealed lines at 320 and 309Å [2.117,118] which subsequently were ascribed to radiative transitions from the doubly-excited He I system to the "normal" 1s $n\ell$ levels. Theoretical analyses by WU [2.119] also showed that the doubly-excited levels of the type 2p np $^{1,3}P$, 2p nd $^{1,3}D$, i.e., levels with ℓ=L, are metastable against autoionization via the electrostatic interaction. The 320Å-line was classified as the 1s 2p ^3P-$2p^2\ ^3P$ transition [2.119].

The doubly-excited system of He I has been the subject of extensive experimental and theoretical studies. For a thorough discussion we refer to the review article by BERRY [2.115].

Much interest in this system came in connection with a study of the doubly-excited He-levels by absorption techniques. Using synchrotron radiation as a continuous light source, MADDEN and CODLING [2.120] excited several ^1P levels. Such levels also appear as resonances in scattering experiments. These techniques permit studies of strongly-autoionizing levels (typical decay probabilities of 10^{14} - 10^{15} s^{-1}).

By means of beam-foil experiments it has been possible to complement such studies and investigate levels that are metastable against autoionization. The first beam-foil study [2.121] of doubly-excited He showed six transitions in the vicinity of the He II resonance line (303.8Å) that were assigned to the He I doubly-excited system (we use the notation He I^{**} for such levels). Later the work was refined by DROUIN and KNYSTAUTAS [2.122].

Figure 2.6 shows an example of their results. All the He I lines in the figure are

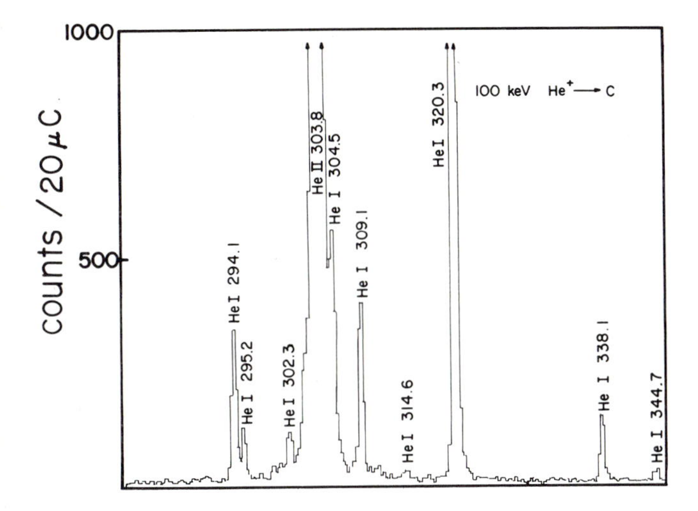

Fig. 2.6 Beam-foil spectrum of He in the vacuum ultraviolet [2.122], observed with a 2.2 m grazing-incidence monochromator. All He I lines in this region are due to transitions from the He I^{**} levels

transitions from 2p $n\ell$ levels in He I into 1s $n'\ell'$ levels. (These lines can therefore be considered as satellites to the 1s-2p transition in He II). In addition to these levels, BERRY et al. [2.123] have also observed transitions connecting the doubly-excited levels. The beam-foil spectra of He between 2200 and 3500Å showed 12 transitions that were assigned to He I^{**}.

These detailed classifications have been made possible by theoretical calculations of level energies and lifetimes for this doubly-excited system. The calculations have been based on several approaches, e.g., the variational method [2,124,125] and 1/Z expansion procedure [2.126,127]. Correlation effects have been included by, for example, MACEK [2.128] and SINANOĞLU and coworkers [2.129,130].

On the basis of theory and beam-foil experiments it is now possible to construct a detailed level diagram for doubly-excited He I, as shown in Fig.2.7.

In Fig.2.7, all transitions observed in beam-foil spectra are indicated by wavelength. BERRY et al. [2.123] found additional weak lines that were ascribed to transitions from higher-lying 2p nd $^3D^o$ states, however. The ^1S and ^3S states mainly decay by autoionization. Also the $^{1,3}P^o$ states are coupled to adjacent continua. These states show the effects of strong configuration interactions. The configurations of the

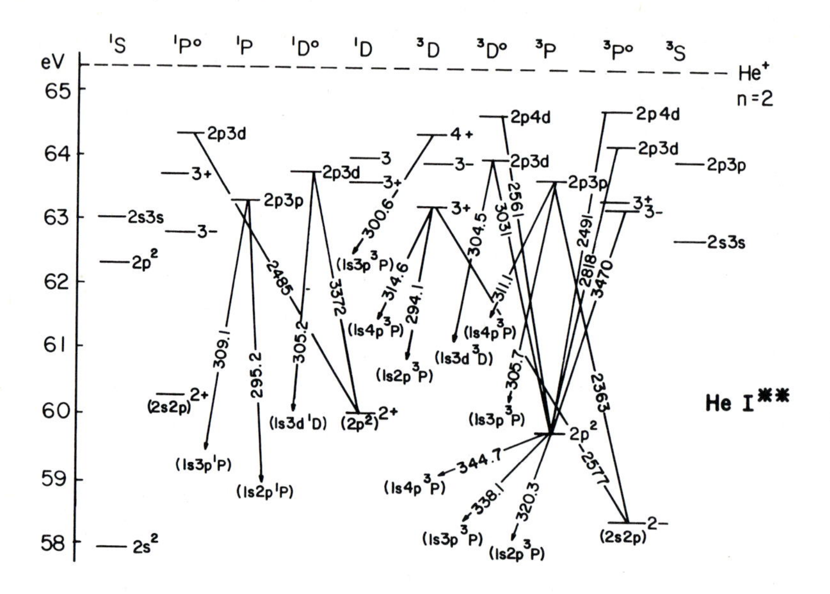

Fig.2.7 Energy level diagram for doubly-excited helium. All transitions seen in beam-foil spectra [2.121-123] are indicated by wavelength, in Angstrom units. Only the lines at 309, 320 and 3013Å have been observed with other light sources

type 2s np, 2p ns and 2p nd, all of which will give rise to $^1P^o$ and $^3P^o$ terms, converge to the same limit (n=2 in He II) and this explains qualitatively the strong configuration mixing which leads to new eigenstates of the + and - type. The + states are strongly autoionizing whereas the - states have much longer lifetimes, and their de-excitation by photon emission is sometimes observable. Similar configuration-mixing also affects the D-states, which are mixtures of 2s nd and 2p np configurations. Of all the photon transitions shown in Fig.2.7, only the lines at 309 and 320Å have been observed with other light sources than BFS. In some cases the classifications were facilitated by lifetime measurements and comparisons with theoretical decay probabilities. However, theory shows that the levels have very similar lifetimes, typically 0.1 ns, largely caused by the far UV transitions in the 300Å region. Additional uncertainties in estimating lifetimes are caused by autoionization processes, the probabilities of which are often more difficult to calculate than those of the radiative decays. BERRY and coworkers [2.123] have here found an interesting way of measuring autoionization probabilities. Certain He I** lines in their beam-foil spectra appeared broader than adjacent lines and this fact was related to very short lifetimes of the lower levels of the corresponding transitions.

The 2577Å line (2s2p ^3P- 3+ ^3D) showed such an effect, and the line profile was also approximately Lorentzian. The width corresponded to a 2s 2p ^3P autoionization lifetime of 7.0 x 0.5 x 10^{-14}s, in good agreement with the value 6.2 x 10^{-14}s found theoretically by BURKE and MCVICAR [2.131].

We have discussed the He I** system at some length because this is the most fundamental of all doubly-excited systems. The two-electron system has also been studied for higher Z. BUCHET et al. [2.132] observed the Li II 1s 2p ^3P - $2p^2$ ^3P transition at 140Å (the corresponding He I line lies at 320Å), together with a number of additional Li II lines from the doubly-excited 2p 3p and 2p 3d configurations. The classifications in Li II demand theoretical calculations of energies and probabilities for radiative and non-radiative transitions. The wavelength accuracy in BFS does not usually permit resolution of all possible satellites, ten of which may sometimes occur within 1Å. In comparison to He I, the Li II doubly-excited system more strongly needs additional experimental studies. At present, several transitions have only been tentatively identified. BERRY et al. [2.60] and SCHLAGHECK et al. [2.133] both suggest observation of the Li II 2s 2p ^3P - $2p^2$ ^3P transition, but unfortunately at different wavelengths, 5510 ± 5 and 5539 ± 2Å, respectively. Both lines show a broadening which might reflect the autoionization rate of a lower level. No beam-foil studies of doubly-excited Be III have been reported. In B IV two beam-foil investigations [2.134] have shown transitions from 2p 3p and 2p 3d levels. The identifications were here based on comparisons with the 1/Z calculations of DRAKE and DALGARNO [2.126] and DOYLE et al. [2.127]. Investigations of systems with even higher Z demand spectroscopic studies with X-ray spectrometers, e.g., of the bent-crystal type. Here MATTHEWS et al. [2.135] observed several transitions from doubly-excited levels in helium-like oxygen, O VII, as satellites to the O VIII Ly-α line at 18.969Å. It is interesting to note that such satellite lines in two- and three-electron spectra were already in 1939 studied by EDLÉN and TYRÉN [2.136]. Transitions from such doubly-excited levels have also been observed in the sun's spectrum, using rocket-borne spectrographs [2.137,138]. Laboratory observations, e.g., using the beam-foil method, serve as valuable tests of the astronomical observations.

Doubly-excited levels in the Li I sequence involve the promotion of one of the 1s electrons to orbits with higher quantum numbers. In Li I, where the ionization energy is 5.4 eV, the lowest doubly-excited level, 1s 2s 2p ^4P, is situated at 57.8 eV. The inner-shell excited configurations in Li I give rise to doublet and quartet terms; the latter cannot autoionize by the Coulomb mechanism because of the $\Delta S=0$ rule. This statement is valid for all quartets lying between 57.8 eV (1s 2s 2p ^4P) and 66.6 eV (the Li II 1s 2p ^3P limit), since here the continua do not contain quartet states. The quartets, such as 1s $2p^2$ ^4P, 1s 2s 3d ^4D, instead make radiative transitions to the metastable 1s 2s 2p ^4P term which autoionizes via the spin-orbit or spin-spin interactions. A few such radiative transitions have been known for quite some time [2.139-141] but they were usually left unclassified until theory [2.142,143] explained their origin. In the early beam-foil studies of Li-spectra, BICKEL et al. [2.144] and BUCHET et al. [2.145] were able to show that the earlier observed Li I** transitions (2337, 2934 and 3714Å) were quite prominent in beam-foil spectra. Evidence of additional, previously-unknown, lines belonging to the Li I quartet system was also found in those two papers, which were the first to demonstrate the superiority of the beam-foil method in studying weakly-autoionizing ("exactly quantized", according to HOLØIEN's [2.124] definition) inner-shell excited levels.

Additional beam-foil studies of the Li I quartet system have followed and the latest results are summarized by BERRY et al. [2.60]. More than 20 transitions between the quartet terms have been classified. For this work, the calculated energies of HOLØIEN and GELTMAN [2.143] and WEISS [2.146] have been essential. The theoretical energies of the 1s 2s ns ^4S and 1s 2p np ^4P series seem to be very reliable, whereas difficulties arise in comparing theory and experiments for the 1s 2s np ^4P levels, probably because of the configuration mixing with 1s 2p ns ^4P levels. Here the situation is still unclear and additional beam-foil measurements, preferably at highest possible resolution, are necessary before the question can be solved. In contrast, WEISS' calculations for the ^4D states (1s 2s nd and 1s 2p nd) are in good accord with beam-foil data. In analyzing the doubly-excited systems in the Li I sequence, it is essential to have access to a very reliable energy for the lowest quartet term, 1s 2s 2p ^4P. The energies of the series limits 1s 2s ^3S and 1s 2p ^3P in Li II are accurately known from spectroscopic data, and a reliable 1s 2s 2p ^4P energy is therefore essential for defining the region of quartet levels and analyzing the quantum defects for these levels. BERRY et al. [2.60] have shown that the theoretical 1s 2s 2p ^4P energies can be experimentally checked by applying the Ritz formula to transitions from ^4D levels observed in beam-foil spectra. It is finally interesting to note that certain inner-shell excited doublet terms, e.g., 1s 2p np ^2D and 1s 2p nd ^2D (n=3,4 ...) also decay by radiative transitions in the Li I system.

The Be II case has been studied by HONTZEAS et al. [2.111]. The energy level diagram obtained from these beam-foil studies is shown in Fig.2.8. The transitions between ^4S and ^4P levels were identified using the calculated values of HOLØIEN and GELTMAN [2.143]. The 2s ^3S and 2p ^3P are known for Be III and by assuming that the ^4D levels have quantum defects of 0.1 or less, the latter can be identified. It is interesting to note that none of the lines shown in Fig.2.8 has been reported in the spark spectra of Be.

Transitions between doubly-excited quartet terms have also been observed in beam-foil studies of multiply-ionized members of the Li I sequence. In the case of B III, several authors [2.61,62,66] report transitions from a few ^4S and ^4D terms. More extensive material is now available for C IV and N V and in both of these spectra, radiative transitions (in the 100-400Å range) from about ten ^4S, ^4P and ^4D terms have been

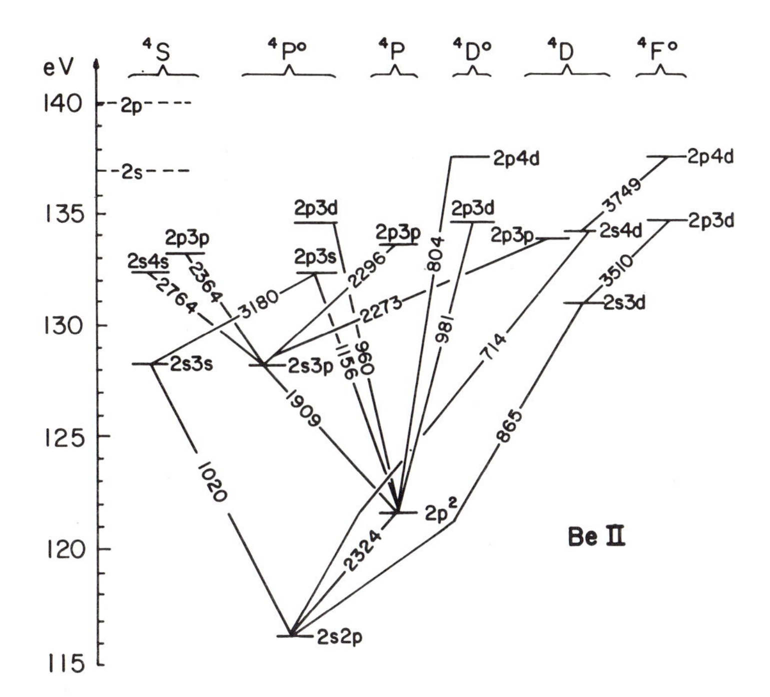

Fig.2.8 Energy level diagram for doubly-excited Be II (1s 2s nℓ and 1s 2p nℓ ^4L configurations). All lines have been observed in beam-foil spectra [2.111], while none of the transitions appear with the spark light source. Transition wavelengths are given in Angstrom units

established [2.147,148]. While these experiments were performed with 2 MV Van de Graaffs, higher beam energies become necessary for further exploration of the Li I sequence. Using a tandem Van de Graaff, MATTHEWS et al. [2.135] were able to extend the work to O VI, where they obtained additional evidence for quartet levels in this sequence. In addition, SELLIN et al. [2.149] have observed autoionization electrons from such quartets in O VI and higher spectra. The autoionization increases in importance as Z increases, because of the strong Z-dependence of the relativistic autoionization processes (spin-orbit, spin-other orbit and spin-spin mechanisms).

Doubly-excited configurations have also been studied in the Na I isoelectronic sequence. Here one of the 2p electrons is promoted to higher orbit and the configurations $2p^5$ 3s ns and $2p^5$ 3p nℓ which lead to quartet terms may be excited. In a beam-foil study of Na BERRY et al. [2.150] observed a number of satellites surrounding the Na II resonance lines $2p^6\ ^1S$ - $2p^5$ 3s ^1P at 372.07Å and $2p^6\ ^1S$ - $2p^5$ 3s ^3P at 376.38Å. These satellites were interpreted as 2p-3s transitions in doubly-excited Na I. The authors further found transitions between doubly-excited Na I quartet terms, e.g., the 3s 3p $^4D_{7/2}$ - 3s 3d $^4F_{9/2}$ transition at 3882.8Å; this line had not been observed previously. In the Na I sequence there is a strong mixing between doubly-excited doublet and quartet terms, explainable by the open 2p shell. The doublets are apt to autoionize quickly, preventing observation of radiative decays of the doubly-excited levels. However this mixing does not occur for the quartet levels with J quantum numbers higher than is possible for states in the doublet continuum. Such beam-foil work complements photo-absorption measurements, in which doubly-excited autoionizing Na I doublet levels can be observed as resonances in absorption spectra [2.151,152].

Beam-foil data for doubly-excited Mg II were reported by LUNDIN et al. [2.153]. The authors observed a line at 3480Å which does not fit into the well-known energy-level diagrams of Mg I - Mg III. By studying the intensity variation of this line with beam energy, it was found that the line showed a behavior lying between those for known Mg II and Mg III transitions and this suggests the previously-unexplored Mg II doubly-excited quartet system as a possible origin. BERRY et al. [2.150] have tentatively classified this Mg line as the Mg II 3s 3p $^4P_{7/2}$ - 3s 3d $^4F_{9/2}$ transition.

The existence of such doubly-excited levels in the Na I sequence has also been found for higher members. PEGG et al. [2.154] thus found evidence for core-excited levels in Cℓ VII.

In a recent study of the beam-foil spectrum of Ca, EMMOTH et al. [2.155] observed three previously-unlisted spectral lines in the region 3800-4000Å. Excitation functions here suggested that these lines are transitions in the doubly-excited quartet levels of Ca II. HANSEN [2.156] has calculated the energies for levels belonging to the $3p^5$ 3d 4s configuration of Ca II and he also found much less doublet-quartet mixing than in the Na I sequence. The beam-foil study of [2.155] has probably also excited the $3p^5$ 3d 4p levels of Ca II, the radiative decay of which to $3p^5$ 3d 4s was observed.

Several authors [2.157-159] have discussed the fact that beam-foil source profusely populates doubly-excited levels. It is believed that the high collision rate within a time much shorter than the relaxation times plays a decisive role. VEJE and coworkers [2.157-159] have measured excitation functions for doubly-excited levels in Li I and Be II and they find that the Li I** and Be II** levels show intensity-energy variations similar to Li II and Be III lines, respectively. A closer analysis yields a simple theoretical model (based on probability calculus) according to which the Li I** excitation functions show a behavior proportional to the product of Li I and Li II excitation functions, whereas the Be II** functions vary with energy in a way similar to the product of Be II and Be III functions.

Acknowledgments. I am grateful to Professors S. Bashkin, W. H. Smith, and J. O. Stoner, Jr., for critical reading of the manuscript and valuable suggestions.

References

2.1 G. L. Withbroe, Menzel Symposium on solar physics, atomic spectra and gaseous nebulae, NBS Spec. publ. 353, ed. by K. B. Gebbie (U.S. Govt. Printing Office, Washington, D.C. 1971)p.127

2.2 E. Hinnow: J. Nucl. Materials 53, 9 (1974)

2.3 L. Kay: Phys. Letters 5, 36 (1963)

2.4 S. Bashkin: Nucl. Instr. and Meth. 28, 88 (1964)

2.5 S. Bashkin, A. B. Meinel: Astrophys. J. 139, 413 (1964)

2.6 W. S. Bickel: Appl. Opt. 6, 1309 (1967)
More detailed discussions of the properties of the beam-foil light source can be found in several review articles, e.g. [2.6-9]

2.7 S. Bashkin: Appl. Opt. 7, 2341 (1968)

2.8 H. G. Berry, J. Bromander, R. Buchta: Nucl. Instr. and Meth. 90, 269 (1970)

2.9 S. Bashkin: In *Progress in Optics*, Vol XII, ed. by E. Wolf (North-Holland Publ. Co., Amsterdam 1974)p.289

2.10 I. Martinson, A. Gaupp: Phys. Reports 15C, 113 (1974)

2.11 H. G. Berry, W. S. Bickel, S. Bashkin, J. Desesquelles, R. M. Schectman: J. Opt. Soc. Am. 61, 947 (1971)

2.12 G. Sørensen: *Experimental Studies of Atomic Transitions* (University of Aarhus 1973)

2.13 S. Doorn, C. Foster, T. Hoogkamer, H. Roukens, F. Saris: Nucl. Instr. and Meth. 120, 371 (1974)

2.14 H. G. Berry, I. Martinson, J. Bromander: Phys. Letters 31A, 521 (1970)

2.15 J. O. Stoner, Jr., L. J. Radziemski, Jr.: J. Opt. Soc. Am. 60, 1108 (1970)

2.16 S. Bashkin: Nucl. Instr. and Meth. 90, 3 (1970)

2.17 R. Hallin, A. Marelius, R. Sjödin, J. Pihl, J. Lindskog: (to be published)

2.18 N. Andersen, G. W. Carriveau, K. Jensen, E. Veje: Phys. Letters 35A, 19 (1971)

2.19 J. B. Marion, F. C. Young: *Nuclear Reaction Analysis* (North-Holland Publ. Co, Amsterdam 1968)

2.20 H. D. Betz: Revs. Mod. Phys. 44, 465 (1972)

2.21 A. B. Wittkower, H. D. Betz: Atomic Data 5, 113 (1973)

2.22 H. D. Betz, L. Grodzins: Phys. Rev. Letters 25, 211 (1970)

2.23 L. Brown, K. Ford, V. Rubin, W. Trächslin, W. Brandt: In *Beam-Foil Spectroscopy*, ed. by S. Bashkin (Gordon and Breach, New York 1968)p.45

2.24 G. S. Bakken, A. C. Conrad, J. A. Jordan, Jr.: J. Phys. B 2, 1378 (1969)

2.25 M. Dufay: Nucl. Instr. and Meth. 90, 15 (1970)

2.26 L. Kay, B. Lightfoot: Nucl. Instr. and Meth. 90, 289 (1970)

2.27 L. Kay, B. Lightfoot, G. Harding: Nucl. Instr. and Meth. 110, 435 (1973)

2.28 J. O. Stoner, Jr., J. A. Leavitt: Appl. Phys. Letters 18, 477 (1971)

2.29 J. A. Leavitt, J. W. Robson, J. O. Stoner, Jr.: Nucl. Instr. and Meth. 110, 423 (1973)

2.30 J. O. Stoner, Jr., J. A. Leavitt: Optica Acta 20, 435 (1973)

2.31 M. Druetta: unpublished work, quoted by BERRY and SUBTIL [2.63]

2.32 G. W. Carriveau, M. H. Doobov, H. J. Hay, C. J. Sofield: Nucl. Instr. and Meth. 99, 439 (1972)

2.33 F. L. Roesler, J. O. Stoner, Jr.: Nucl. Instr. and Meth. 110, 465 (1973)

2.34 W. Chow, A. D. Maio, M. O. Scully: Nucl. Instr. and Meth. 110, 469 (1973)

2.35 J. O. Stoner, Jr., L. J. Radziemski, Jr.: Nucl. Instr. and Meth. 90, 275 (1970)

2.36 L. Meyer: Phys. Stat. Sol. 44B, 253 (1971)

2.37 L. Bridwell, L. M. Beyer, W. E. Maddox, R. C. Etherton: Nucl. Instr. and Meth. 90, 187 (1970)

2.38 L. M. Beyer, W. E. Maddox, L. B. Bridwell, D. D. Duncan, L. L. Bingham, J. C. Asbell: Nucl. Instr. and Meth. 110, 61 (1973)

2.39 L. M. Beyer, W. E. Maddox, L. B. Bridwell: J. Opt. Soc. Am. 63, 365 (1973)

2.40 J. O. Stoner, Jr., L. J. Radziemski, Jr.: Appl. Phys. Letters 21, 165 (1972)

2.41 P. R. Malmberg, S. Bashkin, S. G. Tilford: Phys. Rev. Letters 15, 98 (1965)

2.42 U. Fink: Appl Opt. 7, 2373 (1968)

2.43 G. W. Carriveau, S. Bashkin: Nucl. Instr. and Meth. 90, 203 (1970)

2.44 S. Bashkin, J. A. Leavitt: private communication

2.45 L. Kay: Proc Phys. Soc. 85, 1963 (1965)

2.46 B. Edlén: In *Handbuch der Physik*, Vol.27, ed. by S. Flügge (Springer, Berlin 1964)p.80

2.47 B. Edlén: Physica Scripta 7, 93 (1973)

2.48 S. Bashkin, D. Fink, P. R. Malmberg, A. B. Meinel, S. G. Tilford: J. Opt. Soc. Am. 56, 1064 (1966)

2.49 U. Fink, G. N. McIntire, S. Bashkin: J. Opt. Soc. Am. 58, 213 (1968)

2.50 M. R. Lewis, T. Marshall, E. H. Carnevale, F. S. Zimnoch, G. W. Wares: Phys. Rev. 164, 94 (1967)

2.51 M. R. Lewis, F. S. Zimnoch, G. W. Wares: Phys. Rev. 178, 49 (1969)

2.52 I. Martinson, H. G. Berry, W. S. Bickel, H. Oona: J. Opt. Soc. Am. 61, 519 (1971)

2.53 A. Denis, J. Desesquelles, M. Dufay: J. Opt. Soc. Am. 59, 976 (1969)

2.54 A. Denis, P. Ceyzeriat, M. Dufay: J. Opt. Soc. Am. 60, 1186 (1970)

2.55 S. Lindeberg: private communication

2.56 J. P. Buchet, M. C. Buchet-Poulizac, G. DoCao, J. Desesquelles: Nucl. Instr. and Meth. 110, 19 (1973)

2.57 M. Dufay, A. Denis, J. Desesquelles: Nucl. Instr. and Meth. 90, 85 (1970)

2.58 M. Dufay, M. Gaillard, M. Carre: Phys. Rev. A 3, 1367 (1971)

2.59 L. Minnhagen, H. Nietsche: Physica Scripta 5, 237 (1972) and (to be published)

2.60 H. G. Berry, E. H. Pinnington, J. L. Subtil: J. Opt. Soc. Am. 62, 767 (1972)

2.61 I. Martinson, W. S. Bickel, A. Ölme: J. Opt. Soc. Am. 60, 1213 (1970)

2.62 J. P. Buchet, M. C. Buchet-Poulizac: J. Opt. Soc. Am. 63, 243 (1973)

2.63 H. G. Berry, J. L. Subtil: Physica Scripta 9, 217 (1974)

2.64 Y. Accad, C. L. Pekeris, B. Schiff: Phys. Rev. A 4, 516 (1971)

2.65 B. Schiff, Y. Accad, C. L. Pekeris: Phys. Rev. A 8, 2272 (1973)

2.66 K. X. To, E. J. Knystautas, R. Drouin: Can J. Spectrosc. 19, 72 (1974)

2.67 M. Eidelsberg: J. Phys. Z 7, 1476 (1974)

2.68 T. N. Chang: J. Phys. B 7, L108 (1974)

2.69 D. J. Pegg, P. M. Griffin, H. Haselton, R. Laubert, J. R. Mowat, R. S. Peterson, I. A. Sellin: Phys. Rev. A 10, 745 (1974)

2.70 P. K. Kabir, E. E. Salpeter: Phys. Rev. 108, 1256 (1957)

2.71 A. M. Ermolaev: Phys. Rev. A 8, 1651 (1973)

2.72 A. M. Ermolaev, M. Jones: J. Phys. B 7, 199 (1974)

2.73 B. Edlén: Arkiv Fysik 4, 441 (1952)

2.74 H. G. Berry, R. Bacis: Phys. Rev. A 8, 36 (1973)

2.75 H. G. Berry, R. M. Schectman: Phys. Rev. A 9, 2345 (1974)

2.76 M. Leventhal: Nucl. Instr. and Meth. 110, 343 (1973)

2.77 H. G. Berry, J. L. Subtil, E. H. Pinnington, H. J. Andrä, W. Wittmann, A. Gaupp: Phys. Rev. A 7, 1609 (1973)

2.78 J. O. Ekberg: Physica Scripta 4, 101 (1971)

2.79 B. C. Fawcett: J. Phys. B 3, 1732 (1970)

2.80 B. C. Fawcett, R. D. Cowan, R. W. Hayes: J. Phys. B 5, 2143 (1972)

2.81 L. J. Curtis, I. Martinson, R. Buchta: Physica Scripta 3, 197 (1971)

2.82 H. G. Berry, R. M. Schectman, I. Martinson, W. S. Bickel, S. Bashkin: J. Opt. Soc. Am. 60, 335 (1970)

2.83 S. Bashkin, I. Martinson: J. Opt. Soc. Am. 61, 1686 (1971)

2.84 M. C. Poulizac, J. P. Buchet: Compt. Rend. Acad. Sci. Paris 274B, 699 (1972)

2.85 J. P. Buchet, M. C. Buchet-Poulizac, A. Denis, J. Desesquelles, G. DoCao: Physica Scripta 9, 221 (1974)

2.86 K. Bockasten: Phys. Rev. A 9, 1087 (1974)

2.87 A. Dalgarno: Adv. in Phys. 11, 281 (1962)

2.88 K. Vogel: Nucl. Instr. and Meth. 110, 241 (1973)

2.89 K. Bockasten, K. B. Johansson: Arkiv Fysik 38, 563 (1968)

2.90 M. Druetta, P. Ceyzeriat, M. C. Poulizac: Compt. Rend. Acad. Sc. Paris 271B, 846 (1970)

2.91 W. N. Lennard, R. M. Sills, W. Whaling: Phys. Rev. A 6, 884 (1972)

2.92 B. Edlén, B. Löfstrand: J. Phys. B 3, 1380 (1970)

2.93 R. Hallin, J. Lindskog, A. Marelius, J. Pihl, R. Sjödin: Physica Scripta 8, 209 (1973)

2.94 B. I. Dynefors, I. Martinson (to be published)

2.95 S. Bashkin, W. S. Bickel, B. Curnutte: J. Opt. Soc. Am. 59, 879 (1969)

2.96 H. G. Berry: J. Opt. Soc. Am. 61, 983 (1971)

2.97 S. Bashkin, J. Bromander, J. A. Leavitt, I. Martinson: Physica Scripta 8, 285 (1974)

2.98 J. O. Ekberg: unpublished material

2.99 S. N. Bhardwaj, H. G. Berry, T. Mossberg: Physica Scripta 9, 331 (1974)

2.100 W. N. Lennard, C. L. Cocke: Nucl. Instr. and Meth. 110, 137 (1973)

2.101 J. Bromander: Nucl. Instr. and Meth. 110, 11 (1973)

2.102 B. Löfstrand: Physica Scripta 8, 57 (1973)

2.103 I. Martinson: unpublished material

2.104 S. Bashkin: private communication

2.105 S. Bashkin: In Proceedings of 4th Intl. Conf. on Beam-Foil Spectroscopy 1975 (to be published)

2.106 H. G. Kuhn: *Atomic Spectra* (Longmans, London 1968)

2.107 I. Bergström, J. Bromander, R. Buchta, L. Lundin, I. Martinson: Phys. Letters 28A, 721 (1969)

2.108 T. Andersen, K. A. Jessen, G. Sørensen: Phys. Rev. 188, 76 (1969)

2.109 A. W. Weiss: Phys. Rev. A 6, 1261 (1972)

2.110 H. G. Berry, J. Bromander, I. Martinson, R. Buchta: Physica Scripta 3, 63 (1971)

2.111 S. Hontzeas, I. Martinson, P. Erman, R. Buchta: Physica Scripta 6, 55 (1972)

2.112 L. Johansson: Physica Scripta 10, 236 (1974)

2.113 A. Ölme: Physica Scripta 1, 256 (1970)

2.114 H. G. Berry, J. L. Subtil: Physica Scripta 9, 217 (1974)

2.115 H. G. Berry, Physica Scripta (to be published)

2.116 J. A. Kernahan, A. Denis, R. Drouin: Physica Scripta 4

2.117 K. T. Compton, J. C. Boyce: J. Franklin Inst. 205, 497 (1928)

2.118 P. G. Kruger: Phys. Rev. 36, 855 (1930)

2.119 T. Y. Wu: Phys. Rev. 66, 291 (1944)

2.120 R. P. Madden, K. Codling: Astrophys. J. 141, 364 (1965)

2.121 H. G. Berry, I. Martinson, L. J. Curtis, L. Lundin: Phys. Rev. A 3, 1934 (1971)

2.122 E. J. Knystautas, R. Drouin: Nucl. Instr. and Meth. 110, 95 (1973)

2.123 H. G. Berry, J. Desesquelles, M. Dufay: Phys. Rev. A 6, 600 (1972)
H. G. Berry, J. Desesquelles, M. Dufay: Nucl. Instr. and Meth. 110, 43 (1973)

2.124 E. Holøien: Nucl. Instr. and Meth. 90, 229 (1970)

2.125 T. F. O'Malley, S. Geltman: Phys. Rev. 137, A 1344 (1965)

2.126 G. W. F. Drake, A. Dalgarno: Phys. Rev. A 1, 1325 (1970)

2.127 H. Doyle, M. Oppenheimer, G. W. F. Drake: Phys. Rev. A 5, 26 (1972)

2.128 J. H. Macek: J. Phys. B 2, 831 (1968)

2.129 O. Sinanoğlu, W. Luken: Chem. Phys. Letters 20, 407 (1973)

2.130 D. R. Herrick, O. Sinanoglu: Phys. Rev. A (to be published)

2.131 P. G. Burke, D. D. McVicar: Proc. Phys. Soc. 86, 989 (1965)

2.132 J. P. Buchet, M. C. Buchet-Poulizac, H. G. Berry, G. W. F. Drake: Phys. Rev. A 7, 922 (1973)

2.133 W. Schlagheck, D. Schürmann, D. Haas, H. v. Buttlar: Phys. Letters 45A, 433 (1973)

2.134 H. G. Berry, M. C. Buchet-Poulizac, J. P. Buchet: J. Opt. Soc. Am. 63, 240 (1973)

2.135 D. L. Matthews, W. J. Braithwaite, H. W. Wolter, C. F. Moore: Phys. Rev. A 8, 1397 (1973)

2.136 B. Edlén, F. Tyrén: Nature 143, 940 (1939)

2.137 N. J. Peacock, R. J. Speer, M. G. Hobby: J. Phys. B 2, 798 (1969)

2.138 A. H. Gabriel: Mon. Not. R. Astron. Soc. 160, 99 (1972)

2.139 H. Schüler: Ann. Physik 76, 292 (1925)

2.140 S. Werner: Nature 118, 154 (1926)

2.141 G. Herzberg, H. R. Moore: Can. J. Phys. 37, 1293 (1959)

2.142 J. D. Garcia, J. E. Mack: Phys. Rev. 138, A 987 (1965)
J. D. Garcia, J. E. Mack: Phys. Rev. 139, AB 4 (1965)

2.143 E. Holøien, S. Geltman: Phys. Rev. 153, 81 (1967)

2.144 W. S. Bickel, I. Bergström, R. Buchta, L. Lundin, I. Martinson: Phys. Rev. 178, 118 (1969)

2.145 J. P. Buchet, A. Denis, J. Desesquelles, M. Dufay: Phys. Letters 28A, 529 (1969)

2.146 A. W. Weiss: unpublished work

2.147 H. G. Berry, M. C. Buchet-Poulizac, J. P. Buchet: J. Opt. Soc. Am. 63, 240 (1973)

2.148 J. P. Buchet, M. C. Buchet-Poulizac: J. Opt. Soc. Am. 64, 1011 (1974)

2.149 I. A. Sellin: Nucl. Instr. and Meth. 110, 477 (1970), and references therein

2.150 H. G. Berry, R. Hallin, R. Sjödin, M. Gaillard: Phys. Letters 50A, 191 (1974)

2.151 J. P. Connerade, W. R. S. Garton, M. W. D. Mansfield: Astrophys. J. 165, 203 (1971)

2.152 H. W. Wolff, K. Radler, B. Sonntag, R. Haensel: Z. Physik 257, 353 (1972)

2.153 L. Lundin, B. Engman, J. Hilke, I. Martinson: Physica Scripta 8, 274 (1973)

2.154 D. J. Pegg, I. A. Sellin, P. M. Griffin, W. W. Smith: Phys. Rev. Letters 28, 1615 (1972
D. J. Pegg, I. A. Sellin, P. M. Griffin, W. W. Smith: Nucl. Instr. and Meth. 110, 489 (1973)

2.155 B. Emmoth, M. Braun, J. Bromander, I. Martinson: Physica Scripta (to be published)

2.156 J. E. Hansen: private communication

2.157 N. Andersen, G. W. Carriveau, A. F. Glinska, K. Jensen, J. Melskens, E. Veje: Physica Scripta 3, 255 (1971)

2.158 N. Andersen, G. W. Carriveau, A. F. Glinska, K. Jensen, J. Melskens, E. Veje: Z. Physik 253, 53 (1972)

2.159 B. Dynefors, I. Martinson, E. Veje: Physica Scripta (to be published)

2.160 B. Engman, A. Gaupp, L. J. Curtis, I. Martinson (to be published)

2.161 L. Barette, R. Drouin: Physica Scripta 10, 213 (1974)

3. Lifetime Measurements

Larry J. Curtis

With 7 Figures

The beam-foil light source has a number of unique features which permit many new types of experiments. Some of these features are rather subtle, but the time-resolved nature of the decay process is so conspicuous that it is apparent why the first and most widely-applied usage of this technique should be in the measurement of atomic lifetimes. Because of the nearly monoenergetic properties of the beam, the time t since excitation directly corresponds to the distance from the foil x and is given by

$$t = x/v \quad , \tag{3.1}$$

where the beam velocity v is calculable from the beam energy E after the beam emerges from the foil and the atomic mass M of the ion from

$$v[\mathrm{mm/ns}] = 13.9\ (E/M)^{1/2}\ [\mathrm{MeV/amu}]^{1/2} \quad . \tag{3.2}$$

Thus the decrease in light intensity with distance from the foil for a spectrally resolved emission line is a measure of the rate of relaxation of the parent level, and directly leads to its mean life. In the absence of repopulation by cascading transitions from higher levels, this mean life is proportional to the negative inverse of the logarithmic derivative of the decay curve of the emitted light.

In addition to this obvious time-resolved nature, the beam-foil source possesses a number of other properties which are advantageous in lifetime determinations. Mass analysis of the ion beam assures that it is isotopically pure and free of contaminants. It has a very low particle density (typically 10^5 ions/cm^3) and thus exhibits no self-absorption, no collisional de-excitation and no inter-ionic field effects. Nearly any charge state of any element can be excited in this manner and studied using emitted optical, UV, or X-ray radiation, or, in some cases, electrons ejected through autoionization processes. The technique permits the study of multiply-excited states, which seem to be much more copiously populated in beam-foil excitation than in other sources. Coupled with these advantages are, of course, a number of disadvantages. The low particle density leads to low light levels (although the light per atom is probably very high compared to other sources). Doppler broadening of the in-flight emitted radiation makes line blending a serious problem. The foil can undergo damage and change its properties slightly with time. The population is not selective, which complicates the decay-curve interpretation with cascade repopulation effects. Although some of these factors did introduce substantial uncertainties in some of the early beam-foil measurements, they have now been studied in great detail and methods have been developed to reduce or circumvent their effects. Thus recent beam-foil measurements must in general be considered as among the most extensive and most reliable lifetime determinations presently available.

As is evidenced by other chapters in this book, the beam-foil source often exhibits both coherent and anisotropic excitation. This has made possible many types of experiments involving fine and hyperfine structure and Stark and Zeeman effects. Improvements in spectral resolution have made possible term identifications and the measurement of Lamb shifts. The source can also be useful in the study of atomic collision and autoionization mechanisms. Thus beam-foil excitation is a very versatile technique, applicable to a broad range of studies. However, the in-flight decay of an excited ion is such a powerful technique for generating time-resolved decay curves, and foil collisions are so unique in their excitation properties, that lifetime determinations must remain one of the areas of primary emphasis in beam-foil spectroscopy.

In this chapter we shall describe the measurement of decay curves by the beam-foil technique (primarily in terms of optical photon emission), the extraction of mean lives from these decay curves, and the procedures which are utilized to assure reliability of the mean lives. The emphasis will always be on beam-foil measurements, but the techniques used for analysis are equally valid whether the decay curve is generated by in-flight decay of a beam excited by a foil, a gas stripper, or a laser beam, or by excitation of a gas cell by a pulsed or modulated electron beam, photon beam, or heavy-ion beam.

One word of caution is in order. A number of examples of possible pitfalls in beam-foil measurements are described in this chapter. The point of including these (often unlikely) troublesome cases is to indicate the degree to which beam-foil measurements can extract reliable mean lives even under difficult circumstances. The reader should recognize that modern beam-foil measurements routinely account for the points raised, and are free of such systematic errors.

3.1 Lifetime Studies as a Basic Area of Atomic Physics

Measurements of atomic lifetimes have become a very active research area in a field which was commonly treated as closed only a little over a decade ago. Although many aspects of this development are more thoroughly treated elsewhere in this book, it is fitting that a discussion of lifetime measurements begin with a brief account of their importance and sudden emergence as a basic research area.

3.1.1 The Need for Lifetime Measurements

Lifetime measurements are needed both to improve fundamental knowledge of atomic structure and to meet specific needs in areas such as astrophysical abundance determinations, plasma diagnostics and optical excitation processes. Solar abundance determinations have been perhaps the most dramatic use of lifetime data, and estimates of solar abundances of iron-group and rare-earth-group elements have been changed by factors up to ten by beam-foil measurements. These measurements are particularly valuable because of the difficulties which the iron and rare-earth groups present to theoretical calculations. The current status of lifetime measurements and solar abundances is described in a recent review article by SMITH [3.1]. Lifetime measurements have also been useful in the determination of abundances in the interstellar medium from satellite data [3.2,3]. Sophisticated theoretical techniques have been developed for the calculation of atomic transition probabilities; these techniques require experimental determinations in order to select among various approximation techniques. These theoretical techniques have been reviewed by several authors [3.4-7], and are extensively treated in this volume also (see chapters 4 and 5). Lifetime data are particularly useful when they can be extrapolated along isoelectronic or homologous sequences [3.8,9], so that a well-placed measurement in one element can provide increased reliability for neighboring elements.

Most applications for atomic lifetime measurements are to specify emission and absorption of radiation under steady-state or dynamical circumstances, quite unlike the free-decay conditions under which a lifetime manifests itself so clearly. Therefore the lifetime results must (except for unbranched decays) be combined with branching-ratio data and presented in terms of spontaneous transition probabilities (A-values) and absorption oscillator strengths (f-values). These quantities are defined and their inter-relations presented in Subsection 3.2.2. It is possible to measure A-values and f-values directly by techniques such as emission from arcs and shock tubes, absorption by a vapor or slow atomic beam, and anomalous dispersion. However, these measurements require a knowledge of the density of the radiating atoms, which has been a source of considerable error, and are generally restricted to the neutral species and also to resonance transitions. A number of extensive review articles which compare the various methods for measuring lifetimes, A-values, and f-values are available [3.10-16]. Relative A-values and f-values can be measured more reliably than absolute values, and provide the necessary complement to lifetime measurements. Since a lifetime measurement compares only the intensity of a given transition with itself at different times, it requires a knowledge neither of the density of radiating particles nor of the efficiency of the detection system. Thus lifetime measurements provide a highly reliable and indispensable overall normalization for A-value and f-value determinations.

3.1.2 Lifetime Measurements Prior to the Development of the Beam-Foil Technique

It is not possible to separate the development of atomic lifetime measurements from that of nuclear lifetime measurements, for it involves a succession of adaptations of nuclear discoveries and techniques to an atomic context.

Although the mathematical concepts of exponential growth and decay are as ancient as the geometric progression, and many examples of exponential and multiexponential processes exist in nature, measurement inaccuracies and the influence of external conditions prevented rate constants from being interpreted as fundamental properties of statistical processes until the discovery of radioactivity [3.17]. The nuclear exponential decay law was discovered by RUTHERFORD [3.18], when a gaseous decay product migrated from a natural radioactive source and decayed to lay a deposit on a nearby surface, thus producing a separated radioactive source with a lifetime of a few hours. In 1900 RUTHERFORD and SODDY [3.19-21] chemically separated a decay product with a lifetime of a few days from its long-lived parent and observed the recovery, or "growing in", within the parent. This established the importance of the cascade repopulated decay curve. RUTHERFORD and SODDY also introduced the concepts of a "half-life" and a "mean life" into the terminology in 1904 [3.22,23]. In 1905 VON SCHWEIDLER [3.24] showed that the exponential decay law can be deduced from the laws of chance with the assumptions that the probability of decay is constant in time and the same for all members of the same species. It is significant that it was in lifetime measurements that this first clear encounter with a process which is not accessible to causality occurred [3.25]. In 1910 BATEMAN [3.26] solved the set of coupled differential equations which describes the unbranched case of sequentially-cascaded decay, and these solutions were applied to the study of the natural radioactive series. It is interesting to note that the possibility of multiple direct cascading was ignored, since the natural radioactive chain members have at most one α-decay and one β-decay feeder level, which differ by many orders of magnitude in lifetimes. (This is to be contrasted with the application to atomic lifetimes, in which the earliest assumptions were that all cascading is direct. The subtleties of indirect cascading described in Subsection 3.6.1 have only recently been considered). Thus lifetimes were among the first nuclear properties to be systematically studied.

The use of lifetime measurements as an indicator of atomic structure has developed at a much slower pace than its nuclear counterpart. Atomic term-value analysis began with Kirchhoff and Bunsen in 1859, and by 1913 had been refined to the level of the Bohr atom. A few quantitative attempts to measure post-excitation radiation from free atoms had been made by this time, such as the 1908 canal-ray studies of WIEN [3.27] and the 1913 work of DUNOYER [3.28], the latter of whom sent sodium atoms through a beam of sodium light and attempted unsuccessfully to photograph light from the emerging atoms. However, it was not until 1916 that the concept of an atomic disintegration constant was formally introduced by EINSTEIN [3.29,30], in an analysis of the atomic radiation process which was constructed in close analogy with nuclear radioactivity. A dynamical calculation of an atomic mean life was performed in 1919 by WIEN [3.31] which combined classical radiation theory with the Bohr atom. It is perhaps indicative of the general underestimation of the importance of atomic mean lives that the WIEN mean-life formula was forgotten, while the Bohr term-value formula appears in nearly all elementary atomic physics textbooks. WIEN's work has been reviewed and the terminology updated [3.32], and is worth recalling here. WIEN assumed that a quantum of energy becomes available after an electron passes from one allowed Bohr orbit to another, and is thus radiated from the latter orbit at constant centripetal acceleration. For circular orbits this yields

$$\tau = \frac{3\lambda^3\zeta^2}{16\pi^3\,\alpha c a_0^2 n_f^4} = \left(\frac{\lambda}{1004.33}\right)^3 \frac{\zeta^2}{n_f^4}\ [\mathrm{ns}] \quad , \tag{3.3}$$

where λ is the radiated wavelength measured in angstroms, ζ the net charge of the nucleus and core electrons, α the fine-structure constant, a_0 the Bohr radius, c the speed of light, and n_f the principal quantum number of the final orbit. This classical formula gives surprisingly accurate results [3.32] and can often provide a good first approximation. The quantum mechanical theory of transition probabilities was developed by DIRAC [3.33,34] in 1926, but mean-life measurements were not then considered a crucial test of quantum theory. However a number of measurements of mean lives were made, and it is noteworthy that in-flight decays of atomic beams were among the first methods used. WIEN [3.27,31,35] performed a series of experiments using in-flight decay of canal rays, which were the precursors of beam-foil and beam-gas spectroscopy. Similar canal-ray mean-life experiments were also performed by DEMPSTER [3.36,37] and others [3.38-42]. WALLERSTEIN [3.43] applied electric and magnetic fields to canal rays and observed the first excited beam quantum beats, but this technique fell dormant with the death of WIEN in 1928. In 1932 KOENIG and ELLETT [3.44] optically excited a thermal atomic beam and observed its in-flight decay in a geometry very much like that of a modern beam-laser experiment. Although this work was qualitatively repeated by SOLEILLET [3.45], this technique also fell into disuse. Early forms of pulsed-electron-beam gas excitation and Kerr-cell chopped-photon-beam excitation methods were also developed in the 1920's. A review of mean-life experiments performed prior to 1933 is given by MITCHELL and ZEMANSKY [3.10], and a perusal of these results attests to the lack of nanosecond time resolution so vital to atomic mean-life measurements. Thus the modern era of atomic mean-life measurements did not begin until after the development of nanosecond response phototubes and coincidence circuits and their application to nuclear-physics measurements after World War II [3.46].

The first application of delayed-coincidence measurements to atomic mean lives was by HERON et al. [3.47,48], in a pulsed electron-beam experiment performed in 1954. However, they used a single-channel method which viewed only one fixed delay window at a time, and hence suffered from low detected intensity. A variation of this experiment was performed in 1955 by BRANNEN et al [3.49] who also used electron excitation, but with delayed coincidences between the cascade and primary photons. Again, single-channel techniques were used, and intensity problems were encountered. In 1961 BENNETT [3.50] adapted multichannel-analyzer techniques, developed in the mid-1950's for nuclear-physics applications, to pulsed electron-beam delayed-coincidence measurements. This provided an instrument which is equally sensitive to photons over a wide range of times after excitation, and greatly increased the usable intensity. In this form, this technique came into fairly wide use at the same time that beam-foil spectroscopy was being developed by KAY [3.51] and by BASHKIN [3.52], which ushered in the period of intense activity in mean-life measurement which is still in progress.

3.2 Definitions of Basic Quantities

Since the primary purpose of a mean-life measurement is to obtain atomic transition probabilities and absorption oscillator strengths, we shall list the relationships among these various quantities with particular emphasis upon their connection to a mean-life measurement. We use the nomenclature of CONDON and SHORTLEY [3.53], and denote an atomic level as corresponding to a given value of total angular momentum J, which in the absence of external fields consists of (2J+1) degenerate states (or sublevels), each specified by a different projection M of the angular momentum along some prescribed axis (for our purposes, the beam direction). All other quantum numbers are denoted simply by γ, and the set of levels which have the same γ is called a term. A transition between levels gives rise to a line, a transition between states is called a component, and a set of transitions between terms is a multiplet. For compactness, we shall generally denote a set of quantum numbers by a single subscript, and indicate in the text whether this refers to a state or to a level, or to both. An exception will occur in discussions of alignment and polarization in which the quantum numbers (γ J M) will appear explicitly to avoid confusion.

3.2.1 Instantaneous Populations

It is standard practice in mean-life studies to speak of the instantaneous populations of the various decaying states and levels. This description is not always relevant for beam-foil excitation since the eigenstates which are populated by the source are not necessarily the same as those of the radiative decay process. Thus the excitation Hamiltonian may not be diagonal in the representation which diagonalizes the decay Hamiltonian. The decaying states are then said to be coherently excited, and can exhibit quantum beats. Since we are primarily concerned here with measuring mean lives it is important to seek conditions which preclude quantum beats. We shall therefore restrict our discussions to situations in which (a) fine-structure levels are so distinct in energy that coherences between them are unlikely to take place, (b) the foil is axially symmetric about the beam so that there are no coherences between different M states [3.54], (c) there are no external fields to mix the states, and (d) hyperfine interactions are negligible. Under these circumstances there can be no coherences between states quantized with respect to the beam, and their individual state populations, as well as their combined level populations completely describe the excitation without recourse to density-matrix methods.

3.2.2 Transition Probabilities and Oscillator Strengths

In an emission process, the instantaneous rate of spontaneous photon emission $I_{u\ell}$ between an upper state or level u, to which N_u atoms are excited, and a lower state or level ℓ, is given by [3.29,30]

$$I_{u\ell} = N_u A_{u\ell} \quad , \tag{3.4}$$

where $A_{u\ell}$ is the spontaneous transition probability. In an absorption process, N_ℓ atoms in a lower state or level ℓ are equivalent, in radiative absorption to the state or level u, to n classical harmonic oscillators given by [3.55]

$$n = N_\ell f_{\ell u} \quad , \tag{3.5}$$

where $f_{\ell u}$ is the absorption oscillator strength. For transitions between levels, the quantities are related by

$$g_\ell f_{\ell u} = \frac{mc}{8\pi^2 e^2} \lambda_{u\ell}^2 g_u A_u = \left(\frac{\lambda_{u\ell}}{2582.7}\right)^2 g_u A_{u\ell} \quad , \tag{3.6}$$

where $\lambda_{u\ell}$ is the transition wavelength in Angstrom units, g_u and g_ℓ are the degeneracies of the upper and lower levels, respectively, and $A_{u\ell}$ is measured in nanoseconds. For components, the expressions are the same, except the degeneracies do not appear. The mean life τ_u and its inverse α_u are defined according to

$$1/\tau_u \equiv \alpha_u \equiv \sum_\ell A_{u\ell} \quad . \tag{3.7}$$

Thus f-values can be computed from inverse mean lives through the relationship

$$g_\ell f_{\ell u} = \left(\frac{\lambda_{u\ell}}{2582.7}\right)^2 g_u \left(\frac{A_{u\ell}}{\alpha_u}\right) \alpha_u \quad , \tag{3.8}$$

where $A_{u\ell}/\alpha_u$ is the relative branching ratio, which cannot be determined in a mean-life measurement and must be obtained through some other source. Inverse mean lives can be computed from f-values through the relationship

$$\alpha_u = \sum_\ell \left(\frac{2582.7}{\lambda_{u\ell}}\right)^2 \frac{g_\ell f_{\ell u}}{g_u} \quad . \tag{3.9}$$

In theoretical calculations, it is convenient to define the line strength S, which is symmetric in emission and absorption. It is related to the transition probability by

$$S \propto (\lambda_{u\ell})^{2L+1} g_u A_{u\ell} \quad , \tag{3.10}$$

where L is the multipolarity of the emitted radiation. The proportionality constants for E1, M1, and E2 radiation are presented by SHORE and MENZEL [3.56,p.445]. Under the assumption of spin-orbit coupling, the line strength is the same for all levels in a multiplet [3.57], and if the A-value or f-value for one line in a multiplet is known, the values for all other lines in that multiplet can be obtained by appropriate wavelength and degeneracy corrections.

3.3 Measurement of Beam-Foil-Excited Decay Curves

Subsequent sections will describe refined techniques for extracting mean lives from decay curves. However, the most important part of any mean-life measurement lies not in the analysis, but in the experimental measurement itself, which must be designed to obtain the most reliable and informative decay-curve information possible.

3.3.1 Strengths and Limitations of the Beam-Foil Technique

Beam-foil excitation is by far the most widely used method of direct mean-life measurement, and more mean-life measurements have been made by this technique than by all other direct techniques combined. For example, a summary of transition probabilities for atomic absorption lines formed in interstellar clouds has recently been compiled by MORTON and SMITH [3.58], who cite 159 mean-life measurements of 101 multiplet transitions. Of these, 96 are beam-foil, 44 are modulated electron-beam phase-shift, 8 are pulsed electron-beam delayed-coincidence, 6 are modulated resonance-radiation phase-shift, 3 are high-field level-crossing and 2 are Hanle measurements. Further, the transitions in this compilation are not even particularly well suited to beam-foil methods, since only 18 of the 101 transitions were in atoms more than once ionized. Similarly, LAUGHLIN and DALGARNO [3.59] have recently presented a comparison between theoretical calculations and experimental measurements of transition probabilities along isoelectronic sequences of Be, B and N, and cite 67 mean-life measurements of 24 transitions. Only 5 of these measurements were by methods other than beam-foil.

Thus, because of its wide use, beam-foil excitation is often applied to systems for which it is not the optimum technique available. For example, decay curves of transitions from singly-excited levels in neutral or singly-ionized atoms could in general be measured by pulsed-electron beam-gas delayed-coincidence techniques with much better wavelength resolution (hence less blending) and higher signal-to-noise ratios than are possible using beam-foil methods. One factor which militates against the beam-foil technique for the important case of neutral atoms is that very few neutrals emerge from a foil. Of course, the yield of neutrals is enhanced by working at the smallest possible particle velocity, but this introduces two serious handicaps. For one, beam scattering becomes especially serious. For another, foil life becomes quite short, particularly for heavy ions.

The alternative of the pulsed electron-beam method has been developed to a high degree of sophistication by a number of workers [3.60-63] which makes it an extremely attractive technique to complement beam-foil methods. This is particularly true in view of the developments introduced by ERMAN [3.16] in which high-frequency deflection techniques, combined with kilovolt electron beams, make possible nanosecond pulses at megahertz repetition rates with average currents of several milliamps during the pulse. The technique also has the advantage of a variable time window, so that long-lived components can be followed on one time scale, short-lived components on another, and a combined analysis performed. Thus, any transition which can be strongly excited by pulsed-electron beam-gas techniques can probably be more favorably measured by these techniques than by beam-foil methods, if appropriate equipment is available.

The mean lives which can be extracted from beam-foil decay curves are generally limited to the range from 10^{-6}s to 10^{-11}s, although values as short as 3.7×10^{-12}s have recently been reliably extracted by BARRETTE and DROUIN [3.64] using new high spatial resolution techniques. The beam velocity must lie between the speed of light and the slowest speed with which ions can emerge from the foil in a collimated beam. The measurable flight paths are between the shortest beam length resolvable by the optical system and the furthest downstream distance which can be viewed without beam spreading distortions or loss of the signal in the background. It must be kept in mind that a proper determination of a decay curve requires that one make measurements at a number of points. Thus a reasonable, practical limitation on the shortest beam length one can use is approximately 10 times the shortest resolvable length. These considerations have been examined by BROMANDER [3.65], who devised the plot shown below in

Fig.3.1 to illustrate the measurable mean-life ranges using various types of accelerators.

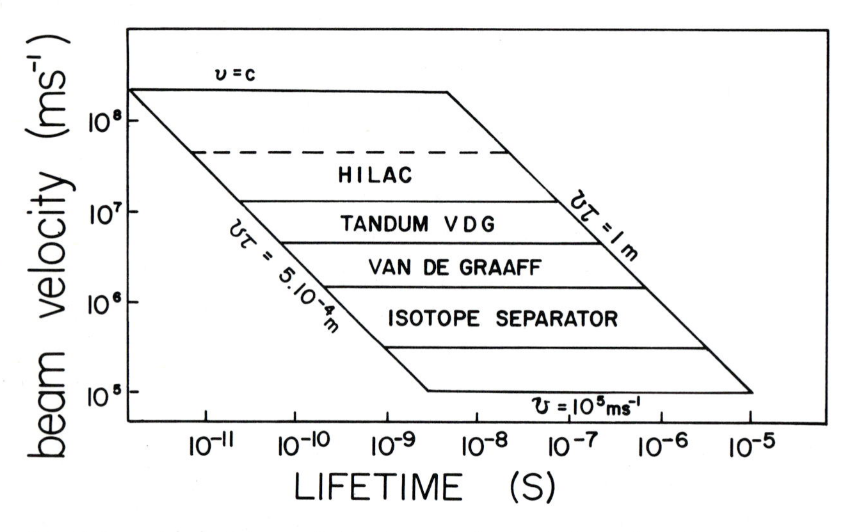

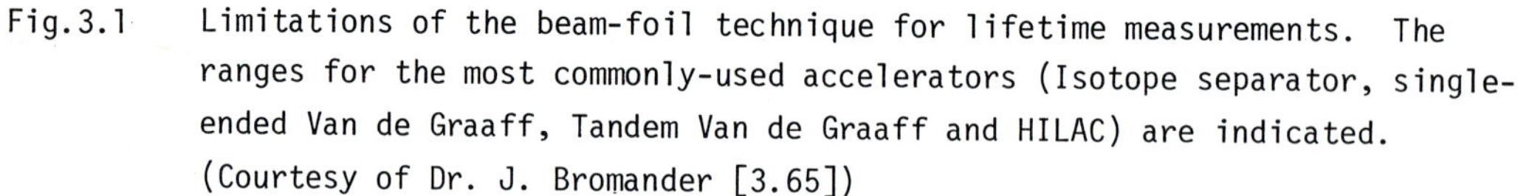
Fig.3.1 Limitations of the beam-foil technique for lifetime measurements. The ranges for the most commonly-used accelerators (Isotope separator, single-ended Van de Graaff, Tandem Van de Graaff and HILAC) are indicated. (Courtesy of Dr. J. Bromander [3.65])

Most of the weaknesses of the beam-foil technique are traceable either directly or indirectly to the low light levels involved, and could be eliminated if higher resolving power could be used. The strengths of the technique lie in the unparalleled ability of the beam-foil source to excite all types of ions to high ionization degrees, high excitation states, and multiply-excited states. For detailed descriptions of these areas, the reader is referred to recent review articles on beam-foil spectroscopy by MARTINSON and GAUPP [3.66], by BASHKIN [3.67], and by MARTINSON [3.68,69], and to a recent review article on multiply-excited states by BERRY [3.70].

3.3.2 Details of Beam-Foil Apparatus and Measurement Procedures

The widespread use of the beam-foil technique is probably due in part to the reasonably simple nature of the basic experimental apparatus, which is relatively inexpensive if a suitable accelerator facility is available. However, in the past several years rather sophisticated instrumentation and refined data-acquisition procedures have been incorporated into beam-foil lifetime measurements, which have substantially improved the accuracy of the data. Many of these techniques were developed primarily in efforts to resolve short-period quantum beats in fine- and hyperfine-structure measurements, and exceed normal requirements for mean-life work. Since quantum beats and mean-life decay curves are often measured in the same laboratory on the same apparatus, this has led to a marked increase in the precision to which lifetime measurements are now routinely performed.

Modern beam-foil measurements are in sharp contrast to those of a decade ago, when the paucity of mean-life data justified order-of-magnitude estimates. Thus, in the early 1960's, a beam-foil study could provide valuable information by classifying level mean lives as "short", "medium", or "long" according to the length of a wavelength-dispersed photograph of the beam. As required accuracies have risen, measurement techniques have been refined, and the uncertainties in the earlier measurements must not be allowed to compromise the high reliability of modern beam-foil measurements.

Today, beam-foil decay-curves are usually obtained by stepwise translating the (5-20 $\mu g/cm^2$) foil [3.71] along the beam axis, referenced relative to a fixed point in the evacuated (10^{-6} torr) foil chamber which is viewed by a high-speed optical system. The foil motion is achieved by a trolley which is driven by a precision machine screw with low backlash and high positional accuracy and reproducibility [3.72], often to within $\pm$ 0.05 mm. From 20 to 40 foils are mounted on a rotatable turret, and a given foil can be reproducibly selected by an external control dial and indicator. The foil-translation mechanism is usually driven by a stepping motor which is shaft encoded to provide a channel-advance signal to a multichannel analyzer operated in the multiscaling mode. The output of the optical detection system is thus accumulated in a separate scaling channel for each position of the foil relative to the viewed position. The step size (as short as 0.1 mm) and the total number of steps are programmable (in some laboratories they are controlled by an on-line computer), and the travel is recycled many times to signal-average any instrumental drifts, beam fluctuations, or foil aging. If the signal averaging is done over many cycles, the stepping motor can be advanced after equal amounts of time at each foil position. However, in order to remove the effects of beam fluctuations, the stepping-motor advance is more often gated by collection of a fixed amount of some monitoring quantity indicative of the instantaneous excitation to the level studied. One type of system uses the total beam charge collected in a Faraday cup, but this is sensitive to the average charge state leaving the foil; that average charge can change with foil aging. A more commonly-used system which avoids this problem collects light from the transition studied at a fixed distance from the foil, either directly by a foil-fixed phototube, or through a fiber optics link to a second monochromator and phototube. In either case, the signal either averages to the same total accumulation time in each channel, or a channel-by-channel dark-count correction is made.

The details of the optical system vary with the wavelength region, but single-photon detection techniques are used almost exclusively. For the optical wavelength region, a lens focuses light onto the entrance slit of a monochromator from a segment of the beam often 0.1 mm or shorter in length. This resolution can be checked by quantum-beat measurements, and focus is often achieved by translating the foil past the optical system and minimizing the travel necessary for the light to rise from near extinction on the upstream side to its full value on the downstream side. When one views closer than 1 mm to the foil, the foil holder begins to vignette the collected light, so the sharpness of the rise does not indicate the actual spatial resolution. Since the light is not strong, a fast (low dispersion) monochromator is used, with low resolution to admit the Doppler-broadened spectral lines. The spectrometer is refocused to a moving source to minimize this broadening [3.73-76]. A low-noise photomultiplier tube is placed at the exit slit. Some photomultiplier tubes are cooled to reduce the dark-count rate to a few per second. In the vacuum UV wavelength region, a lens can not be used, but grating masks are used to limit the beam length viewed, and channel electron multipliers or sodium-salicylate-coated photomultiplier tubes serve as photon detectors. The optical system usually views at an angle 90° with the beam, but occasionally an oblique angle is used to follow a very short mean life into the foil.

The system described above is typical of most modern beam-foil laboratories and, although the specific details may vary somewhat, it should provide a general guide to the procedures presently used. A representative modern beam-foil decay-curve is shown in Fig.3.2.

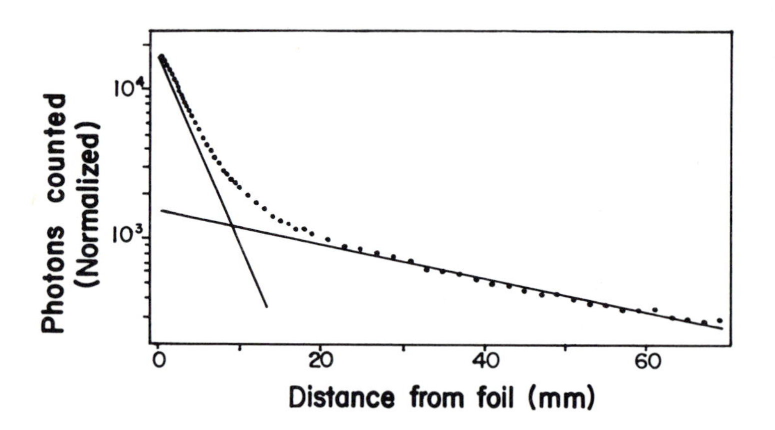

Fig.3.2 Representative beam-foil intensity decay-curve for the 2348Å Be I line ($2s^2\ ^1S$ - $2s2p\ ^1P$) measured at 180 keV ion energy [3.144]

A number of possible pitfalls exist which are carefully examined as part of any measurement. One such problem involves the fact that the foil characteristics change slightly during bombardment. It was observed by CHUPP et al. [3.77] that the energy thickness of the foil increases with bombardment. This was investigated by BICKEL and BUCHTA [3.78], who determined that for an 80 keV Li^+ ion beam incident upon a carbon foil, the foil increased its thickness at 2.7 $\mu g/cm^2$/hour. At energies for which the energy loss in the foil is a reasonable fraction of the total beam energy, foils are therefore replaced often and energy analyses are made frequently. The foil also broadens the beam velocity distribution. Errors introduced into mean lives by beam velocity straggling and beam spreading due to interaction with the foil have been investigated by CURTIS et al. [3.79] and by KAY [3.80], who present calculations which indicate that these effects are negligible, even at rather low beam velocities. The possibility that beam spreading could cause ions to escape from the viewing volume has been studied by ETHERTON et al. [3.81]. They found that by placing a chamber-fixed beam collimator just upstream of the spectrometer slit and monitoring on Faraday cup current, these effects became negligible even at distances corresponding to more than 70 ns from the foil. The most important foil-dependent quantity requiring accurate determination is the average velocity of the foil-emergent particles. Several methods for its determination are available. A post-foil electrostatic energy analyzer [3.82] is one common method. Another method utilizes the Doppler shift of the in-flight emitted spectral line when viewed at a skew angle to the beam [3.83]. It is also possible to determine the velocity by a measurement of a precisely known quantum-beat frequency. Some beam-foil laboratories have facilities for beam pulsing [3.84] which permit the use of delayed-coincidence time-of-flight measurements to determine foil-emergent particle velocities. Another common approach is to calibrate the accelerator energy, using various nuclear resonances and thresholds [3.85]. The thickness of the foil can then be determined by optical transmission [3.86] or proton energy-loss measurements and the energy loss of the ion in the foil computed from theoretical estimates [3.87-89]. However, there is some indication that the theoretical estimates are not directly applicable to the beam-foil case, and may overestimate losses due to ions scattered in nuclear encounters, which tend to be removed from the forward-going beam [3.90]. This is particularly crucial in the case of very heavy ions, where the energy loss in the foil may be an appreciable fraction of the total energy, and researchers usually choose one of the direct velocity measurement techniques for such work.

Another problem involves line blending. Since the beam-foil source requires fast optics and wide acceptance angles viewing a moving source, severe Doppler broadening is unavoidable. Thus there is a chance, particularly for rich spectra, that the tails of neighboring lines will overlap with the line measured, adding unrelated components to the decay curve of interest. A number of approaches to this problem are available. The spectrum can be examined for blending under conditions which can reduce Doppler widths [3.66,Table 3] even though these conditions are not themselves suitable to lifetime determinations. If the decay is branched, or is part of a multiplet, the various branches or related lines can be separately measured and compared. The decay curve can also be measured for equivalent times both on and off the line peak; slow-varying backgrounds should vanish from the difference between these measurements [3.91]. Thus the effects of line blending may occasionally lead to increased uncertainties in lifetimes, but that is probably not the most serious limitation brought about by blending. It is unfortunate that the possibilities of blending generally restrict the application of the beam-foil source to strong lines, which stand up above the tails of neighboring lines, and to lines which are well separated from their neighbors.

One of the most widely discussed problems in beam-foil excitation involves cascade repopulation of the levels, which introduces the mean lives of higher-lying levels into the time dependence of the decay of each level. The analysis of these effects has been greatly aided by improved measurement techniques. Sub-millimeter spatial resolution and closely-spaced data points, together with low-noise detection equipment which permits measurement far out onto the decay-curve tails, have made possible very elegant numerical analyses. The general problem of cascade effects is described in detail in the next subsection, where we shall see that an adequate allowance for cascading in beam-foil measurements can be made in all but the most unusual circumstances. These rare cases where cascade effects can be serious and methods for handling them are discussed in Section 3.6.

Thus the quality of beam-foil decay-curve measurements has undergone a vast improvement in recent years, both in terms of the sensitivity of its instrumentation and in the procedures which are routinely performed to assure reliability. The low light intensities do not permit signal-to-noise ratios as favorable as in sources such as electron beam excitation, but with sufficient data accumulation time, decay curves of comparable quality can be obtained.

3.3.3 Cascade Repopulation - A Tractable Problem

As we saw in subsection 3.1.2, the problem of cascade repopulation has been familiar since the discovery of radioactivity, but for practical reasons the measurement of lifetimes in atomic systems greatly enlarges cascade contributions. This can be seen by observing that atomic mean lives usually lie between the nanosecond and the millisecond range, while the mean lives of the natural radioactive series range from half a microsecond to 20 billion years. Thus mean lives of cascade-coupled atomic levels are much more likely to overlap on a given time scale. Further, atomic systems have more closely spaced levels with fewer transitions forbidden by selection rules than are generally found in nuclei. Thus, multiple direct cascading, while common in atomic systems, is virtually non-existent in natural radioactivity.

The atomic cascade problem was discussed by WIEN [3.35,p.435] in 1927 and is present in essentially all generally applicable atomic excitation sources, and efforts to eliminate it often introduce problems much more serious than cascading. For example, the pulsed-electron beam-gas source can be made cascade-free by the use of threshold excitation, which excites no levels above the level of interest. This has a number of drawbacks. In particular, excitation cross sections become vanishingly small near threshold, and light outputs are not sufficient unless high gas pressures are used, which introduce the possibilities of collisional de-excitation and radiation trapping. BENNETT and KINDLMANN [3.92] have made such measurements for neon, and DONNELLY et al. [3.93] have used this technique with noble-gas ions, but it has not been generally applied to other elements. Another technique to eliminate cascade effects from electron beam-gas excitation studies is to measure delayed coincidences between the cascade photon and the photon from the level of interest [3.94,95]. Again the cascades are eliminated at great cost, with the true coincidence rates down by 3 to 4 orders of magnitude from the singles rates. Still a third technique measures delayed coincidences between the inelastically-scattered electron beam and the decay photon [3.96]. This technique, although promising, requires an experimental system of considerable complexity, and has had limited application, primarily in neutral gases and molecules.

In modulated electron-beam phase-shift measurements, the cascades are studied through the frequency dependence of the phase shift. Analysis often treats the system as if only a single direct cascade were present, which is not a probable situation. The general relationship for the phase shift as a function of frequency for an arbitrarily cascaded and blended level has been calculated by CURTIS and SMITH [3.97], but computer simulations indicate that present experimental accuracies are not sufficient to extract more than one or two cascades from phase-frequency curves.

Excitation by optical radiation can eliminate cascades, but this generally provides access only to levels optically connected to the ground state or to metastable states. When resonance fluorescence techniques such as the Hanle effect and high-field level-crossing are extended to study non-resonance transitions through electron-beam or beam-foil methods, they also acquire cascade repopulation effects. DUFAY [3.98] has formally computed the cascade contributions to the Hanle effect. However, to evaluate the extent to which cascades affect these techniques requires a knowledge of the alignment and transfer of alignment of the cascade levels. It is sometimes conjectured that the transfer of alignment in these cases is slight, but the situation is extremely difficult to analyze quantitatively.

Despite the fact that cascades are a comparable problem in nearly every type of mean-life measurement, the words "cascade errors" have become synonymous with the beam-foil source. This is unfortunate in view of the high accuracy of modern beam-foil measurements, and is partially due to the extreme care with which beam-foil experimenters have approached the cascade problem. Other measuring techniques are often beset with so many problems (pressure dependence, sample purity, rf pickup, etc.), that the cascades become of secondary importance. In beam-foil work, these problems are absent, and only the cascades remain, so they are carefully discussed in each paper.

There are a few cases of beam-foil measurements made prior to 1971 which lacked a sufficient dynamic time range and missed subnanosecond components or very long-lived cascades. Some values were thus reported which substantially overestimated the true mean lives. Most of these values have now been revised by subsequent remeasurements, so some care must be taken as to the origin of a given mean life to be sure that it is the currently accepted result. However, there are certain small trends which are important to consider.

In a report of their recent calculations, LAUGHLIN and DALGARNO [3.59] present a comparison of experimental and theoretical mean lives for a number of isoelectronic sequences, and report that, "The experimentally-determined transition probabilities are invariably smaller than our theoretical values." They attribute this to possible beam-foil cascade effects, so it is valuable to investigate the size, statistical significance, and possible origins of any such trend. PINNINGTON et al. [3.99] have examined the ratios of experimental to theoretical mean-life values for beam-foil measurements made in the University of Alberta Laboratory during the past four years. Their results are presented in Table 3.1, along with a similar set of ratios obtained

Table 3.1 Ratio of beam-foil mean-life measurement to theoretical mean-life computation for two unselected samples. The uncertainties are the sample standard deviations; sample size is given in parentheses

Source	$\tau_{BFS}/\tau_{Theor.}$
PINNINGTON et al. [3.99]	
(a) All values	1.24 ± 0.88 (47)
(b) Highest ratio excluded	1.11 ± 0.18 (46)
LAUGHLIN and DALGARNO [3.59]	
(a) All values	1.23 ± 0.51 (24)
(b) Highest ratio excluded	1.12 ± 0.16 (23)

from the compilation of LAUGHLIN and DALGARNO. Although all ratios are in fact greater than one, their values are within one standard deviation of that value. Further, most of the individual ratios were within 10% of unity, with only a few cases (2 of 71) in major disagreement. Therefore we must distinguish between two types of possible experimental errors: gross errors due to ambiguities in identifying mean lives (see

Subsection 3.6.1), which are difficult to typify in quoted uncertainties, and small errors, which are correctly typified by quoted uncertainties (usually 10%). In the vast majority of beam-foil decay curves today the cascade mean lives can either be resolved by a properly designed experiment, or else are too weak to affect the decay curve significantly. While the total neglect of cascading would certainly overestimate primary mean lives, there is no reason to believe that an exponential decomposition of a decay curve is more likely to undercorrect for cascading than to overcorrect for it. Thus, although cascading increases uncertainties, other sources of systematic error must be considered if the results in Table 3.1 constitute a meaningful discrepancy. For example, beam-foil measurements do contain a selection bias, since they can usually be performed only for strong transitions. Thus any theoretical approximation techniques which are especially applicable to strong transitions should be closely examined, as should trends in lifetimes measured by techniques equally applicable to strong and weak transitions. Also, DALGARNO [3.100] has pointed out that theoretical first-order 1/Z extrapolations along isoelectronic sequences uniformly underestimate mean lives, and gradient corrections have caused upward revisions for the higher members of the sequences [3.101]. WEINHOLD [3.102] and ANDERSON and WEINHOLD [3.103] have suggested a technique for calculating upper and lower bounds to theoretical transition probabilities, which, although generally very broad, could provide a useful guide. Such estimates, together with critical evaluations of the exciting mean-life measurements, could resolve the question of whether the small trends in Table 3.1 have significance. However, it must be stressed that we can probably be 95% confident that a mean-life measurement obtained from the beam-foil method in its modern form is reliable to within its statistical error or 10%, whichever is greater [3.99].

3.4 Time Dependence of the Measured Decay Curves

The mean-life determinations discussed in this chapter involve a measurement of the time-dependence of the radiation emitted by a relaxing atomic level. The mean life of the level is directly determined from this decay curve through analytical expressions which will be developed in this section. These expressions are applicable not only to beam-foil measurements, but also to decay curves generated under a wide variety of excitation conditions.

3.4.1 Solution of the Driven Coupled Linear Rate Equations

Most techniques for the direct measurement of atomic mean lives involve a study of the time-dependence of the radiation emitted by a sample during or subsequent to an external current density, $Q(t)$, which generates the excitation. Usually, low densities are used so that collisional effects and radiation trapping within the sample are negligible. In such cases, the instantaneous population $N_n(t)$ of a state or level n is governed by the equation

$$dN_n/dt = \sigma_n Q(t) + \sum_j N_j(t)A_{jn} - N_n(t)\alpha_n \quad , \tag{3.11}$$

where σ_n is the excitation cross section, A_{jn} is the transition probability for a cascade from the states or levels j, and α_n is the inverse mean life of the level n. This equation has an integrating factor, $\exp(\alpha_n t)$, which permits conversion to the form

$$\frac{d}{dt}\,[N_n(t)\,\exp(\alpha_n t)] = \exp(\alpha_n t)\,[\sigma_n Q(t) + \sum_j N_j(t)A_{jn}] \quad . \tag{3.12}$$

This differential equation can be changed to a convenient integral equation by integrating both sides between the limits minus infinity (when it is assumed that none of the excited states n is populated) and an arbitrary time t, shifting the integration variable and exchanging orders of integration and summation to obtain

$$N_n(t) = \sigma_n L_n(t) + \sum_j A_{jn}\int_0^\infty dt'\,\exp(-\alpha_n t')N_j(t-t') \quad , \tag{3.13}$$

where the quantity $L_i(t)$ is the Laplace Transform Convolution of the excitation stimulus

$$L_i(t) \equiv \int_0^\infty dt'\,\exp(-\alpha_i t')Q(t-t') \quad , \tag{3.14}$$

which has the very useful reduction property [3.104,105]

$$\int_0^\infty dt'\,\exp(-\alpha_j t')L_i(t-t') = [L_i(t)-L_j(t)]/(\alpha_j-\alpha_i) \tag{3.15}$$

For i=j, ℓ'Hôpital's rule yields $-\partial L_i/\partial\alpha_i$ for the right-hand side of (3.15).

A relationship of the form of (3.13) holds for each state or level of the system, and the simultaneous solution of the coupled differential equations of a given state with those of its cascades can be performed to specify the instantaneous populations completely. The solution can conveniently be written in a closed-form series decomposition [3.104,105], in which the individual terms can be identified with cascades which contribute to the primary state or level through a specific number of intermediate states or levels. This series can be very neatly generated through an iterative process. As a zeroth approximation we neglect all cascades in (3.13) and obtain $N_n^{(o)}(t) = \sigma_n L_n(t)$. We then use this result to make the first approximation, which neglects cascades into cascades, and sets $N_j^{(o)}(t) = \sigma_j L_j(t)$ in (3.13). The integrals of $L_j(t)$ are easily reduced through the properties of (3.15), and we are ready to substitute this first approximation into (3.13) to obtain a second approximation. This process is repeated until cascading up to a desired order is included, thus generating an expression, which, containing linear terms in the $L_i(t)$ quantities as its only time-dependence, can be written in the form

$$N_n(t) = \sigma_n L_n(t) + \sum_j \{j\to n\} + \sum_{kj}\sum \{k\to j\to n\} + \sum_{\ell kj}\sum\sum |\ell\to k\to j\to n\}$$
$$+ \ldots + \sum \ldots \sum \{m\to(r\text{ steps})\to n\} \quad , \qquad (3.16a)$$

where the quantities in braces are diagrammatic symbols for the various functions which are generic to cascades of a given order. One can show that [3.104,105]

$$\{j\to n\} \equiv \sigma_j A_{jn}\left[\frac{L_n(t)}{(\alpha_j-\alpha_n)} + \frac{L_j(t)}{(\alpha_n-\alpha_j)}\right]$$

$$\{k\to j\to n\} \equiv \sigma_k A_{kj} A_{jn}\left[\frac{L_n(t)}{(\alpha_j-\alpha_n)(\alpha_k-\alpha_n)} + \frac{L_j(t)}{(\alpha_k-\alpha_j)(\alpha_n-\alpha_j)} + \frac{L_k(t)}{(\alpha_n-\alpha_k)(\alpha_j-\alpha_k)}\right]$$

$$\{\ell\to k\to j\to n\} \equiv \sigma_\ell A_{\ell k} A_{kj} A_{jn}\left[\frac{L_n(t)}{(\alpha_j-\alpha_n)(\alpha_k-\alpha_n)(\alpha_\ell-\alpha_n)} + \frac{L_j(t)}{(\alpha_k-\alpha_j)(\alpha_\ell-\alpha_j)(\alpha_n-\alpha_j)}\right.$$
$$\left. + \frac{L_k(t)}{(\alpha_\ell-\alpha_k)(\alpha_n-\alpha_k)(\alpha_j-\alpha_k)} + \frac{L_\ell(t)}{(\alpha_n-\alpha_\ell)(\alpha_j-\alpha_\ell)(\alpha_k-\alpha_\ell)}\right] \quad , \qquad (3.16b)$$

which can be generalized to the form

$$\{m\to ..(r\text{ steps})..\to n\} \equiv \sigma_m A_{mb}\cdots A_{jn} \sum_{i=n}^{m} \left[L_i(t)/\prod_{j\neq i}(\alpha_j-\alpha_i)\right] \quad .$$

Here i and j range over the (r+1) cascade and primary states or levels and the product of transition probabilities is over the r-step cascade chain. Equation (3.16a) can be refactored into the form

$$N_n(t) = \sum_j \beta_{nj} L_j(t) \quad , \qquad (3.17)$$

where the sum includes one term for the primary level n, and one term for every level which cascades, either directly or indirectly, into it.

In the following discussion, the variables are redefined in dimensionless form. For the beam-foil source, the excitation may be considered an impulse $Q(t) \propto \delta(t)$, so if we make the following replacements in (3.16)

$$L_i(t) \to \exp(-\alpha_i t)$$

$$\sigma_i \to N_i(0) \quad , \qquad (3.18)$$

we obtain a specific expression for the population of any arbitrarily-cascaded beam-foil-excited level.

Three-level examples in which there are two repopulating cascades are often used for illustrative purposes, so we shall present them explicitly here. For a primary level labeled 1, repopulated by cascades from levels labeled 2 and 3, (3.16) and (3.17) yield a relationship

$$N_1(t) = [N_1(0)-\beta_{12}-\beta_{13}]\exp(-\alpha_1 t) + \beta_{12}\exp(-\alpha_2 t) + \beta_{13}\exp(-\alpha_3 t) \quad . \quad (3.19)$$

There are two possible schemes by which this can occur:

Case 1 Direct cascades - both 2 and 3 have transitions to 1, but are not themselves repopulated

$$\beta_{12} = N_2(0)\,A_{21}/(\alpha_1-\alpha_2) \quad ,$$

$$\beta_{13} = N_3(0)\,A_{31}/(\alpha_1-\alpha_3) \quad . \quad (3.20)$$

Case 2 Indirect cascades - 3 cascades into 2, which cascades into 1

$$\beta_{12} = [N_2(0) + N_3(0)A_{32}/(\alpha_3-\alpha_2)]A_{21}/(\alpha_1-\alpha_2)$$

$$\beta_{13} = N_3(0)A_{32}A_{21}/[(\alpha_1-\alpha_3)(\alpha_2-\alpha_3)] \quad . \quad (3.21)$$

As noted in Subsection 3.1.2, the indirect cascading case (with $N_1(0) = N_2(0) = 0$) was first deduced by RUTHERFORD and SODDY in their study of radioactive decay and recovery curves. The implications of direct and indirect cascading will be contrasted in Subsection 3.6.1.

The condition of (3.18) also describes the decay curves for all excitation sources which have a finite shut-off time after which $Q(t)=0$ (for example, a pulsed beam-gas source) provided $t=0$ is interpreted to be after shut-off. Equation (3.16) can also be applied to extract mean lives from driven excitation, with the only difference being that $L_i(t)$ becomes a more complicated function. For example, for a unit Gaussian driving stimulus $Q(t) = (h/\sqrt{\pi})\exp(-h^2t^2)$, the function becomes [3.104-106]

$$L_i(t) = \exp(-\alpha_i t+\alpha_i^2/4h^2)\,[1+\mathrm{erf}(ht-\alpha_i/2h)]/2 \quad . \quad (3.22)$$

Similarly, the empirical shape of a measured excitation pulse $Q(t)$ (the "prompt" curve) can be numerically convoluted to form $L_i(t)$, which can then be inserted into

(3.16). This generalizes to arbitrary cascading a technique used by LAWRENCE [3.107] to extract short mean lives in pulsed-electron beam-gas excitation, and by HELMAN [3.108] in the analysis of fast luminescence behavior. For a unit step function, $L_i(t)$ is given by [3.104,105]

$$L_i(t) = [1 - \exp(-\alpha_i t)]/\alpha_i \quad . \tag{3.23}$$

Insertion of this relationship into (3.16) generalizes to arbitrary cascading the solutions developed by ANKUDINOV et al. [3.109,110] to describe excitation of a beam by a gas cell. This type of analysis is also extended in [3.104,105] to specify the phase shift for a modulated beam-gas excitation source, with arbitrarily complicated cascading included. Thus by evaluation of the single representative integral of (3.14), a surprisingly simple description of nearly any atomic relaxation process is given by (3.16).

3.4.2 A Quantitative Indicator of Level Repopulation - The Replenishment Ratio

Since cascade repopulation is a source of much concern in nearly all mean-life measurements, it would be useful if a reported mean-life determination could be accompanied by some estimate of the degree to which cascading was present. A convenient indicator is provided by the replenishment ratio [3.79], which is defined as the ratio of the cascade repopulation rate to the decay depopulation rate. From (3.11) this can be seen to be

$$R(t) \equiv \left(\sum_j N_j(t) A_{jn}\right) / \left(N_n(t)\alpha_n\right) \quad . \tag{3.24}$$

It is a positive quantity: $R(t) \ll 1$ implies there is little cascading; $R(t) > 1$ indicates a growing-in behavior. For an undriven case, $Q(t)=0$, and (3.11) can be written

$$-d(\ln N_n)/dt = \alpha_n[1-R(t)] \quad , \tag{3.25}$$

clearly exposing the fact that α_n is the upper limit to the negative logarithmic derivative of the decay curve (see subsection 3.6.1). In terms of the exponential sum of (3.17) the initial value of the replenishment ratio is given by

$$R(0) = \sum_j (1-\alpha_j/\alpha_n)\, \beta_{nj}/\sum_k \beta_{nk} \quad , \tag{3.26}$$

which is a convenient form for computation from curve fits.

3.4.3 Intensity Relationships for an Aligned Source

The intensity of radiation which is emitted into all 4π steradians in a given spontaneous transition is proportional to the instantaneous population of the upper state or level, as shown in (3.4). However, in practice, a detector samples the radiation in only some limited solid angle, and has a sensitivity to the radiation's polarization. Thus, if the states are not equally populated, the differing angular patterns of the polarized component radiation can cause the time-dependence of the sampled line radiation and that of its corresponding level population to differ.

In the discussion of the relationship between component radiation and line radiation, most authors (e.g., [3.53,p.97] and [3.56,p.443] assume "natural excitation", or equal populations among the sublevels. This is a poor assumption for beam foil work, but fortunately not a necessary one since all sublevels must have the same mean life. Thus the line intensity emitted into all 4π steradians

$$I_{\gamma J,\gamma' J'} = \sum_{M}\sum_{M'} N_{\gamma JM} A_{\gamma JM,\gamma' J' M'} \tag{3.27}$$

can be written as a product of the level population,

$$N_{\gamma J} = \sum_{M} N_{\gamma JM} \quad , \tag{3.28}$$

and the line transition probability,

$$A_{\gamma J,\gamma' J'} = \sum_{M'} A_{\gamma JM,\gamma' J' M'} \quad , \tag{3.29}$$

since the sum of component transition probabilities taken over lower states is independent of the upper state. Thus (3.4) is valid for both lines and components if all radiation is included.

To describe a realistic experimental situation, let us consider the electric dipole radiated intensities $\Delta I^{\parallel}$ and $\Delta I^{\perp}$ detected at an angle θ to the beam through linear polarizers set parallel and perpendicular to the beam-photon plane. These intensities can be written in terms of the Zeeman components $\Delta M=0$ and $\Delta M=\pm 1$ as

$$\Delta I^{\parallel} = \eta_{\lambda}^{\parallel} \sum_{M} N_{\gamma JM}[A_{\gamma JM,\gamma' J' M}\,(2-3\cos^2\theta) + A_{\gamma J,\gamma' J'}(\cos^2\theta)] \tag{3.30a}$$

$$\Delta I^{\perp} = \eta_{\lambda}^{\perp} \sum_{M} N_{\gamma JM}[A_{\gamma J,\gamma' J'} - A_{\gamma JM,\gamma' J' M}] \quad , \tag{3.30b}$$

where $\eta_{\lambda}^{\parallel}$ and $\eta_{\lambda}^{\perp}$ are the detection efficiencies at the wavelength λ. Sum rules were used to write the $\Delta M=\pm 1$ components in terms of the line and $\Delta M=0$ A-values. The $\Delta M=0$ A-value has the useful property

$$\sum_{M} A_{\gamma JM,\gamma' J' M} = (2J+1)\, A_{\gamma J,\gamma' J'}/3 \quad , \tag{3.31}$$

from which one can easily recover $\Delta I^{\parallel} = \Delta I^{\perp}$ for the natural excitation case, in which $N_{\gamma JM} = N_{\gamma J}/(2J+1)$. Thus the total radiation detected in a standard beam-foil arrangement at $\theta=90°$ is given by the sum of (3.30a) and (3.30b)

$$\Delta I = \sum_M N_{\gamma JM}[(2\eta_\lambda^{\parallel} - \eta_\lambda^{\perp})A_{\gamma JM,\gamma' J' M} + \eta_\lambda^{\perp} A_{\gamma J,\gamma' J'}] \quad . \tag{3.32}$$

Thus, unless $\eta_\lambda^{\perp} = 2\eta_\lambda^{\parallel}$, the intensity is not proportional to the total level population, but rather to a weighted sum of the sublevel populations.

Since all states of a level have the same mean life, the same mean lives will be contained in all component decay curves of a line. However, due to differences in sublevel populations, the admixtures of exponentials in the various components may differ, and the admixture of (3.32) will in general not correspond to that of the total radiation of any component, or to that of the line. This fact has implications which we shall discuss in the next subsection and in succeeding sections.

3.4.4 Distortions Which Preserve the Mean-Life Content of a Decay Curve

In Subsection 3.3.2 methods were described which can eliminate instrumental effects which could distort the decay curves from their multiexponential form. There are other classes of distortions inherent in all decay-curve measurements which do not affect the mean-life content of the curve, but do alter the admixture of exponentials. These distortions are important to consider, for although some analysis techniques (see Section 3.5) are sensitive only to the individual exponential components, others (see Section 3.6) rely on the detailed shape of the decay curve. It is possible to estimate the magnitude of these distortions and correct for them, as well as to adjust experimental conditions so as to eliminate them.

a) *Finite Time Window* The measured data are instrumentally integrated over some finite resolution time t. For a beam-foil source, this corresponds to the beam segment viewed by the optical system divided by the beam velocity; for a delayed coincidence measurement, it is the channel width (for a beam-foil source, Δt can be less than the channel width ΔT, which is the step size). The average value of the population over this interval is

$$\overline{N_n(t)} = \int_{t-\Delta t/2}^{t+\Delta t/2} dt' \; N_n(t')/\Delta t \quad . \tag{3.33}$$

Using (3.17) and (3.18) this becomes [3.111]

$$\overline{N_n(t)} = \sum_j \beta_{nj}[\sinh(\tfrac{1}{2}\alpha_j\Delta t)/(\tfrac{1}{2}\alpha_j\Delta t)] \exp(-\alpha_j t) \quad . \tag{3.34}$$

The admixture of exponentials is altered if the resolution width approaches any of the contributing mean lives.

b) *Finite Solid Angle* If the source is aligned, the intensity ΔI which enters a solid angle at 90° to the beam direction is given by (3.32). If we insert (3.17) and (3.18) we obtain

$$\Delta I = \sum_j \left\{ \sum_M \beta_{(\gamma JM)j} \left[(2\eta_\lambda^{\parallel} - \eta_\lambda^{\perp}) A_{\gamma JM,\gamma' J' M} + \eta_\lambda^{\perp} A_{\gamma J,\gamma' J'} \right] \right\} \exp(-\alpha_j t) \quad , \qquad (3.35)$$

so, unless every β is independent of M, we again have an exponential admixture which is different from that describing the level population [3.112]. Since the coefficient in braces involves the transition probabilities, the admixture can differ among the various branches of the decay, due to differing angular distributions of the same total intensity. This dependence can be eliminated by forcing $\eta_\lambda^{\perp} \to 2\eta_\lambda^{\parallel}$. This can be achieved by measuring each decay curve both in parallel and in perpendicular polarized light, correcting for the instrumental polarizations, and forming $\Delta I^{\parallel}/\eta_\lambda^{\parallel}$ + $2\Delta I^{\perp}/\eta_\lambda^{\perp}$. The same result can be attained by viewing the source through a linear polarizer with its axis inclined at the "magic angle" $\cos^{-1}(1/\sqrt{3}) = 54.7°$ to the beam-photon plane [3.113]. The polarizer will then pass 1/3 of the parallel and 2/3 of the perpendicularly-polarized light, which gives the desired result for a detection system which has no instrumental polarization. Instrumental polarization can be reduced to less than 1% by use of a polarization scrambler, which creates pseudo-depolarized light after polarization analysis [3.114].

With both polarizations viewed at an angle θ to the beam direction by a detection system with no instrumental polarization, we obtain, adding (3.30a) to (3.30b), with $\eta_\lambda^{\parallel} = \eta_\lambda^{\perp}$, and the substitution of (3.17) and (3.18),

$$\Delta I = \sum_j \left\{ \sum_M \beta_{(\gamma JM)j} \left[A_{\gamma JM,\gamma' J' M}(1-3\cos^2\theta) + A_{\gamma J,\gamma' J'}(1+\cos^2\theta) \right] \right\} \exp(-\alpha_j t) \quad . \qquad (3.36)$$

Thus even without instrumental polarization, the geometrical detection efficiencies cause the admixture to differ from that of $N_n(t)$ in a manner dependent upon alignment, branch, and angle. An exception occurs if the radiation is viewed at the "magic angle" to the beam, defined as above to be $\theta_M = 54.7°$, for which the component transition probability drops out and the M sum over β becomes unweighted.

It should be emphasized that the distortions discussed above change only the admixture of the exponentials, and not the exponential mean lives themselves. Thus an analysis which simply fits exponentials will not be affected by these distortions, but they must be considered in more sophisticated techniques which utilize cascade relationships between jointly-analyzed decay curves.

3.5 Mean-Life Extraction by Exponential Fits to Individual Decay Curves

The customary method for determining mean lives consists of approximating the decay curve of the desired level through the adjustment of the C_j and α_j parameters in the fitting function $\bar{I}$

$$\bar{I} \equiv \sum_j C_j \exp(-\alpha_j t) \quad . \tag{3.37}$$

As we have seen, an atomic decay curve is usually a multiexponential sum with a very large number of terms. However, in practice it is often sufficient to include only one or two cascade exponentials to obtain a good approximation of the primary mean life. The cascade mean lives obtained in such a fit should be considered averaged effective values, which do not necessarily correspond to any specific level. However, experiences with detailed cascade analyses (see section 3.6) have shown that there are cases in which one or two cascades completely dominate the repopulation, and are reliably recoverable from the decay-curve fit.

Nearly any analysis of exponential mean lives begins with a graphical plot on semilog paper, with the mean-life components resolved by repeated subtraction of straight lines. This curve-peeling approach can in some cases yield reliable results, but its limitations in accuracy due to cumulative error buildup should be apparent. However, the graphical method is still commonly used to obtain reasonable starting values for the computer programs which will be described here.

3.5.1 Maximum Likelihood Method

The maximum likelihood method utilizes the likelihood function L [3.115] defined as

$$L \equiv \prod_{i=1}^{N} \left[\bar{I}(C_1, C_2, \ldots \alpha_1, \alpha_2, \ldots t_i) / \int_{(t)} dt \, \bar{I} \right] \quad , \tag{3.38}$$

where $\bar{I}$ is defined in (3.37), the product is over i individual counts, each registered at a time t_i, and the integration extends over the entire measurable time window. The fit is accomplished by varying the parameters in such a way as to maximize L. Since the equations are non-linear, this requires a brute-force search. Computer programs have been developed which achieve maximization by the MALIK [3.116] stepping procedure, as well as by an alternative iteration procedure [3.117]. Provisions have been made to accept grouped multichannel data by correcting for the fact that the decay shifts the count centroid away from the channel centroid [3.118-120]. A modification has also been suggested [3.121] in which (3.37) is rewritten in terms of an alternative parametrization $C_j \rightarrow C'_j \alpha_j$. The correlation between C'_j and α_j is weaker than between C_j and α_j, which speeds the convergence. Use of these maximum-likelihood programs has been limited to one or two exponentials with constant background terms.

3.5.2 Non-Linear Least Squares Method

The most frequently-used technique is the method of least squares, which involves a minimization of the quantity [3.122]

$$S^2 = \sum_i [I(t_i) - \overline{I}\,(C_1,C_2,\ldots\alpha_1,\alpha_2,\ldots t_i)]^2\, W_i \quad . \tag{3.39}$$

Here $I(t_i)$ is the number of counts in channel i, t_i is the time centroid of channel i, $\overline{I}$ is the fitting function of (3.37), W_i is the uncertainty weight in the measurement of $I(t_i)$, and i is summed over channels. Minimization of S^2 is equivalent to maximization of L if the measured values are gaussian distributed about the parent distribution. For Poisson statistics, the best fit is not obtained by minimizing this mean square deviation from the mean since the Poisson distribution is always skewed below the mean, particularly so for counts less than around 100. Thus it is sometimes desirable to pool several channels on the decay curve tails in making this analysis.

Again, since (3.37) is not linear in the α_j parameters, neither L nor S^2 can be optimized by analytical differentiation and inversion (unless only a single exponential contributes, in which case the logarithmic intensities can be linearly fitted). Thus minimization must be achieved through numerical search methods, such as the method of steepest descent or the Gauss-Seidel linearization (Taylor Series). Since each of these techniques has a characteristic efficiency which depends upon the proximity to minimization, they are often combined through a scheme such as Marquardt's Algorithm [3.123] (the maximum-neighborhood method). If a true minimum is achieved, the resulting fitting parameters are independent of how the minimum was located and various programs should differ only in the time and number of iterations required for convergence. Weighted non-linear least-squares fitting programs are contained in the systems libraries of most large computer installations (e.g., [3.124-126]), and most packages already include an exponential sum as an optional function. Special programs of this type have also been developed specifically for use in mean-life data reduction (e.g., [3.127,128]). The amount of programming required is substantial, and most laboratories adapt an existing library program.

Although the fits should be independent of the specific search program, they are often quite sensitive to the weights chosen. For convenience, the weights are based on the measured data rather than the parent distribution and include a number of different sources of uncertainty. Since a decay curve begins with a high number of counts and often ends when the signal becomes comparable with fluctuations in the background, the local accuracies over the decay curve are expected to vary greatly. Statistical fluctuations are often not sufficient to account for the observed spread of data points, particularly in the high count region, where statistical inaccuracies may be only 0.1%, and non-statistical errors must be considered. A possible weighting which includes an incoherent sum of statistical and non-statistical errors is given by

$$W_i = \left\{I(t_i) + [\varepsilon I(t_i)]^2 + 2B_i\right\}^{-1} \tag{3.40}$$

Here ε is a small number, chosen so that $(W_i)^{-1/2}$ will match the spread of data points for large $I(t_i)$ while contributing a negligible amount for small $I(t_i)$, and B_i denotes any measured backgrounds which have been pre-subtracted to give $I(t_i)$. (The background may arise from several separate Poisson-distributed contributions, since such a merger is also Poisson-distributed with a mean equal to the sum of the separate means). If the fit is found to be sensitive to ε, a detailed study of non-statistical errors is necessary. In a modern beam-foil measurement, such as was described in subsection 3.3, non-statistical errors have largely been removed and statistical weights are usually sufficient. If weights are correctly chosen and properly normalized, the sample variance can be compared with that of a parent gaussian distribution of the same number of degrees of freedom (the number of data points minus the number of fitting parameters) by the χ^2 test [3.122]. The number of exponential terms which must be included for an adequate but not overparameterized fit can be selected by comparing the values of S^2 for various inclusions with a table of corresponding χ^2 probabilities [3.122]. The uncertainties in the individual parameters (for either the least-squares or maximum-likelihood methods) can be estimated by examining χ^2 as a function of each parameter in the neighborhood of the minimum.

It is seldom that beam-foil decay curves can be fitted to more than three exponentials (six free parameters). One limitation is imposed by the finite time-window (as shown in Fig.3.1; it is difficult to resolve times shorter than 0.05 ns or longer than 100 ns within a standard beam-foil chamber). Thus, only under unlikely circumstances could four exponential mean lives all differ from each other by at least an order of magnitude, and thus have a region of dominance within the measured decay curve. In pulsed beam-gas excitation, it is possible to vary the duty cycle and examine the decay curve in various time domains [3.16], but even here it is difficult to extract the same four exponentials from two separate but equivalent measurements of a decay curve. LANCZOS [3.129] has studied the numerical implications of the exceedingly non-orthogonal properties of the exponential functions, and has noted extraordinary sensitivity of the fitted parameters to small changes in the data. This causes difficulties in applications to physical problems, where the aim of the fits is not merely to approximate the data closely by a mathematical function, but to determine accurately the physically meaningful parameters. LANCZOS estimates that to fit four or five exponentials would require measurement accuracies of from 6 to 8 significant figures, which is quite unrealistic when compared to actual measurement accuracies. However, in normal mean-life determinations, the situation is not as hopeless as this analysis would imply, since we desire accuracy in only one fitted parameter, the primary level's mean life. If this mean life dominates the decay curve, the fact that nonphysical values are recovered for the other parameters is not serious. The LANCZOS *caveat* must certainly be heeded in attempts to extract cascade mean lives or relative initial populations from decay-curve fits.

The knowledge that the decay curve is a sum of exponentials, which have special analytical properties, is not incorporated into the basic fitting process. It is possible to utilize these properties, either in conjunction with a traditional fit, or in alternative fitting procedures, and partially counteract the disadvantages of non-orthogonality.

3.5.3 Differentiation and Integration of Decay Curves

One aspect of the exponential function which is not contained in the basic fitting procedure but which can sometimes increase the reliability of a fit involves its very simple properties under differentiation and integration. The derivative of (3.37) is

$$d\bar{I}/dt = -\sum_j (C_j\alpha_j)\exp(-\alpha_j t) \quad , \tag{3.41}$$

while its integral from to to infinity is

$$\int_t^\infty dt'\bar{I}(t') = \sum_j (C_j/\alpha_j)\exp(-\alpha_j t) \quad . \tag{3.42}$$

Therefore the differentiated and integrated decay curves should contain the same mean lives, but with coefficients which differ in a predictable way. This can be coupled with the fact that the early and the late portions of a decay curve are usually quite different in information content. The early portions contain short-lived exponentials and have high statistical accuracy, while the late portions contain long-lived exponentials and have low statistical accuracy. If the early portion possesses sufficient statistical accuracy to permit numerical differentiation [3.130], the process will further enhance the short-lived components and eliminate entirely the necessity for a constant background subtraction. Similarly, if the late portion can be integrated (this requires an accurate background subtraction and an extrapolation of the integration beyond the end of the decay curve, which can be done iteratively [3.50,p.28]) the long-lived exponentials will be further enhanced, and the integration will reduce the statistical fluctuations. The fitting procedure can be performed three times: once for the raw decay curve, once for the early portion of the differentiated curve, and once for the late portion of the integrated decay curve. An example [3.131] is shown in Fig.3.3. Agreement among the coefficients and mean lives in the various fits indicates the exponential content has been correctly described, and disagreement can indicate that an inappropriate number of exponential terms was used.

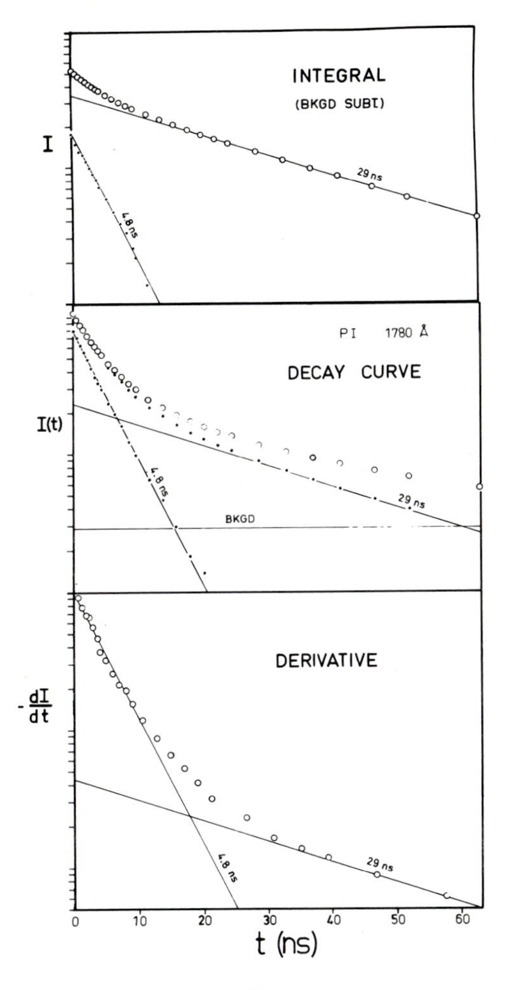

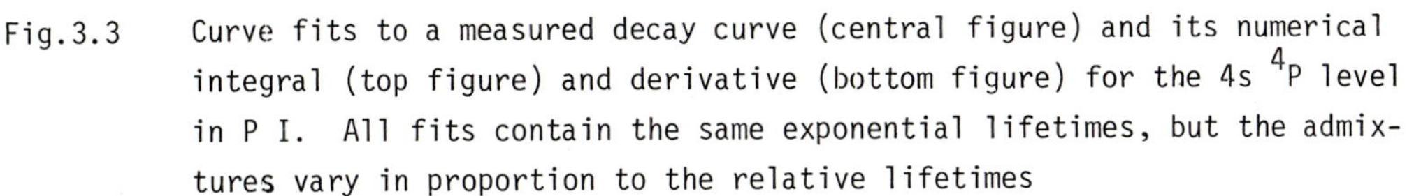

Fig.3.3 Curve fits to a measured decay curve (central figure) and its numerical integral (top figure) and derivative (bottom figure) for the 4s ^{4}P level in P I. All fits contain the same exponential lifetimes, but the admixtures vary in proportion to the relative lifetimes

3.5.4 Expansion About a Close-Lying Mean Life

In some cases, beam-foil decay curves appear to be relatively free of cascades. The question then arises of whether they really are cascade-free, or merely an effective sum of close-lying values. As we shall show in Subsection 3.6.1, there are situations in which this question is virtually impossible to answer. However, there are fitting techniques which are especially sensitive to small lifetime differences which can be used in such cases. As was pointed out by DENIS et al. [3.132], if (3.16) is Taylor-expanded about $\alpha_j = \alpha_n + \Delta\alpha$ for a contribution from a single direct cascade j, the equation becomes

$$N_n(t) = \exp(-\alpha_n t)\ [N_n(0) + N_j(0)A_{jn}(t-\Delta\alpha t^2/2) + \ldots] \tag{3.43}$$

and a polynomial linear fit to the quantity $I_{nf}(t)\ \exp(\alpha_n t)$ (using the value α_n from a single exponential fit) can very sensitively expose small differences in mean life. ERMAN [3.16] has studied the problem in the context of pulsed-electron beam-gas excitation and has concluded that (barring the unlikely situations described in Subsection 3.6.1) it can introduce at worst a few percent error. If the cascade lifetime differed from the primary by more than this amount, the decay data would exhibit measurable curvature. If the lifetimes differed by less, the fitted value would be between the primary and cascade mean lives.

3.5.5 Fourier-Transform Methods

An alternative procedure has been developed by GARDNER et al. [3.133] in which the exponential sum over a discret set of mean lives is re-expressed as a Laplace integral over a continuous set of mean lives

$$I(t) = \sum_j C_j \exp(-\alpha_j t) \rightarrow \int_0^\infty d\alpha\ g(\alpha) \exp(-\alpha t) \quad , \tag{3.44a}$$

where

$$g(\alpha) \equiv \sum_j C_j \delta(\alpha-\alpha_j) \quad , \tag{3.44b}$$

and the function $g(\alpha)$ takes the form of a frequency spectrum with peaks of height C_j centered on the values α_j. Here $g(\alpha)$ can be computed through the inversion of this integral equation, and since $g(\alpha)$ has entirely different properties from $I(t)$, this approach may avoid some of the problems created by the non-orthogonality of the exponential sum. Since inverse Laplace transform calculations for empirical functions present practical problems [3.108], a method of inversion using Fourier transforms was employed [3.133], which has been extended to permit the use of modern Fast Fourier Transform Algorithms [3.134] by SCHLESINGER [3.135]. Unfortunately this method requires the observation of all decays for many mean lives, and is seriously limited when more restricted ranges must be used. Another disadvantage is that the uncertainties in the parameters are difficult to estimate. The technique has not generally been applied to atomic mean-life work, but at the least it could provide useful starting values for an exponential search program.

3.5.6 Method of Moments

Another technique which transforms exponential data into an alternative form has been developed by BAY [3.136,137] and extended to include several exponentials by DYSON and ISENBERG [3.138], ISENBERG and DYSON [3.139], and SCHUYLER and ISENBERG [3.140]. Moments of the decay curve are numerically computed over the range of measured data

$$u_k \equiv \int_0^T dt\ t^k I(t) \quad , \tag{3.45}$$

where k runs from 0 to one less than twice the number of exponentials in the fitting function (so as to match the number of parameters). The fitting function (3.37) is also integrated and becomes (corrected for truncation)

$$\overline{u_k}\ (C_1,C_2,\ldots\alpha_1,\alpha_2,\ldots k) = \sum_j C_j \left[k!/\alpha_j^{k+1} - \int_T^\infty dt\ t^k \exp(-\alpha_j t)\right] \quad , \tag{3.46}$$

and a least-squares fit is made to the moments

$$S^2 = \sum_k [u_k - \overline{u_k}\ (C_1,C_2,\ldots\alpha_1,\alpha_2,\ldots k)]^2 \quad . \tag{3.47}$$

The infinite integrals in the fitting function can be generated recursively, but this correction usually contributes only a small perturbation in the fit. In order to integrate the data reliably, a special numerical filter was developed [3.138-140] to remove the high-frequency noise components. The technique has been successfully used to analyze simulated three-component decay curves, and an extensive review of the method and its application to biological systems is given by YGUERABIDE [3.141].

3.6 Mean-Life Extraction by Joint Analysis of Cascade-Related Decay Curves

In the previous section, analysis techniques were discussed by which mean lives are extracted from individual decay curves. These methods are in general highly reliable and provide a conclusive determination in most cases. However, there are a few situations in which the extraction is either particularly difficult, or for which there is an ambiguity in the assignment of the fitted mean lives. In such cases, it is possible to utilize additional conditions between cascade-related decay curves to sharpen the determination and to remove the ambiguities. We shall therefore discuss a few of the most difficult possible situations, and describe the techniques by which even these examples become tractable.

3.6.1 Ambiguities in the Assignment of Fitted Mean Lives

It is known [3.142] that substitution of certain combinations of decay rates and initial populations into (3.16) will remove the exponential term corresponding to the primary mean life. For example, if $N_3(0):N_2(0):N_1(0) = 5:3:1$ and $\alpha_3:\alpha_2:\alpha_1 = 3:8:27$ (unbranched) then (3.21) yields $\beta_{12} = 0$ and $\beta_{13} = N_1(0)$ and a single exponential results, although three different mean lives are involved. We mention this to show that a single exponential does not necessarily preclude cascades nor unambiguously determine mean lives.

The primary mean life can be extracted even if it is not among the fitted exponentials. This mean life may not be exhibited explicitly as an exponential, but it is nonetheless manifested implicitly through the relative coefficients of the cascade and primary admixtures, if both decay curves are measured. The decay curve $N_1(t)$, containing only the exponentials present in its direct cascade, and presenting a less drastic example which clearly illustrates the above point, is shown in Fig.3.4.

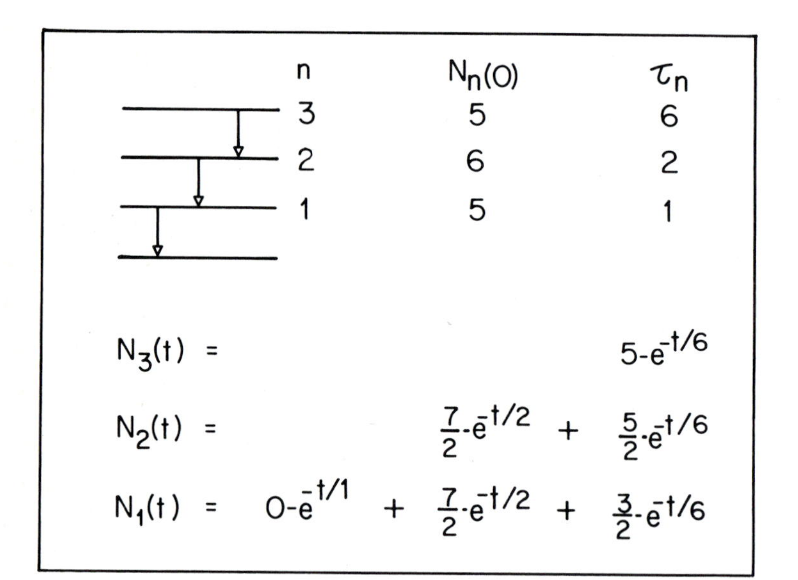

Fig.3.4 Example of an indirectly-cascaded level with a decay curve which does not involve the lifetime of that level. Although that lifetime does not appear as an exponential in any of the decay curves, it can be extracted from a comparison of the admixtures

In such a two-level example, the differential equation,

$$dN_1/dt = N_2(t)A_{21} - N_1(t)\alpha_1 \quad , \tag{3.48}$$

couples the solutions of (3.17) and (3.18) to yield the relationship

$$\frac{\beta_{22}}{\beta_{23}} = \frac{\alpha_1-\alpha_2}{\alpha_1-\alpha_3} \, \frac{\beta_{12}}{\beta_{13}} \quad , \tag{3.49}$$

and α_1 can be evaluated from the ratios of the coefficients and the cascade mean lives. Such implicit information can be utilized to analyze a wider class of situations in which the primary exponential term, although not fortuitously zero, is small, due either to heavy cascading or to a short mean life. Generalized techniques to extract mean lives from cascade-coupled sets of decay curves will be described in the following subsections.

A frequently-cited example in which ambiguities in the assignment of mean lives occur is the "growing-in" decay curve. In this situation, one or more of the exponential coefficients is negative, and a maximum may occur in the decay curve or in one of its derivatives. If we examine the possibilities for such an occurrence in the absence of indirect cascades, (3.19) and (3.20) admit two possibilities: a short-lived primary level of low initial population, or a cascade level which is shorter-lived than the primary. Selection between these possibilities is called the direct cascade "growing-in ambiguity". Unfortunately, such ambiguities are not restricted to growing-in decay curves.

If a level were repopulated only by direct cascades, it would follow from the above that the primary mean life would be equal to or shorter than the shortest fitted mean life which has a positive coefficient. Thus in the absence of growings-in, the primary would always be the fastest contributor. Although this has been found generally to be the case, it is not a rigorously valid condition if indirect cascading is present. An example is given in Fig.3.5 of an indirect cascade scheme in which the primary level is of intermediate mean life, but exhibits no growing-in.

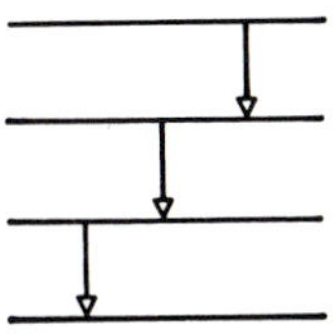

n	N_n	τ_n
3	7	1
2	0	9
1	5	2

$$N_3(t) = 7\cdot e^{-t/1}$$

$$N_2(t) = \frac{63}{8}\cdot e^{-t/9} - \frac{63}{8}\cdot e^{-t/1}$$

$$N_1(t) = 1\cdot e^{-t/2} + \frac{9}{4}\cdot e^{-t/9} + \frac{7}{4}\cdot e^{-t/1}$$

Fig.3.5 An example of a decay curve in which a cascade is the fastest contributing lifetime, but no negative "growing-in" coefficients are exhibited

Again, the ambiguity can be resolved if the cascade decay curves are also measured. Despite this counter-example to the "fastest non-growing-in" argument, it is still possible to establish a rigorous upper limit to the level mean life by examination of the logarithmic derivative of the decay curve. Since the replenishment ratio cannot be negative, (3.25) implies that

$$-d(\ell n\ N_n)/dt \leq \alpha_n \quad . \tag{3.50}$$

KAY [3.143] has shown that this inequality is valid, regardless of cascading conditions, instrumental time-resolution, detection efficiency, data-point spacing, or velocity-dispersion effects. Although the situations described above may seem formidable, they are handled rather easily, if the decay curves of the cascade levels are also accessible to measurement, by techniques which are described in the subsections to follow.

3.6.2 Constrained Fits

One technique for incorporating measured cascade information into the analysis is the constrained fit. The decay curves of contributing cascades are measured and fitted by non-linear least-squares methods. Decay constants from the strongest lines observed in the cascades are then forced into the fitting function of the primary decay-curve least-squares fit, but the cascade mean-life values are fixed and only their coefficients are allowed to vary freely in the fit. Since these coefficients are linear parameters in the fit, they do not display the non-orthogonal behavior of the mean-life fits, and their use can account for a higher complexity of cascading than would otherwise be possible. MARTINSON et al. [3.144] have used this technique to fix up to five cascade mean lives in Be I decay curves. TIELERT and BUKOW [3.145] have developed a technique which minimizes the joint χ^2 for two measured decay curves which contain common mean lives, and have thus analyzed hydrogen decay curves including the effects of five mean lives, only one of which was free to vary. Although their analysis was primarily intended to determine the coefficients for population determinations, it recovered a highly reliable value for the free mean life.

3.6.3 Linearly-Fitted Normalizations of Cascade-Related Decay Curves

The constrained-fit methods described above use only the information that cascade exponentials are also contained in the primary decay curve, and ignore the relationships among the coefficients of the various admixtures which are imposed by the population equations, such as (3.49). This does have certain advantages, since mean lives are independent of excitation conditions and coefficients are not, so decay curves measured under quite different circumstances can be combined by a constrained fit. However, if we impose the additional condition that the primary level and all of its significant cascade levels are observed under equivalent excitation conditions, then their populations are coupled by the differential equation

$$dN_n/dt = \sum_j N_j(t)A_{jn} - N_n(t)\,\alpha_n \quad . \tag{3.51}$$

If the decay curves are measured under circumstances such that they are free of all types of distortions including those discussed in Subsection 3.4.4, and are corrected to remove all background and blend contributions, then the decay intensity $I_k(t)$ (observed in any decay branch) is proportional to the instantaneous population of the decaying level k

$$I_k(t) = N_k(t)/\xi'_k \quad , \tag{3.52}$$

where ξ'_k is a constant which depends upon the detection efficiency, the transition probability of the branch observed, the data-accumulation period, the slit width, etc. Thus the various $I_k(t)$ are Arbitrarily-Normalized Decay Curves (ANDC) and $\xi_k = \xi'_j/\xi'_n$ is the undetermined parameter which normalizes the jth cascade relative to the primary. If all ANDC are measured with a common time base and initialization, (3.51) becomes [3.146]

$$dI_n/dt = \sum_j \xi_j I_j(t) - \alpha_n I_n(t) \quad , \tag{3.53}$$

which represents a separate linear relationship connecting α_n and the various ξ_j for each instant of time, with coefficients obtainable directly from the measured ANDC. Here the empirical decay curves themselves, rather than their approximate mathematical representations, are the fitting functions, and cascade mean lives occur only implicitly and are not determined. Equation (3.53) provides as many independent linear relationships as there are data channels which can be satisfied simultaneously by a standard multiple linear regression [3.122], which analytically minimizes S^2 in

$$S^2 = \sum_i (y_i - \sum_k a_k x_{ki})^2 \quad . \tag{3.54}$$

There is a choice in the possible definitions of the variables y_i and x_{ki}, which can provide alternative approaches, depending upon the accuracy of the data. If the statistical accuracy is sufficient to permit a numerical differentiation of the primary decay-curve, the variables may be defined directly from (3.53) as

$$y_i = \sqrt{W_i}\,\dot{I}_n(t_i) \quad , \quad x_{ki} = \sqrt{W_i}\,I_k(t_i), \; k = n, j_1, j_2, \ldots \quad , \tag{3.55}$$

where $\dot{I}_n = dI_n/dt$, t_i is the time coordinate of the ith channel, and W_i is the weight factor (this definition allows the use of general-purpose multiple-regression subroutines, which often are not designed to accept data of explicitly varying weight). Sophisticated numerical differentiation formulae often smooth over many data points, which is neither necessary nor desirable when it is followed by least-squares fitting. Thus a crude but adequate three-point differentiation formula for quasi-exponential data of equal spacing ΔT is given by

$$\dot{I}(t) = I(t)\ \ell n[I(t+\Delta T)/I(t-\Delta T)]/(2\Delta T) \quad . \tag{3.56}$$

If we assume uncorrelated statistical errors (neglecting the error correlation between I_n and $\dot{I}_n$), a satisfactory weight-factor is given by

$$W_i = [(\Delta T^{-2} + \alpha_{no}^2)I_n(t_i) + \sum_j \xi_{jo}^2\ I_j(t_i)]^{-1} \quad , \tag{3.57}$$

where α_{no} and ξ_{jo} are estimated values of α_n and ξ_j, not varied in the fit, but iterated if incorrectly estimated [3.147].

If the data cannot reliably be differentiated, an alternative formulation has been suggested and utilized by KOHL [3.148], CURTIS et al. [3.149], KOHL et al. [3.150], and SCHECTMAN et al. [3.151], in which both sides of (3.53) are integrated between arbitrary limits T_I and T_F, and the variables of (3.54) become

$$y_i \equiv \sqrt{W_i}\ [I_n(T_F) - I_n(T_I)] \quad , \quad x_{ik} \equiv \sqrt{W_i}\int_{T_I}^{T_F} dt\ I_k(t) \quad . \tag{3.58}$$

This formulation has the disadvantage that it combines several data points in the integration, so that the number of degrees-of-freedom is not inferred directly by the number of data points, complicating the interpretation of goodness-of-fit criteria and the definition of weight factors. However, the problems of numerical differentiation can be severe, and this integration approach provides an effective means of pooling data without degrading the time-resolution of the decay-curve.

In either formulation, the number of independent relationships must greatly exceed the number of cascades for any analyzable case. Thus the applicability of this method to a given set of measurements is indicated by the goodness of the fit obtained. If one or more significant cascade decay-curves has been neglected, the analysis should produce a poor fit. Contrarily, if a given cascade is neglected and a satisfactory fit can be obtained, then its contribution to the repopulation is probably small. There are a number of ways to test the goodness-of-fit. One involves reciprocity of the fit under interchange of the role of the independent variables in the fit. In our formulation, we have chosen y_i as the axis along which deviations are minimized. If one of the other variables assumes this role, the minimization process is altered, but the values inferred for α_n and the ξ_i should be the same to within fit variances if the decay curves are truly correlated. The χ^2 test also gives a measure of goodness-of-fit, provided the weights chosen accurately reflect the uncertainties in the measurement. With appropriate weights, the regression also provides an estimate of the uncertainties in α_n and ξ_j.

The goodness-of-fit can be displayed on a graphical plot if the number of parameters can be reduced to two. As an example, consider a level 1 which has cascades from levels 2 and 3. Equation (3.53) becomes

$$dI_1/dt = \xi_2 I_2(t) + \xi_3 I_3(t) - \alpha_1 I_1(t) \quad . \tag{3.59}$$

If the ratio $e = \xi_3/\xi_2$ can be determined, either empirically or by a prior multiparameter regression, this equation can be rewritten for the ith channel as

$$[-d(\ell n I_1)/dt] = \alpha_1 - \xi_2[(I_2+eI_3)/I_1]_i \quad . \tag{3.60}$$

Thus a plot of one bracketed quantity vs. the other will yield α_1 and ξ_2 as an intercept and a slope, respectively, about which the scatter of points is an obvious goodness-of-fit indicator. An example of this is shown in Fig.3.6 for the 4p ^2P level in

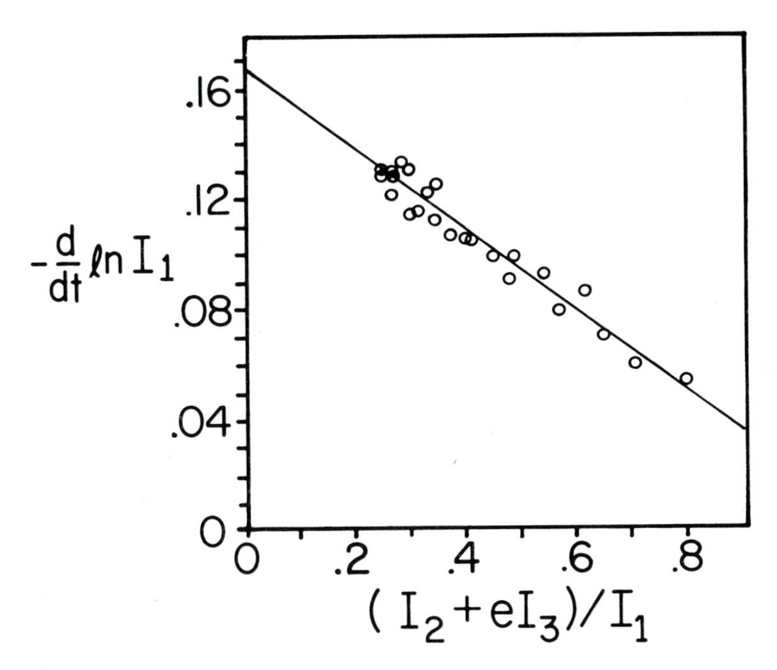

Fig.3.6 Plot of the negative logarithmic derivative of the primary level vs. the adjusted sum of its cascade decay curves for the 4p ^2P level in Ca II. These quantities are linearly related, so the ordinate intercept gives α_{4p} and the slope gives the relative normalization of the cascades to the primary

Ca II. The decay curve of 4p ^2P was heavily repopulated by fast cascades from 5s ^2S and 4d ^2D which gave it a strong growing-in shape. This greatly hampered accurate curve fitting and could have allowed ambiguities in the assignment of mean lives if the cascade decay-curves had not also been measured. A three-parameter regression of the three measured decay-curves was made with (3.59) which yielded values of α_1 = 0.168, ξ_2 = 0.86 and ξ_3 = 0.43. The bracketed quantities in (3.60) were then computed with e set at 0.50, and the goodness-of-fit is clear from the plot.

This method employs exact linear relationships to incorporate cascade decay-curves into the analysis of the primary decay-curve, does not involve exponential curve-fitting, and provides self-consistency checks of its validity. The method utilizes the same decay-curves as standard curve-fitting techniques, so is an additional rather than an alternative approach, which can be very useful in resolving possible ambiguities and extracting mean lives in troublesome cases.

3.7 Cascade-Free Methods

Several methods have been developed which can, in certain circumstances, completely eliminate cascade effects from fast ion-beam mean-life measurements.

3.7.1 Beam-Foil Coincidence Techniques

Although photon-photon coincidence techniques have provided a method for making cascade-free mean-life measurements in nuclear physics and in electron-beam excited-atom studies, low light-levels make its application very difficult in the beam-foil case. However, this technique has been used successfully by MASTERSON and STONER [3.152] to measure one mean life for which a single cascade dominates. In their experiment, the true coincidence rate was only about 1/40 of the chance coincidence rate, but they were able to subtract the chance coincidences by use of a single-channel approach (a multichannel analyzer recorded both the fixed-delay-window peak and surrounding background, permitting an estimation of the background under the peak). In order to obtain a signal 3 times the statistical fluctuations in the chance coincidences, about $(3x40)^2$ = 14400 total coincidences were required per channel, the accumulation of which took approximately 40 hours per point, or 10 days of running time for a 6-point decay curve. Although these times indicate the great difficulties inherent in such a measurement, the results clearly demonstrate that the measurement is possible. The method has a unique advantage in that it can establish a correlation between the cascade and primary transitions, and can therefore be used to verify level schemes.

3.7.2 Use of Alignment to Discriminate Against Cascades

It is well known that beam-foil excitation often produces alignment in substate populations, which (as shown in Subsection 3.4.3) gives rise to linear polarization of the radiation emitted into limited solid angles. For a given level, the population consists of two incoherent non-interacting portions: (a) the "remnant" of the initial foil excitation, and (b) the "repopulation" added by cascades. Since the sublevels all decay with the same mean life, the remnant retains its initial alignment, while the repopulation contains whatever alignment it transfers from the higher-lying states (which in general is reduced by the $\Delta M = 0,\pm 1$ selection rules). NEDELEC [3.153] has obtained general expressions for the transfer of alignment by cascades, which indicate that it is in most cases small. Let us therefore consider a case in which cascade-induced alignment is negligible, either because the cascade levels are not aligned, or because their alignment is not efficiently transferred. In this case (3.17) and (3.18) for the sublevels can be written as

$$N_{\gamma JM}(t) = [N_{\gamma JM}(0) - \sum_k \beta_k] \exp(-\alpha_{\gamma J} t) + \sum_k \beta_k \exp(-\alpha_k t) \quad , \tag{3.61}$$

where k is summed over the cascades, and since the cascade contributions to all sublevels of the primary level are the same, the β_k are M-independent. This expression suggests an analysis technique which can be understood by consideration of (3.30a) and (3.30b). If the light is viewed at $\theta=90°$ through parallel and perpendicular polarizers, the results being corrected for instrumental polarization, the difference in intensities yields

$$\Delta I^{\parallel}/\eta^{\parallel} - \Delta I^{\perp}/\eta^{\perp} = \sum_M N_{\gamma JM}(t)[3A_{\gamma JM,\gamma' J' M} - A_{\gamma J,\gamma' j'}] \quad . \tag{3.62}$$

Substituting in (3.61) and using the sum rule of (3.31), we find that the k terms vanish, leaving the expression

$$\Delta I^{\parallel}/\eta^{\parallel} - \Delta I^{\perp}/\eta^{\perp} = \left\{\sum_M N_{\gamma JM}(0)[3A_{\gamma JM,\gamma' J' M} - A_{\gamma J,\gamma' J'}]\right\} \exp(-\alpha_{\gamma J} t) \quad , \tag{3.63}$$

so the difference between the two multiexponential decay curves should be a single exponential with the mean life of the primary level. This approach has been used by BERRY et al. [3.154]. After subtracting the unpolarized background and correcting for instrumental polarizations (which were made small by use of a scrambler) the two multiexponential polarization decay-curves of the Li II 4787Å transition (indicated by hollow circles) were subtracted to give a single exponential curve (shown by solid circles). Although the polarization was small, tolerable accuracies in the subtracted differences were achieved by keeping statistical uncertainties in the measured decay curves below 1%. They also used this technique to remove blending effects in hydrogenic decay curves. The S-state portions, being unpolarized, cancelled on subtraction, and higher angular momentum state polarizations showed an energy dependence. Thus for the He II 3-5 transition the 5f mean life dominated the subtracted decay curve at

300 keV, while the 5p mean life dominated at 1.5 MeV. For BERRY et al.'s results, see Fig.3.7.

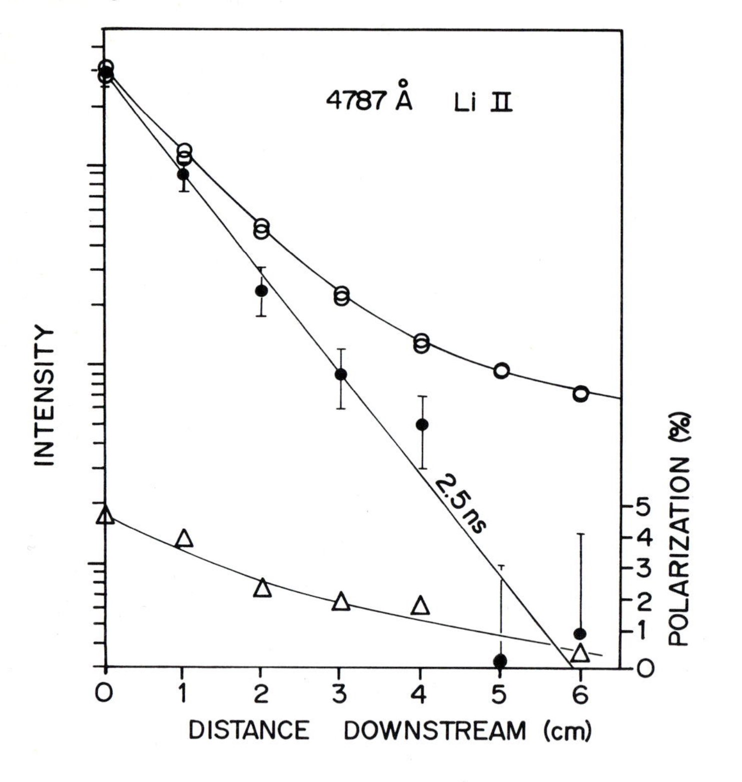

Fig.3.7 Decay curves for the Li II 3p ^1P - 4d ^1D transition measured in polarizations parallel (upper O) and perpendicular (lower O) to the beam axis and their subtracted difference (●), multiplied by 10 for common (arbitrary) normalization. The lower part of the figure (Δ) shows the change of polarization as the cascading fraction increases

An automated system for measuring (3.63) directly, coupling rotating polarization analyzers to a phase-sensitive detector, has been used by SCHECTMAN [3.155], which may make this specialized technique more generally applicable. A similar technique, utilizing the lack of cascade-induced alignment to suppress cascade effects in the beam-foil Hanle effect has been utilized by LIU and CHURCH [3.156] and CHURCH and LIU [3.157].

3.7.3 Laser Excitation

One of the most promising new techniques which has been used for mean-life measurements is selective excitation of a fast ion beam in flight by an intersecting laser beam. This technique provides cascade-free decay curves, and shares with beam-foil excitation its high isotopic purity and low particle density, while avoiding the problems of beam energy spread due to losses in the foil. The first experiment of this type was performed by ANDRÄ et al. [3.158,159] who Doppler-tuned an argon-ion laser through 9Å by variation of the intersection angle between the beams, which then corresponded to a Ba II resonance transition. This technique can routinely yield mean lives good to better than 1%, and is being extended through the use of tunable dye lasers and beam excitation within a laser cavity. However, in this form, the technique cannot be applied to levels which are not optically excitable from the ground state, and does not have the access to highly ionized and multiply-excited states of a beam-foil source.

Some of the limitations of this method can be eliminated by a variation developed by HARDE and GUTHÖHRLEIN [3.160], in which the ion beam is pre-excited in a gas cell, (or foil) before receiving further excitation from a continuous-wave intracavity dye laser, which is chopped at 82 Hz. Although the total emitted intensity is not cascade-independent, the difference between the chopped decay curves, measured with a digital lock-in photon counting system, is, and decays as a single exponential at points downstream from the laser. To verify this, consider a component transition between an upper state n and a lower state k (the laser does not pump the various sublevels equally, so the components must be considered separately). Without laser excitation, the upper state population $N_n^{\,0}(t)$ is governed by

$$dN_n^{\,0}/dt = \sum_i N_i(t)A_{in} - N_n^{\,0}(t)\,\alpha_n \quad . \tag{3.64}$$

With laser excitation, the population becomes $N_n^{\,1}(t)$, governed by

$$dN_n^{\,1}/dt = W_{kn}[N_k^{\,1}(t) - N_n^{\,1}(t)] + \sum_i N_i(t)A_{in} + N_n^{\,1}(t)\,\alpha_n \quad , \tag{3.65}$$

where W_{kn} is the stimulated transition probability [3.161]. Notice that the cascade contributions are the same in either case, so if we subtract (3.64) from (3.65) we obtain

$$\left(\frac{d}{dt} + \alpha_n\right)\left(N_n^{\,1}(t) - N_n^{\,0}(t)\right) = W_{kn}\left(N_k^{\,1}(t) - N_n^{\,1}(t)\right) \quad . \tag{3.66}$$

Thus during the dwell time in the laser when $W_{kn} \neq 0$ there will be a pumping (either up or down) proportional to the difference in population between the upper and lower levels. Thus it is important to place the laser at an appropriate distance downstream from the pre-exciter to optimize that difference. Downstream from the laser, where $W_{kn} = 0$, (3.66) yields a single exponential of inverse mean life α_n, if the upper level populations with and without the laser differ. This method has so far been applied only to ground state and metastable state cases, but it may well be one of the major applications of beam-foil methods in the future.

3.8 Concluding Remarks

In the past few years the measurement and analysis of beam-foil decay curves has been substantially improved and mean lives measured by this technique must be considered to be among the most reliable values presently available. Cascade effects, once a serious problem, are now correctly accounted for in the vast majority of cases. Beam-foil measurements have provided more atomic mean-life information than all other direct measurement techniques combined, even for neutral and singly-ionized atoms for which other higher intensity sources are available. Post-foil optical pumping using tunable dye lasers offers an extremely interesting possibility, which may lead to new types of lifetime measurements. The beam-foil source has a unique ability to produce highly-ionized and multiply-excited states of virtually any atom, and lifetime measurements in such systems present a very fertile area for future research.

Acknowledgments

I am grateful to Dr. J. Bromander for permission to use his published figure and to Dr. G. Sørensen for making available unpublished materials. Many helpful suggestions have been gained from discussions with Professors W. S. Bickel, D. G. Ellis, C. G. Montgomery, H. J. Simon, Wm. Hayden Smith, and with Dr. J. Hansen. I am indebted to Professors H. G. Berry, I. Martinson, E. H. Pinnington and R. M. Schectman for their critical reading of the first draft, and to Deborah MacDonald for her help in the preparation of the final manuscript.

References

3.1 P. L. Smith: Nucl. Instr. and Meth. 110, 395 (1973)

3.2 D. C. Morton, J. F. Drake, E. B. Jenkins, J. B. Rogerson, L. Spitzer, D. G. York: Astrophys. J. 181, L103 (1973)

3.3 D. C. Morton, J. F. Drake, E. B. Jenkins, J. B. Rogerson, L. Spitzer, D. G. York: Astrophys. J. 181, L110 (1973)

3.4 D. Layzer, R. H. Garstang: In *Annual Reviews of Astronomy and Astrophysics*, Vol. 6, ed. by L. Goldberg (Annual Reviews, Inc., Palo Alto 1968)pp.449-494

3.5 R. J. S. Crossley: In *Advances in Atomic and Molecular Physics*, Vol. 5, ed. by D. R. Bates, I. Estermann (Academic Press, New York, London 1969)pp. 237-296

3.6 A. W. Weiss: Nucl. Instr. and Meth. 90, 121 (1970)

3.7 O. Sinanoğlu: Nucl. Instr. and Meth. 110, 193 (1973)

3.8 M. W. Smith, W. L. Wiese: Astrophys. J. Suppl. Ser. 23, 103 (1971)

3.9 T. Andersen, A. Kirkegård Nielsen, G. Sørensen: Nucl. Instr. and Meth. 110, 143 (1973)

3.10 A. C. G. Mitchell, M. W. Zemansky: *Resonance Radiation and Excited Atoms* (Cambridge University Press, London, New York 1961)pp.96-153

3.11 E. W. Foster: Rept. Progr. Phys. 27, 469 (1964)

3.12 W. R. Bennett, Jr., P. J. Kindlmann, G. N. Mercer: Appl. Opt. Suppl. 2, 34 (1965)

3.13 K. Ziock: In *Methods of Experimental Physics*, Vol. 4B, ed. by V. W. Hughes, H. L. Schultz (Academic Press, New York, London 1967)pp. 214-225

3.14 W. L. Wiese: In *Methods of Experimental Physics*, Vol. 7A, ed. by B. Bederson, W. L. Fite (Academic Press, New York, London 1968)pp. 117-141

3.15 A. Corney: In *Electronics and Electron Physics*, Vol. 29, ed. by L. Marton, (Academic Press, New York, London 1970)pp.115-231

3.16 P. Erman, Physica Scripta 11, 65 (1975)

3.17 Procedures almost identical to our modern lifetime determinations were used by A. V. Harcourt, W. Esson: Phil. Trans. Roy. Soc. London 156, 193 (1866). They made very precise measurements of monomolecular chemical reaction rates and fitted their results to one and two exponentials. In an appendix, Esson correctly solved the population differential equation for both a branched cascade and for a blend. However, the extracted rate constants were strongly dependent upon temperature and concentration and could not then be given fundamental meaning.

3.18 E. Rutherford: Phil. Mag. 49, 161 (1900)
Similar qualitative results had been obtained with chemically-separated sources a few months earlier by J. Elster, H. Geitel: Ann. Physik 69, 83 (1899)

3.19 E. Rutherford, F. Soddy: J. Chem. Soc. 81, 321, 837 (1902)

3.20 E. Rutherford, F. Soddy: Phil. Mag. 4, 370, 569 (1902)

3.21 E. Rutherford, F. Soddy: Phil. Mag. 5, 576 (1903)

3.22 E. Rutherford: Phil. Trans. Roy. Soc. London A 204, 169 (1904)

3.23 F. Soddy: Nature 69, 297 (1904)
Soddy later wrote that, "So far as I know, the period of average life was first deduced by Mr. J. K. H. Inglis, to whom I put the problem." [F. Soddy: *The Interpretation of Radium and the Structure of the Atom*, 4th ed. (G. P. Putnam's Sons, New York 1920)p. 113]

3.24 E. von Schweidler: *Congrès Internat. Radiologie* (Liège 1905) Chapt.15, Section 1

3.25 M. von Laue: *History of Physics* (Academic Press, New York 1950)p. 111

3.26 H. Bateman: Proc. Cambridge Phil. Soc. 16, 423 (1910)

3.27 W. Wien: Ann d. Phys. 30, 369 (1909)

3.28 L. Dunoyer: Le Radium 10, 400 (1913)

3.29 A. Einstein: Verhandl. Deut. Phys. Ges. 18, 318 (1916)

3.30 A. Einstein: Phys. Z. 18, 121 (1917)

3.31 W. Wien: Ann. Physik 60, 597 (1919)

3.32 L. J. Curtis: J. Opt. Soc. Am. 63, 105 (1973)

3.33 P. A. M. Dirac: Proc. Roy Soc. A112, 661 (1926)

3.34 P. A. M. Dirac: Proc. Roy Soc. A114, 243 (1927)

3.35 A summary of papers, published during 1909-1927, is given by W. Wien: In *Handbuch der Experimentalphysik*, Vol. 14, ed. by W. Wien, F. Harms (Akademische Verlagsges, Leipzig 1927)

3.36 A. J. Dempster: Phys. Rev. 15, 138 (1920)

3.37 A. J. Dempster: Astrophys. J. 57, 193 (1923)

3.38 J. S. McPetrie: Phil. Mag. 1, 1082 (1926)

3.39 H. Kerschbaum: Ann. Physik 79, 465 (1926)

3.40 H. Kerschbaum: Ann. Physik 83, 287 (1927)

3.41 E. Rupp: Ann. Physik 80, 528 (1926)

3.42 R. d'E. Atkinson: Proc. Roy. Soc. 116, 81 (1927)

3.43 I. Wallerstein: Phys. Rev. 33, 800 (1929)

3.44 H. D. Koenig, A. Ellett: Phys. Rev. 39, 576 (1932)

3.45 P. Soleillet: Compt. Rend. 194, 783 (1932)

3.46 Z. Bay, G. Papp: Rev. Sci. Instr. 19, 565 (1948). (These investigations were carried out in 1943-44, but could not be published because of wartime circumstances).

3.47 S. Heron, R. W. P. McWhirter, E. H. Rhoderick: Nature 174, 564 (1954)

3.48 S. Heron, R. W. P. McWhirter, E. H. Rhoderick: Proc. Roy. Soc. (London) A234, 565 (1956)

3.49 E. Brannen, F. R. Hunt, R. H. Adlington, R. W. Nicholls: Nature 175, 810 (1955)

3.50 W. R. Bennett, Jr.: In *Advances in Quantum Electronics*, ed. by J. Singer (Columbia University Press, New York 1961)

3.51 L. Kay: Phys. Letters 5, 36 (1963)

3.52 S. Bashkin: Nucl. Instr. and Meth. 28, 88 (1964)

3.53 E. U. Condon, G. H. Shortley: *The Theory of Atomic Spectra* (Cambridge University Press, London 1957)

3.54 D. G. Ellis: J. Opt. Soc. Am. 63, 1232 (1973)

3.55 R. Ladenburg, F. Reiche: Naturwiss. 27, 584 (1923)

3.56 B. W. Shore, D. H. Menzel: *Principles of Atomic Spectra* (John Wiley and Sons, New York 1968)

3.57 L. R. Maxwell: Phys. Rev. 38, 1664 (1931)

3.58 D. C. Morton, W. H. Smith: Astrophys. J. Suppl. Ser. 26, 333 (1973)

3.59 C. Laughlin, A. Dalgarno: Phys. Rev. A 8, 39 (1973)

3.60 e.g., W. R. Bennett, P. J. Kindlmann, G. N. Mercer: Appl. Opt. Suppl. 2, 34 (1965)

3.61 J. Z. Klose: Phys. Rev. 141, 181 (1966)

3.62 W. R. Bennett, P. J. Kindlmann: Phys. Rev. 149, 38 (1966)

3.63 G. M. Lawrence: Phys. Rev. 175, 40 (1968)

3.64 L. Barrette, R. Drouin: Physica Scripta 10, 213 (1974)

3.65 J. Bromander: Nucl. Instr. and Meth. 110, 11 (1973)

3.66 I. Martinson, A. Gaupp: Physics Reports 15C, 113 (1974)

3.67 S. Bashkin, In *Progress in Optics*, Vol. XII, ed. by E. Wolf (North-Holland, Amsterdam 1974)pp. 289-344

3.68 I. Martinson: Nucl. Instr. and Meth. 110, 1 (1973)

3.69 I. Martinson: Physica Scripta 9, 281 (1974)

3.70 H. G. Berry: Physica Scripta (to be published)

3.71 Many laboratories evaporate their own foils; others obtain them from suppliers such as the Arizona Carbon Foil Company, Tucson, Arizona, or the Yissum Research Development Company, Hebrew University, Jerusalem. One might remark that foil thickness is often unspecified or poorly determined

3.72 L. Lundin has incorporated much of the precision technology of ruling engines into foil-drive design

3.73 J. O. Stoner, Jr., J. A. Leavitt: Appl. Phys. Letters 18, 368 (1971)

3.74 J. O. Stoner, Jr., J. A. Leavitt: Appl. Phys. Letters 18, 477 (1971)

3.75 J. O. Stoner, Jr., J. A. Leavitt: Optica Acta 20, 435 (1973)

3.76 J. A. Leavitt, J. W. Robson, J. O. Stoner, Jr.: Nucl. Instr. and Meth. 110, 423 (1973)

3.77 E. L. Chupp, L. W. Dotchin, D. J. Pegg: Phys. Rev. 175, 44 (1968)

3.78 W. S. Bickel, R. Buchta: Physica Scripta 9, 148 (1974)

3.79 L. J. Curtis, H. G. Berry, J. Bromander: Physica Scripta 2, 216 (1970)

3.80 L. Kay: Physica Scripta 5, 139 (1972)

3.81 R. C. Etherton, L. M. Beyer, W. E. Maddox, L. B. Bridwell: Phys. Rev. A 2, 2177 (1970)

3.82 D. Halliday: *Introductory Nuclear Physics*, 2nd ed. (Wiley, New York, London 1960)p. 247

3.83 E. H. Pinnington: Nucl. Instr. and Meth. 90, 93 (1970)

3.84 R. M. Schectman, L. J. Curtis, C. Strecker, K. Kormanyos: Nucl. Instr. and Meth. 90, 197 (1970)

3.85 J. B. Marion: Rev. Mod. Phys. 38, 660 (1966)

3.86 J. O. Stoner, Jr.: J. Appl. Phys. 40, 707 (1969)

3.87 J. Lindhard, M. Scharff, H. E. Schiøtt: Kgl. Danske Videnskab. Selskab, Mat.-Fys. Medd. 33, 14 (1963)

3.88 L. C. Northcliffe, R. F. Schilling: Nucl. Data Tables A7, 233 (1970)

3.89 K. B. Winterbon: AECL 3194 (Scientific Document Distribution Office, Chalk River, Canada 1968)

3.90 P. Hvelplund, E. Laegsgård, J. Ø. Olsen, E. H. Pederson: Nucl. Instr. and Meth. 90, 315 (1970)

3.91 C. L. Cocke, B. Curnutte, J. H. Brand: Astron. and Astrophys. 15, 299 (1971)

3.92 W. R. Bennett, Jr., P. J. Kindlmann: Phys. Rev. 149, 38 (1966)

3.93 K. E. Donnelly, P. J. Kindlmann, W. R. Bennett, Jr.: IEEE J. Quant. Electr. QE10, 848 (1974)

3.94 R. D. Kaul: J. Opt. Soc. Am. 56, 1262 (1966)

3.95 G. H. Nussbaum, F. M. Pipkin: Phys. Rev. Letters 19, 1089 (1967)

3.96 R. E. Imhof, F. H. Read: Chem. Phys. Letters 3, 652 (1969)

3.97 L. J. Curtis, W. H. Smith: Phys. Rev. A 9, 1537 (1974)

3.98 M. Dufay: Nucl. Instr. and Meth. 110, 79 (1973)

3.99 E. H. Pinnington, A. E. Livingston, J. A. Kernahan: Phys. Rev. A 9, 1004 (1974)

3.100 A. Dalgarno: Nucl. Instr. and Meth. 110, 183 (1973)

3.101 C. Laughlin, M. N. Lewis, Z. J. Horak: J. Opt. Soc. Am. 63, 736 (1973)

3.102 F. Weinhold: J. Chem. Phys. 54, 1874 (1971)

3.103 M. T. Anderson, F. Weinhold: Phys. Rev. A 10, 1457 (1974)

3.104 L. J. Curtis: J. Opt. Soc. Am. 64, 495 (1974)

3.105 L. J. Curtis: Amer. J. Phys. 36, 1123 (1968)

3.106 N. H. Gale: Nucl. Phys. 38, 252 (1962)

3.107 G. M. Lawrence: Phys. Rev. A 2, 397 (1970)

3.108 W. P. Helman: Int. J. Radiat. Phys. Chem. 3, 283 (1971)

3.109 V. A. Ankudinov, S. V. Bobashev, E. P. Andreev: Zh. Eksper. I. Teor. Fiz.48, 40 (1965)

3.110 V. A. Ankudinov, S. B. Bobashev, E. P. Andreev: Sov. Phys. JETP 21, 26 (1965)

3.111 W. S. Bickel, A. S. Goodman: Phys. Rev. 148, 1 (1966)

3.112 L. J. Curtis, R. M. Schectman, J. L. Kohl, D. A. Chojnacki, D. R. Shoffstall: Nucl. Instr. and Meth. 90, 207 (1970)

3.113 B. L. Moiseiwitsch, S. J. Smith: Rev. Mod. Phys. 40, 238 (1968), with erratum 41, 574 (1969)

3.114 W. Hanle: Z. Instrumentenk. 51, 488 (1931)

3.115 J. Orear: *Notes on Statistics for Physicists*, Report UCRL-8417 (University of California Radiation Lab., Berkeley 1958)

3.116 F. Grard: Nucl. Instr. and Meth. 34, 242 (1965)

3.117 P. H. R. Orth, W. R. Falk, G. Jones: Nucl. Instr. and Meth. 65, 301 (1968)

3.118 A. H. Jaffey: Nucl. Instr. and Meth. 81, 155 (1970)

3.119 A. H. Jaffey: Nucl. Instr. and Meth. 81, 253 (1970)

3.120 A. H. Jaffey: Nucl. Instr. and Meth. 81, 218 (1970)

3.121 G. F. Lee, G. Jones: Nucl. Instr. and Meth. 91, 665 (1971)

3.122 P. R. Bevington: *Data Reduction and Error Analysis for the Physical Sciences* (McGraw-Hill, New York 1969)

3.123 D. W. Marquardt: J. Soc. Ind. Appl. Math. 11, 431 (1963)

3.124 C. Daniel, F. S. Wood, J. W. Gorman: *Fitting Equations to Data - Computer Analysis of Multifactor Data for Scientists and Engineers* (Wiley, New York London 1971)
Program GAUSHAUS 360 D-13.6.007 available from Share Program Library Agency, Research Triangle Park, North Carolina 27709

3.125 R. Fletcher: *A Modified Marquardt Subroutine for Non-Linear Least Squares*, AERE report, R.6799 (Harwell Atomic Energy Res. Estab.)
Program VB01A available Harwell Subroutine Library, Harwell, England

3.126 R. H. Moore, R. K. Ziegler: *The Solution of the General Least-Squares Problem with Special Reference to High Speed Computers*, Report LA-2367 (Los Alamos Scientific Laboratory, Los Alamos, New Mexico)

3.127 D. J. G. Irwin, A. E. Livingston: Computer Phys. Comm. 7, 95 (1974), Program HOMER

3.128 P. C. Rogers: MIT Lab. Nucl. Sci. Tech. Report No. 76, NYO 2303, Program FRANTIC

3.129 C. Lanczos: *Applied Analysis* (Prentice-Hall, Englewood Cliffs, New Jersey 1956)pp. 272-282

3.130 Problems with reliable numerical differentiation of statistically fluctuating data can be severe, but satisfactory results have been obtained using numerical filtering programs, e.g., A. Savitsky, M. J. E. Golay: Analytical Chem. 36, 1627 (1964)

3.131 L. J. Curtis, I. Martinson, R. Buchta: Physica Scripta 3, 197 (1971)

3.132 A. Denis, P. Ceyzériat, M. Dufay: J. Opt. Soc. Am. 60, 1186 (1970)

3.133 D. G. Gardner, J. C. Gardner, G. Laush, W. W. Meinke: J. Chem. Phys. 31, 978 (1959)

3.134 J. W. Cooley, J. W. Tukey: Math. Comp. 19, 297 (1965)

3.135 J. Schlesinger: Nucl. Instr. and Meth. 106, 503 (1973)

3.136 Z. Bay: Phys. Rev. 77, 419 (1950)

3.137 Z. Bay, V. P. Henri, H. Kanner: Phys. Rev. 100, 1197 (1955)

3.138 R. D. Dyson, I. Isenberg: Biochem. 10, 3233 (1971)

3.139 I. Isenberg, R. D. Dyson: Biophys. J. 9, 1337 (1969)

3.140 R. Schuyler, I. Isenberg: Rev. Sci. Instr. 42, 813 (1971)

3.141 J. Yguerabide: In *Methods in Enzymology*, Vol. XXVI, ed. by C. H. W. Hirs, S. N. Timasheff (Academic Press, New York, London 1972)pp. 498-578

3.142 L. J. Curtis: In *Proceedings of the Second European Conference on Beam-Foil Spectroscopy and Connected Topics*, ed. by M. Dufay (University of Lyon, Lyon, France (1971)

3.143 L. Kay: Physica Scripta 5, 138 (1972)

3.144 I. Martinson, A. Gaupp, L. J. Curtis: J. Phys. B 7, L463 (1974)

3.145 R. Tielert, H. H. Bukow: Z. Physik 264, 119 (1973)

3.146 L. J. Curtis, H. G. Berry, J. Bromander,: Phys. Letters 34A, 169 (1971)

3.147 D. R. Barker, L. M. Diana: Am. J. Phys. 42, 224 (1974)

3.148 J. L. Kohl: Phys. Letters 24A, 125 (1967)

3.149 L. J. Curtis, R. M. Schectman, J. L. Kohl, D. R. Shoffstall: Nucl. Instr. and Meth. 90, 207 (1970)

3.150 J. L. Kohl, L. J. Curtis, R. M. Schectman, D. A. Chojnacki: J. Opt. Soc. Am. 61, 1656 (1971)

3.151 R. M. Schectman, L. J. Curtis, D. A. Chojnacki: J. Opt. Soc. Am. 63, 99 (1973)

3.152 K. D. Masterson, J. O. Stoner, Jr.: Nucl. Instr. and Meth. 110, 441 (1973)

3.153 O. Nedelec: J. Physique 27, 660 (1966)

3.154 H. G. Berry, L. J. Curtis, J. L. Subtil: J. Opt. Soc. Am. 62, 771 (1972)

3.155 R. M. Schectman: private communication

3.156 C. H. Liu, D. A. Church: Phys. Rev. Letters 29, 1208 (1972)

3.157 D. A. Church, C. H. Liu: Nucl. Instr. and Meth. 110, 147 (1973)

3.158 H. J. Andrä, A. Gaupp, W. Wittmann: Phys. Rev. Letters 31, 501 (1973)

3.159 H. J. Andrä, A. Gaupp, K. Tillmann, W. Wittmann: Nucl. Instr. and Meth. 110, 453 (1973)

3.160 H. Harde, G. Guthöhrlein: Phys. Rev. A 10, 1488 (1974)

3.161 A. E. Siegman: *An Introduction to Lasers and Masers* (McGraw-Hill, New York 1971)p. 256

4. Theoretical Oscillator Strengths of Neutral, Singly-Ionized, and Multiply-Ionized Atoms

The Theory, Comparisons with Experiment, and Critically-Evaluated Tables with New Results*

Oktay Sinanoğlu

With 4 Figures

In the mid-sixties, the beam-foil spectroscopy (BFS) of BASHKIN [4.1] opened up the way to the accurate determination of atomic lifetimes. The arc and intensity methods turned out then to be mostly in error by factors like two or more. The new accuracy was about ± 20% initially; later, upon successful treatment of cascade effects [4.2,3], about ± 3%. The phase-shift method too supplemented BFS nicely. It could measure lifetimes in neutrals (and molecules), while BFS is largely unsuited to neutral atoms but can go to highly-charged ions [4.4].

This development was paralleled in the theory. Pre-1967 methods turned out often in error by factors of 2 - 3 ($1s^2 2s^2 2p^m \rightarrow 1s^2 2s 2p^{m+1}$ transitions of, for example, A II, III, etc., where A represents an arbitrary species, A({B,C,N,O,F,Ne,Na}) [4.5-8], and sometimes by factors of 10 - 30 (KL $3s^n 3p^m 3d^k$ transitions with $f^{El} < 0.1$) [4.9] compared with the new theory [4.10-16]. In our notation, { } means the set of the included quantities.

Prior to 1967, one had the Coulomb approximation [4.17] (good for Rydberg-like, out-of-shell transitions), the Hartree-Fock method (RHF) [4.18], the Z-expansion [4.19,20] (good for high-Z, highly-charged ions where electron correlation is less and less important), the limited configuration-interaction (C.I. = SOC, superposition of configurations) or the very similar multi-configuration self-consistent field method (MCSCF) [4.21,22], the large C.I. (SOC [4.23], with no way specified as to how to choose from among the infinitely many configurations), and the many-body diagrammatic perturbation theory [4.24] (MBPT). A brief guide to these methods is given in the SINANOĞLU BFS-1973 article [4.25]. They are discussed in more detail in a recent book which includes key papers on these methods [4.26].

Theoretically the in-shell (KL → KL'), (KLM → KL'M') transitions and the nearby "intershell" ones (e.g., KL → KL'[M], $2s^2 2p^2 \rightarrow 2s 2p^2 3p$; $n \rightarrow n' \pm 1$) are the more

---

*This chapter is dedicated to the fond memory of the two great atomic scientists and human beings, Professor Edward U. Condon of the United States and Professor A. Yutsis of the Soviet Union.

difficult (while for many applications, they are also of the greatest interest). They involve sensitively all the non-closed-shell-type correlation effects in both initial and final states. That is why, in 1967, this writer decided to test the new theory (NCMET) he had developed for many-electron systems on such oscillator strengths (f^{El}). Many f^{El}'s were calculated with no prior knowledge of new data. The agreement within experimental error, (on the whole ± 11%) [4.25] with modern data (BFS, Phase-Shift, Hanle effect, etc.) (and more data coming after the theoretical results were published) was therefore a particularly significant confirmation and cross check of theory and experiment. Where, due to cascade effects, BFS laboratories had reported several very different values for a f^{El}, NCMET agreed with one of the values.

This picture of the NCMET accuracy and results has not changed in later papers. The results reported since 1969 are essentially the same as the earlier SINANOǦLU-ÖKSÜZ-WESTHAUS results, all making use of the fully automated program system, "ATOM", of SINANOǦLU-ÖKSÜZ-WESTHAUS with some improvements by SINANOǦLU and LUKEN (which however do not change the essential results), and a KLM-shells version by SINANOǦLU and BECK. All of this NCMET work to 1973 is surveyed in the SINANOǦLU BFS-1973 article [4.25].

In 1973, however, f^{El} results on neutrals and +1 ions were published by NICOLAIDES [4.27,28] using the SINANOǦLU NCMET and the auto-program system "ATOM" of SINANOǦLU made available to NICOLAIDES, which gives all the determinants (or configurations) that are to occur in the NCMET "charge wave function" ψ_C ($\equiv \psi_{CD}$). The latter is the special wave function introduced by SINANOǦLU [4.5-7] to include all the specifically non-closed type correlations hence affecting the charge densities. The NICOLAIDES results showed large discrepancies [4.27,28] from BFS, reminiscent of the pre-NCMET and pre-BFS data.

These discrepancies have now been resolved [4.30,31] showing the difficulty lay not in the NCMET itself, but in its incorrect application by NICOLAIDES. The new results [4.31] of SINANOǦLU and LUKEN (Be I, B I, C I, C II, N I, N II, O I, O II, etc.) are in good agreement with BFS. This confirms [4.30,31] that NCMET itself properly used was capable of good results on low ionization stages as well as on highly-ionized systems.

In the present chapter we shall concentrate on theoretical aspects since BFS-1973 [4.25]. We shall also introduce here the new concepts of pre-Rydberg states, charge supermultiplets, and supertransitions, particularly appropriate for the $N \approx Z$ region (N = no. of electrons, Z = atomic number), and give a spectral interpretation of the NCMET charge wave function. These latter have come up recently in work by the author on the theory of molecular spectra in the vacuum ultraviolet [4.13], but they are particularly pertinent also to neutral-atom spectra.

The present state of calculations of oscillator strengths in the low (N-Z) region displaying the strong dips in the KL → KL' $f^{El}(1/Z)$ curves is shown in Fig.4.1.

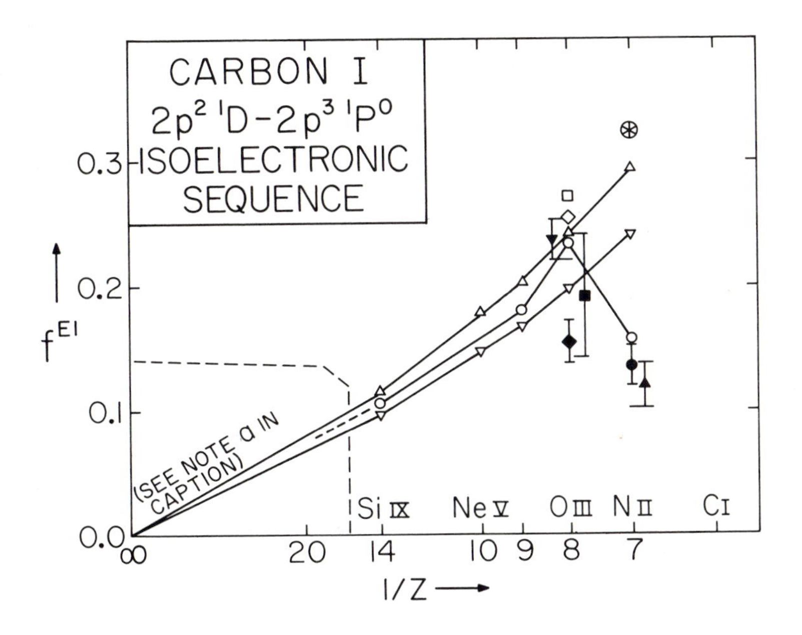

Fig.4.1 Example of oscillator strengths for an isoelectronic series calculated from the Non-Closed-Shell Many-Electron Theory (NCMET) of SINANOĞLU compared with results of BFS experiments and pre-NCMET calculations. (No values are indicated for C I because the $2p^3\ ^1P^o$ state lies in the continuum and autoionizes in this case). ○ NCMET ψ_c calculations using correct roots (this work and [4.31]); ⊛ Incorrect result due to use of wrong root of the SINANOĞLU NCMET charge wave function Hamiltonian matrix [4.27,28], resulting in wave functions with variational collapse (see text); ▽ Orbital theory. These points include both the results of restricted Hartree-Fock (RHF) calculations [4.32], and Z-expansion approximation to RHF [4.33]; △ 2 x 2 C.I. Z expansion (includes $2s^2 2p - 2p^3$ internal correlations in ground state [4.33]; □ [4.34]; ◇ NBS tables [4.35] (when different from other values indicated above); ● BFS [4.36]; ▲ BFS [4.37]; ▼ BFS [4.38]; ■ BFS [4.39]; ◆ BFS [4.40]. Note a: contrary to previous assumptions in the literature, the high-Z portion of these $f^{El}(1/Z)$ curves (enclosed by dotted line) has recently been found to behave very differently from the behavior shown here, due to relativistic/intermediate coupling effects as well as the combined NCMET correlations. Before using extrapolations in this region, the reader should consult [4.41] and [4.25,Table 5]

It was thought until recently that the RHF and the Z-expansion (comparable to RHF for the high (Z/N) region) results and extrapolations in the $f^{E1}(1/Z)$ curves for $(1/Z)\to 0$, would be reliable for highly-stripped heavy ions (down to KL-shells). Since BFS-1973 [4.25] however, it has been found that, though such non-relativistic multiplet f^{E1} values [4.25,Table 1] (including also the correlation effects given by NCMET) are of the correct order of magnitude, new relativistic intermediate-coupling effects [4.41] change the picture drastically in that end of the $f^{E1}(1/Z)$ curves also.

Finally, in this chapter, we present the fully NCMET-evaluated $f^{E1}(1/Z)$ values for essentially all of the $1s^2 2s^n 2p^m \to 1s^2 2s^{n'} 2p^{m'}$ transitions of the Be I, B I, C I, N I, O I and F I isoelectronic sequences, including the correct neutral and +1 ion-transitions f^{E1} values [4.31] and critically evaluating the others. All of these values which have been scrutinized are from SINANOĞLU, WESTHAUS, and LUKEN, and, thus, provide a standard uniform NCMET table, replacing also the incorrect values reported in [4.27].

4.1 The Non-Closed-Shell Many-Electron Theory

For any N-electron state, the exact wave function is given by [4.5-7]

$$\psi_{exact} = \psi_C + \chi_U \quad , \tag{4.1}$$

with $\langle \psi_C | \chi_U \rangle = 0$. The "charge wave function", ψ_C, was introduced by SINANOĞLU for its physical significance, as it contains only and all of the specifically non-closed-shell-correlation and orbital effects [4.5-7]. In (4.1), χ_U represents the remaining, "all external correlations", the only ones that survive in truly closed-shell systems like He and Ne.

The total charge distribution

$$\rho(\vec{r}) = \langle \Psi | \sum_{i>1}^{N} \delta(\vec{r} - \vec{r}_i) | \Psi \rangle \tag{4.2}$$

is determined in large part by ψ_C,

$$\rho(\vec{r}) \approx \rho_C(\vec{r}) = \langle \psi_C | \sum_i \delta(\vec{r} - \vec{r}_i) | \psi_C \rangle \quad . \tag{4.3}$$

Similarly, for a transition density $(0 \to n)$,

$$\rho_{on}(r) \approx \rho_{on}^{C}(\vec{r}) = \langle \psi_C(0) | \sum \delta(\vec{r} - \vec{r}_i) | \psi_C(n) \rangle \quad . \tag{4.4}$$

The ψ_C is rigorously [4.5-7]

$$\boxed{\psi_C = \phi_{RHF} + \chi_{INT} + \chi_F} \quad , \qquad (4.5)$$

consisting of the Hartree-Fock wave function for that state, the "internal" (INT), and the "semi-internal" correlations (F). Then,

i) In NCMET, all states arising from the same atomic shells (e.g., KL-shells: all terms of all $1s^n 2s^m 2p^k$ configurations) are treated as a class. (For discussion's sake we continue with the KL case).

ii) Given the ϕ_{RHF}, the rest of ψ is already mathematically determined [4.5-7]. The ψ_C is made up exactly of the "Hartree-Fock sea", the intravalency $\overline{V}$ set of spin orbitals $\{k\}_{\overline{V}}$

$$\{k\} = \{1s_\alpha, 1s_\beta, 2s_\alpha, \ldots, 2p_{+1\beta}\} \quad , \qquad (4.6)$$

and of the new NCMET-derived "semi-internal orbitals", $\{\hat{f}\}$. In (4.6), α refers, as usual, to "spin-up", β to "spin-down", and +1 to m_ℓ.

iii) The $\{\hat{f}\}$ are also finite in number. (There are just a few $\hat{f}$'s ($f_{ij;k}$ for i, j, k ε $\{k\}_{\overline{V}}$); in most KL-shell states there is just one). The $\hat{f}$, in KL-shells, can be expanded in spherical harmonics. By symmetry, the series can be truncated:

$$\hat{f}(\vec{r}) = f_s(r)\, Y_{00}(\vec{\Omega}) + f_p(r)\, Y_{1m}(\vec{\Omega}) + f_d(r)\, Y_{2m'}(\vec{\Omega}) + f_f(r)\, Y_{3m'}(\vec{\Omega}) \quad . \qquad (4.7)$$

At most four semi-internal radial orbitals of s, p, d, f types need to be determined for the ψ_C of each KL state.

These new orbitals $f_\ell(r)$ are easiest found variationally.* For KL-shells, SINANOǦLU

*An $f_\ell(r)$ is a closed-form individual function, the solution of a particular integro-differential equation [4.42] given by NCMET. It need not be expanded in an infinite basis set, but if it is expanded, say in a hydrogenic basis $R_{n\ell}(r)$, one gets an entire Rydberg ℓ-series:

$$f_{(\ell)}(r) = \sum_{n\geq 3}^{\infty} a_n\, R_{n(\ell)} \quad .$$

What we find in practice is that

$$f_{(\ell)}(r) \simeq R^*_{3*\ell} \quad (\text{for } \ell = 0,1,2) \quad ,$$

i.e., only one term represents the function well, if it is determined variationally with an optimized orbital exponent. (Thus the $3*\ell$ is not like an actual 3ℓ-orbital in its charge distribution, but a much more shrunk new kind of orbital involved in the strong non-dynamical correlations of the L-shell electrons.

and ÖKSÜZ [4.43,44] found them to good approximation to be of the form

$f_s \sim$ 3s-like

$f_p \sim$ 3p-like

$f_d \sim$ 3d-like

$f_f \sim$ 4f-like

i.e., $r^2\ e^{-\xi r}$ and $r^3\ e^{-\xi' r}$. The functions are easily and automatically orthogonalized to 1s, 2s, 2p by the system, "ATOM". More elaborate expansions, with more variational parameters independently varied, etc., do not change the results appreciably. It is remarkable that many charge-like properties are given to few percent accuracy with these simplest of f's (see Section 4.3). Contrasted to the very expensive and large configuration-interaction calculations which could only be carried out on one or two atomic states before, even in the first runs with NCMET we reported 113 states and many f^{El} values with very little computer cost [4.8,43,44]. The reason, of course, is in the NCMET having predicted all the significant configuration-interaction effects (in ψ_c), leaving out the infinitely many configurations occurring in χ_U which, however, do but little for charge-like properties, e.g., oscillator strengths (from $\rho_{on}(\vec{r})$ (4.4)).

iv) From remarks (ii) and (iii) above, we see that ψ_c contains rigorously a finite number of determinants, namely, those from the intravalency set of N-particle spin-orbital wave functions and from the set of semi-internal orbitals. Specifically, there are customarily 10-60 determinants for $1s^2 2s^n 2p^m$ terms; these are all figured out and listed by our program system [4.5-7,43,44,49] "ATOM" just from a specification of Z, N and the spectroscopic state desired.

v) The "dynamical correlations" [4.5-7] (all-external, χ_U) [4.50] of MET, affect mainly the energy quantities. They are and need to be included for calculating term-splittings [4.52], excitation energies, ionization potentials, or electron affinities [4.43,44], etc., by the special ("pair correlations" $\varepsilon_{ij;k\ell}$) techniques of MET [4.4-7,52].

Some charge-like properties we calculated using NCMET are compared in Table 4.1 with the values obtained from standard atomic theory, the RHF (Hartree-Fock).

Table 4.1 Sample charge-like properties as given by the "charge wave function" ψ_c of NCMET of this writer and co-workers compared with values obtained from standard atomic-structure theory (Hartree-Fock, RHF) and experiment

Property	RHF	NCMET (this work)[a]	Experiment	Reference
Electric Dipole Multiplet Oscillator Strength, f^{El}				
(KL → KL')				
N II $2p^2\ ^3P - 2p^3\ ^3D^o$	0.236	0.102	(0.101) ± 0.006	[4.45]
(KL → KL'[M])				
N I $2p^2\ ^4S^o - 2p^2 3s\ ^4P$	0.082	0.271	(0.28) ± 0.01	[4.46]
(KLM → KLM')				
P II $3s^2 3p^2\ ^3P - 3s3p^3\ ^3P^o$	0.292	0.041	(0.040)	[4.47]
Electric Quadrupole Transition Rate A (KL → KL')				
O I $^1D_2 - {}^1S_o$	1.422 [s^{-1}]	1.183 [s^{-1}]		
Generalized Oscillator Strength (function of momentum transfer, q)				
$F_{on}(q)$ at q = 1.0 [a.u.] C I $2s^2 2p^2\ ^3P - 2s2p^3\ ^3P^o$	0.08	0.03		
Hyperfine Structure Magnetic Dipole Constant A_J				
^{9}Be $2s2p\ ^3P_1^o$ (I = 3/2), μ = -1.1776	-119.3 [MHz]	-136.9 [MHz]	-139.4 [MHz]	[4.48]
Electric Field Gradient at the Nucleus q_J				
Be I $2s2p\ ^3P_1^o$	0.0591 [a.u.]	0.0553 [a.u.]		
Electronic Quadrupole Moment Q				
Be I $2s2p\ ^3P_2^o$	2.045 [a.u.]	2.296 [a.u.]		
Charge Density at the Nucleus $\rho(o)$				
Be I $2s2p\ ^3P_2^o$	35.477 [a.u.]	34.120 [a.u.]		
Spin Density at the Nucleus $\rho_s(o)$				
Be I $2s2p\ ^3P_2^o$	0.619 [a.u.]	0.728 [a.u.]	0.739 [a.u.]	[4.16][b]

[a] See references to the work of SINANOĞLU and various co-workers in text and references

[b] Based on data from [4.48] and NCMET hyperfine structure orbital contribution [4.16]

4.2 A Spectroscopic Interpretation of the Charge Wave Function

The low-lying spectroscopic states of KL-shell atoms are of the types:

Intravalency States:

$$\{\bar{V}\} = \{1s^2 2s^2 2p^m \\ 1s^2 2s 2p^{m+1}\} \tag{4.8}$$

or *Rydberg States:*

$$\{R\} = \{1s^2 2s^2 2p^{m-1} ns \\ 1s^2 2s^2 2p^{m-1} nd \\ 1s^2 2s 2p^m np\}, \quad n \geq 3 \quad .$$

We see now that, according to the NCMET ψ_c theorem [4.13]:

i) The χ_{INT} part of the ψ_c of <u>any</u> intravalency state contains (the mixings with) <u>all</u> the intravalency states of the same symmetry.

ii) The χ_F of the ψ_c of a <u>given</u> intravalency state of particular symmetry contains the mixings with <u>all</u> the Rydberg states and series of the same overall symmetry $({}^{2S+1}L^{\pi})$.

Moreover, although entire Rydberg series {R} of the ℓ = s,p,d,f types could mix in, the SINANOĞLU-ÖKSÜZ and SINANOĞLU-LUKEN ψ_c calculations of several hundred KL-states have shown the main mixings to be only with $n\ell$ from among the set {3s,3p,3d,4f} with optimized exponents.

This leads to the Spectral Charge Wave Function Theorem [4.13]:
The strongest mixings, intensity borrowings, level crossings with Z, etc., take place among the same $\{{}^{2S+1}L^{\pi}\}$ spectroscopic terms corresponding to terms of the form

$$\{\bar{V}\} = \{1s^2 2s^m 2p^n; \quad {}^{2S+1}L^{\pi}\} \quad , \tag{4.9}$$

and

$$\{pR\} \equiv 1s^2 2s^q 2p^r \left.\begin{matrix} 3s \\ 3p \\ 3d \\ [4f] \end{matrix}\right\}; \quad {}^{2S+1}L^{\pi} \quad . \tag{4.10}$$

{pR} is defined below.

The NCMET thus delineates (purely from the properties of the N-electron Schrödinger equation) [4.5-7], a special class of states, {pR}, which we shall call the pre-Rydberg states [4.13]. This class stands apart from the true Rydberg states

$$\{R\} \equiv \{1s^2 2s^s 2p^t \; n\ell \; n'\ell'\}, \; n > 3, \; \ell \leq 2, \; n' > 4 \quad , \tag{4.11}$$

which should mix less, in general, with the $\overline{V}$, and thus fit into regular fixed quantum-defect (Δ_ℓ) Rydberg levels, with $\varepsilon_n \propto (n + \Delta_\ell)^{-2}$. The pre-Rydberg levels, on the other hand, should be the main ones deviating from the Rydberg formulas.

Written in the spectroscopic interpretation, (4.1) and (4.5) become [4.9]

$$\psi_{exact} = \psi_c + \chi_U = \psi_{\{\overline{V}\}} + \chi_{\{pR\}} + \chi_F^{\{R\}} + \chi_U \quad . \tag{4.12}$$

Since experience shows that the $\chi_F^{\{R\}} \approx 0$ in our many calculations [4.43,44], we may simplify to

$$\psi_c \cong \psi_c^s \equiv \psi_{\{\overline{V}\}} + \chi_{\{pR\}}$$

($\psi_c^s \equiv$ the "spectral charge wave function").

Thus NCMET indicates: all the intravalency and pre-Rydberg states of the same spectroscopic term designation (e.g., $^1P^o$, etc.) should be treated on the same footing. We thereby view these strongly intermingling states as an <u>NCMET charge supermultiplet</u> [4.13]. Then too, strong intensity borrowings will occur mainly within the component transitions of one big supertransition, supermultiplet oscillator strength

$$\overline{f^{E1}}^{\,S} \equiv (1/n_\gamma) \sum_{\gamma,\gamma'} f^{E1}(\gamma LS \rightarrow \gamma' L' S) \quad , \tag{4.14}$$

where $n_\gamma \equiv$ number of LS terms of differing $\overline{V}$, pR configurations γ in the initial supermultiplet.

The component transitions ($\gamma LS\pi \rightarrow \gamma' L' S\pi'$) of the lowest $^1D_2 \rightarrow {}^1P_1^o$ supertransition are shown in Table 4.2 as an example.

Table 4.2 Two NCMET ψ_c charge supermultiplets, for $\{^1D\}$ and $\{^1P^o\}$, in the C I, N II, O III ... isoelectronic sequences. The $^1D \rightarrow {}^1P^o$ E1 transitions may be viewed as the components of one supertransition between $\{^1D\}$ and $\{^1P^o\}$

Initial supermultiplet $\{^1D_2\}$	Terms observed in	Final supermultiplet $\{^1P^o_1\}$	Terms observed in
$1s^22s^22p^2$	C I - P X	$1s^22s2p^3$	N II - P X
$1s^22p^4$	O III, Na VI	$1s^22s^22p3s$	C I - PX
$1s^22s^22p3p$	C I - O III	$1s^22s^22p3d$	C I - P X
$1s^22s^22p4f$	N II	$1s^22s2p^23p$	O III, F IV
$1s^22s2p^23s$	Na VI - A VIII, P X	$1s^22s2p^24d$	-
$1s^22s2p^23d$	F IV, Na VI, Mg VII, P X	$1s^22p^33s$	-
$1s^22p^33p$	-	$1s^22p^33d$	-

4.3 NCMET Calculations

The accurate calculations of atomic structures and properties, i.e., with the full inclusion of electron correlation in both ground and excited states made possible with NCMET, are fully automated in our computer program system, "ATOM", described in [4.5-7] and in more detail in [4.49]. The NCMET and "ATOM" basically can work in two parts:

i) For charge-like properties (as in Table 4.1), calculation of the charge wave function, ψ_c (and its energy $E_c \equiv E_{RHF} + E_{INT + F}$).

ii) For energy quantities, obtaining also the all-external correlation energy, ΔE_U, (or for extreme accuracy for any property, of χ_U) [4.53].

The methodology for both parts has been given already [4.5-8,43,44,49]. Here a number of comments will be made on the calculation of ψ_c's and f^{E1}'s.

It is expedient but not essential to distinguish between (i) "Type A", which are states lowest of their symmetry (e.g., $2s^22p^2\ {}^3P$ in C I, $2s2p^3\ {}^3P^o$ in N II, $2s^22p^4\ {}^1D$ in O I, $2s2p^4\ {}^4P$ in F III), and (ii) "Type B" states which have others of lower energy but of the same spatial and spin symmetry and the same parity (quantum numbers $LS\pi$), in the actual (exact) spectrum. These we call "non-lowest" states.

Atomic (or +ξ-ionic) states experimentally labeled with a configuration $1s^2 2s^n 2p^m$ (KL-shells $\{\overline{V}\}$ states) but which are of Type B, i.e., non-lowest, are listed in Table 4.3.

Table 4.3 States labeled with a dominant KL - $\{\overline{V}\}$, i.e., $1s^2 2s^n 2p^m$ configuration, but which are not the lowest of their symmetry. Column 1 shows the spectroscopic term being considered. Column 2 shows the number of lower terms having the same LSπ. Column 3 lists the system in which the term occurs. All other spectroscopic $\{\overline{V}\}_{KL}$ terms of any other ξ-ion are the lowest of their symmetry

Spectroscopic term	No. of lower terms of same LSπ	Atom or Ion	Spectroscopic term	No. of lower terms of same LSπ	Atom or Ion
$2p^2\ {}^1S$	∞	Be I	$2s2p^4\ {}^4P$	1	N I
${}^1S$	1	B II, C III N IV, ...	${}^2D$	∞	N I
$2s2p^2\ {}^2S$	4	B I	${}^2S$	∞	N I
$2p^3\ {}^2P^o$	3	C II	${}^2P$	∞	N I
${}^2P^o$	1	N III, O IV F V, ...	${}^2P$	1	O II
$2s2p^3\ {}^3P^o$	1	C I	$2p^5\ {}^2P^o$	3	F III
${}^1D^o$	∞	C I	$2s2p^5\ {}^3P^o$	2	O I
${}^1P^o$	∞	C I	${}^1P^o$	∞	O I
${}^1P^o$	1	N II	${}^1P^o$	1	F II
$2p^4\ {}^3P$	1	O III	$2s2p^6\ {}^2S$	∞	F I
${}^1D$	1	O III			
${}^1S$	2	O III			

All other KL- $\{\overline{V}\}$ states in all isoelectronic sequences are of Type A.

Although our main concern in this chapter is on second row (KL) atoms, we show also some "non-lowest" states occurring among the KLM- (KL $3s^4 3p^s 3d^t$) - intravalency states $\{\overline{V}\}_{KLM}$, in Table 4.4.

Table 4.4 Examples of $\{\overline{V}\}_{KLM}$ spectroscopic terms with dominant KL $3s^r3p^s3d^t$ configuration designations which are not the lowest of their symmetry. Note the interesting changes along the Si I isoelectronic series

Spectroscopic term	No. of lower terms of same $LS\pi$	Atom or Ion	Spectro-scopic term	No. of lower terms of same $LS\pi$	Atom or Ion
$3p^2\ ^1S$	∞	Mg I	$3s^23p3d\ ^1P^o$	1	Si I
	∞	Aℓ II		2	P II
	1	Si III		?	S III
	1	P IV	$3s^23p^33d\ ^1P^o$	1	Cℓ II
$3s3p^2\ ^2S$	∞	Aℓ I		1	K IV
	1	Si II		1	Ca V
	0	P III	$3s^23p^63d^3\ ^4F$	1	Sc I
$3s3p^3\ ^3P^o$	∞	Si I	^2D	3	Sc I
	0	P II			
	0	S III			
$^1P^o$	∞	Si I			
	1	P II			
	0	S III			

Whether of Type A or B, given a $\overline{V}$ state, to be labeled with a dominant KL configuration $(1s^n2s^m2p^q)$ (analogous for other shells), the calculation of ψ_c is carried out in the following systematic way:

i) List all NCMET-predicted determinants or configurations (essentially from symmetry and vector-coupling properties, within χ_{INT} and χ_F). "ATOM" generates these by itself and prints them out (as e.g., in the sample outputs reproduced in [4.5-7,25]. This list is good for an entire isoelectronic series.

ii) Obtain the 1s, 2s, 2p orbitals by the RHF (non-closed-shell Hartree-Fock) method. But from which configuration? This differs in Type A and B (see below).

iii) Obtain first trial semi-internal orbitals f_s, f_p, f_d, f_f. With these, calculate the N-electron, K x K Hamiltonian matrix (K = number of determinants found in step i). This sizable amount of atomic algebra (by Condon-Shortley and/or Racah methods; orthogonalization of the f_ℓ to 1s, 2s, 2p; evaluation of Coulomb integrals; etc.) is also all done automatically by the SINANOĞLU-ÖKSÜZ programs [4.43,44,49] within the system "ATOM".

iv) Diagonalize the K x K large $\mathcal{H}$-matrix. This gives K-roots, N-electron energy levels. Which root(s) are the one(s) desired depends on Type A or B (and on four sub-categories [4.31] within Type B).

v) Step (iii) is repeated by varying the semi-internal trial orbitals $\{f_s, f_p, f_d, f_f\}$ (their exponents, etc.; see 4.3.3 below), recalculating the K x K $\underset{\approx}{\mathcal{H}}$, then doing step (iv) again. These iterations (iii-v) continue until the optimum desired energy level is obtained. What this optimum is, which root is minimized or maximized, etc., what happens to the ordering of roots during the orbital variations, etc., is straightforward in Type A, but requires knowledge of variational theory for Type B (see below).

vi) The final output consists of the coefficients of the determinants (or configurations) and the $\{1s, 2s, 2p, f_s, f_p, f_d, f_f\}$-orbitals, this the full charge wave function ψ_C; also the energy $E_C \equiv E_{RHF} + E_{(INT + F)}$. The term $E_{(INT + F)}$ is the charge (or complete "non-dynamical") correlation energy.

The ψ_C's of the two states can be fed into various sub-programs of "ATOM" in order to calculate transition probabilities, spin densities, hyperfine structure, quadrupole moments, or many other important atomic parameters [4.49]. In these parts, full N-electron matrix elements are evaluated by special algorithms we developed [4.8] for the convenient inclusion of the non-orthogonality of orbitals from different states in the transition matrix elements, etc. Typical computer time for calculation of the ψ_C, E_C of a KL state (on CDC-6600) is 50 seconds, for the f^{El}, about 150 seconds. (This may be compared with the hours that were often taken up in conventional configuration-interaction calculations in other work in the literature).

4.3.1 The L^2, S^2 Symmetry of ψ_C

The ψ_C output of "ATOM" is an eigenfunction of L^2 and S^2. In the original SINANOĞLU-ÖKSÜZ program, this is ensured by applying the raising, lowering operators $L_\pm$, $S_\pm$ to the determinants generated by NCMET. If the determinants are complete, $L_\pm$, $S_\pm$ simply give back determinants in the same set $\{\text{determinants}\}_{NCMET}$. Then when the K x K $\mathcal{H} = (\mathcal{H})_{det-det}$, matrix is diagonalized, (due to $[H,L^2] = 0 = [H,S^2]$), the resulting ψ_C will also be an exact L^2, S^2 eigenfunction. Occasionally $L_\pm$, $S_\pm$ will give one or two additional determinants which then complete the {determinants} - set with regard to L^2, S^2. These new ones are virtual - triple - or more - excitations out of ϕ_{RHF}. SINANOĞLU and ÖKSÜZ tested the inclusion or omission of these and found them to make very little contribution. (These also give zero contribution rigorously to first order in ψ_C [4.54,55]). Nevertheless, we routinely use in "ATOM" the complete L^2, S^2 set.

An equivalent way to ensure the exact L^2,S^2 eigenfunction property of ψ_c is by the application of L^2,S^2 (definite L and S) projection operators to a newly-generated determinants set. This version [4.49,56], programmed by WESTHAUS and SINANOĞLU, is also in the "ATOM" system.

4.3.2 Dipole Length vs. Dipole Velocity

The f^{El} can be calculated from the dipole-length form

$$f_r^{El} = b\,\varepsilon\,S_r \quad , \tag{4.15}$$

or from the dipole-velocity form

$$f_v^{El} = b(1/\varepsilon)S_v \quad , \tag{4.16}$$

with b the combination of physical constants, ε the transition energy, S_r and S_v line-strengths containing the operators

$$\left(\sum_{i\geq 1}^{N} \vec{r}_i\right) \text{ or } \left(\sum_{i\geq 1}^{N} \vec{\nabla}_i\right) \quad .$$

There is a host of other forms too [4.57,58] involving operators resulting from a succession of commutations $[\ldots[[H,\vec{r}],\vec{r}],\ldots]$.

The $f_r^{El} = f_v^{El}$ with exact eigenfunctions of H. This agreement is a necessary but not sufficient condition for the correctness of a calculated f^{El} (see [4.57]). For example, sometimes the RHF method gives f_r^{El} and f_v^{El} in mutual agreement, but both off from the exact value by large factors.

Another useful form for f^{El} we use is

$$f_{\sqrt{rv}} \equiv \sqrt{f_r^{El}\,f_v^{El}} = b\sqrt{S_r S_v} \quad . \tag{4.17}$$

The advantage of this is its lack of dependence on the excitation energy ε.

We report in our work in general both the f_r^{El} and the f_v^{El} values. Table 4.5 shows the typical spread in f_r^{El} vs. f_v^{El} for the transitions calculated by SINANOĞLU and WESTHAUS [4.8] with Type A (lowest-of-symmetry) states. The difference is of the order of 1%, within experimental and NCMET ψ_c accuracy.

For f_r^{El} and f_v^{El} it also does not matter whether ε_{exact} or $\varepsilon_c = \varepsilon_{RHF} + \Delta\varepsilon_c$ is used since $\varepsilon_{exact} = \varepsilon_{RHF} + \Delta\varepsilon_U$ (actually ε_U is also easily given by MET/NCMET [4.52] and

$\varepsilon_c >> \Delta\varepsilon_U$, so that with "$\varepsilon_{theor}$" (i.e., ε_c) or "ε_{expt}", f^{El}'s differ by only of the order of a few percent for Type A [4.62] (see [4.25,Table 1]). This difference is unimportant, being less than the experimental errors.

Table 4.5 The dipole length-dipole velocity spread vs. (theory-experiment) spreads of the SINANOĞLU-WESTHAUS NCMET ψ_c oscillator strengths all involving Type A, i.e., lowest-of-their-symmetry states. Column 3 lists geometric mean values between the dipole length and dipole velocity calculations which do not depend on energy. Column 4 shows the effect of calculating the oscillator strengths with either the calculated or observed level energy, which is seen to be less than the experimental error

Transition $2p\ ^2P^o$ - $2p^2\ ^2S$	$\frac{f_r^{El} - f_\nabla^{El}}{f_r^{El}}$	$f_{\sqrt{r\nabla}}^{El}$	$\frac{f_r^{El}(\varepsilon\text{-theory}) - f_r^{El}(\varepsilon\text{-expt})}{f_r^{El}(\varepsilon\text{-expt})}$	f_{expt}	Reference
C II	0.008	0.121	0.091	0.117 ± 0.008	[4.59]
N III	0.011	0.084	0.070	0.060 ± 0.012	[4.60]
O IV	-0.029	0.070	0.055	0.071 ± 0.008	[4.61]

4.3.3 Semi-Internal Orbital Variations (Type A, Lowest-of-Symmetry, States)

In steps (iii)-(v) above, the $\{f\} \equiv \{f_s, f_p, f_d, f_f\}$ are varied to optimize the energy. For Type A states calculated by SINANOĞLU-ÖKSÜZ [4.43,44], (and f^{El}'s SINANOĞLU-WESTHAUS [4.8]), the energy level desired is the lowest root of the K x K $\underset{\approx}{\mathcal{H}}$-matrix. Therefore this energy is minimized with regard to the {f} orbitals. In the initial SINANOĞLU-ÖKSÜZ calculations after some experimentation with more elaborate variations (several Slater-type orbitals for each $f_\ell(r)$, separate variations of STO exponents ξ_ℓ^*, etc.), it was concluded that good results ($\Delta E \sim 0.1$ eV) were obtained already with single STO's, one for each of f_s, f_p, f_d, f_f (4.7) and with exponents such that $\xi_{3s} = \xi_{3p} = \xi_{3d} = (5/3)\xi_{4f}$. That these simple semi-internals are indeed adequate, because the minima in E for Type A with regard to ξ's are very shallow and insensitive to minor variations in the $f_\ell(r)$ so long as they are 3s, p, d and 4f-like and with the ξ's pulling them into the electron-dense regions of the $2s^n2p^m$-shells, is seen in Table 4.6.

Table 4.6 compares the SINANOĞLU-ÖKSÜZ ψ_c results with our more elaborate ones in which ξ_{3s}, ξ_{3p}, ξ_{3d} and ξ_{4f} are separately varied (also in some cases several d-like or other STO's are included). All of these calculations we perform on our same program system "ATOM" which allows the more elaborate variations to be carried out using more interations, but requiring hardly any more computer time. (Such runs have been made by SINANOĞLU and LUKEN).

Table 4.6 Comparison of f-value calculations for transitions between Type A (lowest-of-symmetry) states. The SINANOĞLU-ÖKSÜZ-WESTHAUS NCMET results were calculated ("simpler variation") using NCMET charge wave functions constrained to a single non-linear parameter $\xi = \xi_{3s} = \xi_{3p} = \xi_{3d} = (5/3)\xi_{4f}$. The later LUKEN, NICOLAIDES, TÖZEREN, SINANOĞLU runs in our laboratory made use of charge wave functions obtained from more variations ("fully varied") in which each ξ $\{\xi_{3s},\xi_{3p},\xi_{3d},\xi_{4f}\}$ was varied independently, and a fifth non-linear parameter ξ'_{3d} was also added to the wave function. The results are seen to be still the same

			NCMET (SINANOĞLU)			
Ion	Transition	RHF	Simpler variation [4.43]	Fully varied [4.27,28]	Experiment	Reference
C II	$2p\ ^2P^o - 2p^2\ ^2D$	0.263	0.125	0.121	0.101	[4.63]
					0.108	[4.64,65]
					0.114	[4.66]
					0.127	[4.64]
N III	$2p\ ^2P^o - 2p^2\ ^2D$	0.213	0.114	0.111	0.103	[4.45]
					0.112	[4.67]
N II	$2p^2\ ^3P - 2p^3\ ^3D^o$	0.236	0.100	0.097	0.094	[4.67]
					0.101	[4.45]
					0.105	[4.46]
					0.109	[4.66]
O III	$2p^2\ ^3P - 2p^3\ ^3D^o$	0.200	0.100	0.097	0.102	[4.61,68]
N II	$2p^2\ ^3P - 2p^3\ ^3S^o$	0.334	0.218	0.211	0.189	[4.45]
					0.209	[4.67]
O III	$2p^2\ ^3P - 2p^3\ ^3S^o$	0.272	0.183	0.178		

We conclude that for KL-shells, Type A states, which include all the KL-states of +2 and higher ions, all A I, A II ground states and a few excited A I, A II ones, the f^{E1}-values of SINANOĞLU-WESTHAUS-ÖKSÜZ are still the main accurate and systematic theoretical (i.e., NCMET) values available in the literature.

4.4 States Not Lowest of Their Symmetry

We now show that the proper treatment of A I and A II states often requires that one take a closer look at the physical situation, and that an uncritical application of the formalism may lead to large errors. Those errors can, however, be avoided and reliable values obtained.

4.4.1 Neutral and Singly-Ionized Atoms

The $\overline{V} \to \overline{V}'$ intravalency transitions' multiplet oscillator strengths f^{El} (LS $\to$ L'S) between terms of $1s^2 2s^n 2p^m \to 1s^2 2s^{n'} 2p^{m'}$ for all isoelectronic sequences are given by SINANOĞLU and WESTHAUS [4.5-8] (and for the higher Z's in the sequences, from the $(1/Z) \to 0$ extrapolations [4.25] of the SINANOĞLU-WESTHAUS [4.8] values in the straight line regions). In [4.25,Table 1] these values are compiled. It will be noted that in both [4.8] and [4.25], the A I and A II values are largely missing. This omission was deliberate, as we shall see now.

In the doubly- and higher-ionized members of the KL-shell atom isoelectronic sequences, the first El-allowed (absorption) transitions are of the $\overline{V} \to \overline{V}'$ intravalency type ($1s^2 2s^2 2p^m \to 1s^2 2s 2p^{m+1}$; $\Delta L = 0, \pm 1$; $\Delta S = 0$; $\pi \cdot \pi' = -1$). The $\overline{V}'(1s^2 2s 2p^{m+1};\ {}^{2S+1}L'{}^{\pi'})$ terms excited are the lowest energy levels of their (L'Sπ') symmetry. Thus the variational minimum principle applies to both the initial ($\overline{V}$) and the final state ($\overline{V}'$) calculations for such f^{El}'s. (Lowest root of the K x K ψ_c calculation for each case is minimized (step (iv), Section 4.3) for the ψ_c' of ($\overline{V}'$) as well as for the ψ_c of ($\overline{V}$)).

At the lower ξ = Z-N members of the KL-isoelectronic series, however, the first El transitions are often of the $\overline{V} \to$ pR, pre-Rydberg type

$$\overline{V}(1s^2 2s^2 2p^m) \to pR \begin{Bmatrix} 1s^2 2s^2 2p^{m-1}\ 3s,3d \\ \\ 1s^2 2s 2p^m 3p \end{Bmatrix} \quad . \tag{4.18}$$

The reason for this is easily seen in Fig.4.2, where we plot the energies (relative to the ground term) of the terms $2s2p^3\ {}^1P^o$ (this is our $\overline{V}'$), $2p3s\ {}^1P^o$ and $2p3d\ {}^1P^o$, the latter being pRs and pRd, respectively. In Fig.4.2, x indicates the ionization

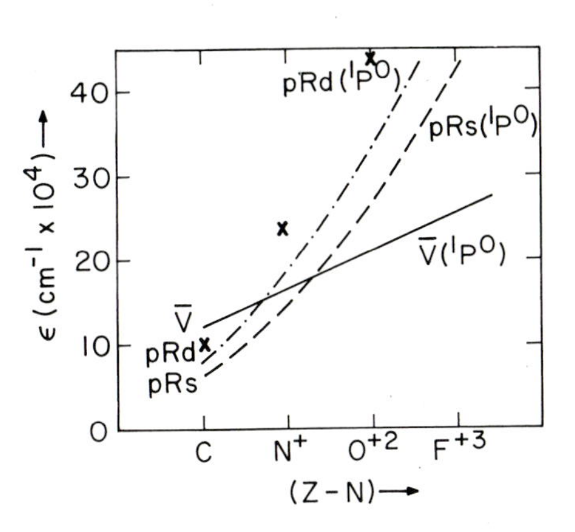

Fig.4.2 The lowest ${}^1P^o$ terms in the C I isoelectronic sequence, plotted relative to the $2p^2\ {}^3P$ ground term. The symbol x indicates the ionization energy. Where not shown, the ionization energy is above the top of the diagram

$$\varepsilon(\overline{V}) = E(2s2p^3\ {}^1P^o) - E(2p^2\ {}^3P)$$

$$\varepsilon(pR_s) = E(2p3s\ {}^1P^o) - E(2p^2\ {}^3P)$$

$$\varepsilon(pR_d) = E(2p3d\ {}^1P^o) - E(2p^2\ {}^3P)$$

energy. At C I, $\overline{V}'$ is in the continuum, and the transitions which occur are of the type $\overline{V} \rightarrow$ pR. However, already at N II, $\overline{V}'$ is bound and able to compete with the pR terms as the final term in the transition, and, for higher values of ξ, $\overline{V}'$ dominates as the final term. In Fig.4.3, we see still another case in which complete crossover has been reached at the neutral member.

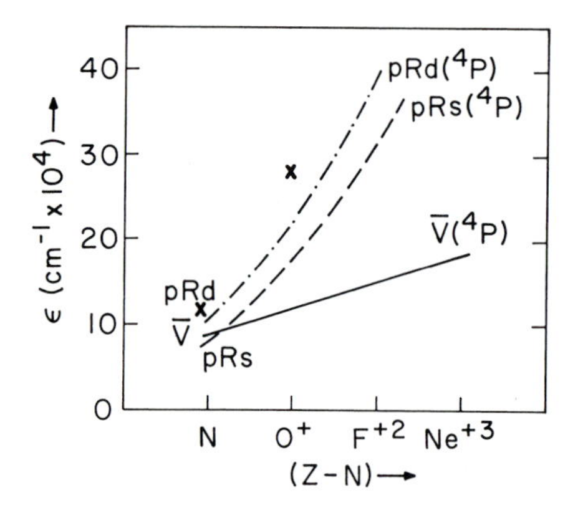

Fig.4.3 The lowest ^4P terms in the N I isoelectronic sequence, plotted relative to the $2p^3\ ^4S^o$ ground term. The symbol x is as in Fig.4.2

$$\varepsilon(\overline{V}) = E(2s2p^4\ ^4P) - E(2p^3\ ^4S^o)$$
$$\varepsilon(pR_s) = E(2p^2 3s\ ^4P) - E(2p^3\ ^4S^o)$$
$$\varepsilon(pR_d) = E(2p^2 3d\ ^4P) - E(2p^3\ ^4S^o)$$

In the upper state ^{2S+1}L supermultiplets, the $\overline{V}'(1s^2 2s2p^{m+1})$ states are not the lowest of that symmetry for the cases that were listed in Table 4.3. These cases all occur for the neutral and singly-ionized species.

There is a simple physical explanation of crossover. The hydrogenic degeneracy of 2s, 2p is split in N-electron atoms due to Coulomb repulsions and the resulting difference in shielding, with the more "penetrating orbital" 3s seeing the nuclear charge more than 2p. This difference diminishes again along an isoelectronic sequence at high Z as Coulomb repulsions become less important relative to nuclear attraction. At high Z, (same N), while the 2s-2p excitation energy difference is decreasing, at the same time the 2ℓ-$3\ell'$ difference is increasing (as Z^2). Thus at the higher Z, large ξ-ions, the $1s^2 2s2p^{m+1}(\overline{V}')$ configuration is lower than the $1s^2 2s^2 2p3\ell$(pR) configuration. The opposite is true near $\xi = 0$ or 1.

When the state to be calculated, for example, N I $2s2p^4\ ^4P$, is thus not the lowest ^4P, one cannot injudiciously minimize some root of the K x K $\psi_c(^4P)\underset{\approx}{\mathcal{H}}$-matrix in step (iii), Section 4.3. The root will collapse past the exact energy of the desired level towards the lowest, hence incorrect, ^4P state.

Aware of this problem, SINANOĞLU-ÖKSÜZ avoided such Type B states. However, they needed at least one or two such 2s-hole states, to be able to extract all of the

$\varepsilon^U_{ij;k\ell}$ all-external pair correlations which they were trying to evaluate from atomic data (from $E_U = E_{exp} - E_c$). They calculated only a few Type B cases, getting reasonable (but rougher than in Type A cases) Type B ψ_c's by "freezing" the 3s, 3p, 3d, 4f exponents at their values from some nearby Type A states. No minimization, hence no collapse. Not being optimized, however, these few special ψ_c's are not very good for the more sensitive oscillator strengths.

While SINANOĞLU and LUKEN were studying the Type B problem, NICOLAIDES in the same laboratory ran off the neutrals and singly-ionized cases using the SINANOĞLU-ÖKSÜZ-WESTHAUS-LUKEN programs. It appears that he had taken the root of the K x K $\mathcal{H}$-matrix, initially corresponding to the largest $\overline{V}'$ RHF determinant coefficient, and kept varying the semi-internal exponents, minimizing his energy. SINANOĞLU and LUKEN questioned these results at the time and later, noting the variational collapse and the abnormality of the resulting determinantal coefficients and renormalization factors $(1 + \langle\chi|\chi\rangle^{-1}$, suggesting an incorrect root was being calculated.

NICOLAIDES' results [4.27,28] show large discrepancies from experiment and from the new calculations by SINANOĞLU and LUKEN [4.31], as seen in Table 4.7. HIBBERT [4.30] was the first to note in a paper that NICOLAIDES-BECK results [4.73] suffer from variational collapse towards incorrect roots. The LUKEN-SINANOĞLU [4.31] analysis of the problem agrees with the conclusions of HIBBERT.

4.4.2 Variational Collapse and Its Avoidance

Consider the ψ_c calculation for a $\overline{V}'$ state which is not the lowest of its symmetry, for example, N I $2s2p^4\ ^4P$ (see Fig.4.3). The NCMET ψ_c determinants generated by "ATOM" are shown in Table 4.8. For compactness, we show these lumped into configurations.

In the Z-N > 1, multiply-ionized atoms for this N I isoelectronic sequence, the $1s^2 2s2p^4\ ^4P$ would be the lowest ^4P (Fig.4.3). Then we would get the 1s, 2s, 2p orbitals from the $1s^2 2s2p^4\ ^4P$ RHF calculation (an optional subroutine attached to "ATOM"). Next we would vary the $\xi_{3s}, \xi_{3p}, \xi_{3d}$ and ξ_{4f} exponents in the STO form semi-internals one at a time (or tied together; see Section 4.3) and keep minimizing the lowest root of the $\underset{\approx}{\mathcal{H}}$. The $\overline{V}'(1s^2 2s2p^4)$ being the correct configurational assignment for this state, it remains dominant during these variations, i.e., $c_1 \gg c_K (K \neq 1)$ in the ψ_c. Then $\langle\chi_c|\chi_c\rangle \ll 1$, as seen from the renormalization factor $(1 + \langle\chi_c|\chi_c\rangle)^{-1}$ in Table 4.8 for the more highly charged ions A" III, A"' IV, ... in the A I isoelectronic sequence.

In A I and A' II too, the RHF $\overline{V}'$ configuration $1s^2 2s2p^4\ ^2P$ has at first (before any mixings) lower energy than the pR single configuration $1s^2 2s^2 2p^2 3s$ with $\xi_{3s} = 2.3$. Just as the actual $\overline{V}'$ and pR levels cross with Z in the actual ion sequence in Figs. 4.2,3, however, as the ξ_{3s} is varied for A II, the single-configuration energies $E^0_{\overline{V}'}$,

Table 4.7 Oscillator strength results from the NCMET of SINANOĞLU for the $\overline{V} \rightarrow \overline{V}'$ ($2s^2 2p^m \rightarrow 2s2p^{m+1}$) intravalency transitions of neutral (A I) and singly-ionized (A II) atoms calculated [4.31] by the proper application of the variational principles for states not lowest of their symmetry. Earlier calculations, listed under C.A.N., are incorrect, due to variational collapse into lower roots. Where two calculated values appear, the top one is from the dipole length formulation, the other from dipole velocity

Ion	Transition ($\overline{V} \rightarrow \overline{V}'$)	λ [Å]	RHF	C.A.N. [4.27,29]	[4.31] NCMET ψ_C proper variational	Experiment	Reference
C I	$2s^2 2p^2\ ^3P - 2s2p^3\ ^3P^o$	1329	0.169	0.092	0.027	(0.039)	[4.69]
			0.115		0.045		
N II	$2s^2 2p^2\ ^1D - 2s2p^3\ ^1P^o$	660	0.240	0.324	0.157	0.13	[4.36]
			0.096		0.160	0.12	[4.37]
	$2s^2 2p^2\ ^1S - 2s2p^3\ ^1P^o$	746	0.847	0.195	0.255	0.25	[4.36]
			0.441		0.340	0.22	[4.37]
N I	$2s^2 2p^3\ ^4P^o - 2s2p^4\ ^4P$	1134	0.490	0.286	0.035	0.058	[4.63]
			0.557		0.077	0.078	[4.70]
						0.080	[4.66]
						0.083	[4.46]
						0.13	[4.71]
O II	$2s^2 2p^3\ ^2D^o - 2s2p^4\ ^2P$	538	0.329	0.311	0.158	0.166	[4.61]
			0.143		0.184	0.153	[4.39]
						0.142	[4.38]
	$2s^2 2p^3\ ^2P^o - 2s2p^4\ ^2P$	581	0.304	0.007	0.086	0.099	[4.61]
			0.155		0.120	0.102	[4.39]
						0.185	[4.38]
O I	$2s^2 2p^4\ ^3P - 2s2p^5\ ^3P^o$	792	0.344	0.190	0.088	autoionizes	[4.72]
			0.328		0.147		

and E^o_{pR} cross, as shown in Fig.4.4 at $\xi_{3s} \cong 1.4$. After the configuration-interaction $\underset{\approx}{\mathcal{H}}$-mixing occurs in the ψ_C, however, because the symmetry of $\overline{V}'$ and pR is the same, the crossing is avoided (non-crossing rule). In fact, what started out as the dominant $\overline{V}'_o$ configuration ($1s^2 2s2p^4$) follows along the dotted curve as the lowest root to become a predominantly pR^o ($1s^2 2s^2 2p^2 3s$) configuration. Similarly, the second root turns into a dominantly $\overline{V}'$ configuration-containing state, although it had started as a pR^o determinant.

This explains, as noted by HIBBERT [4.30], the NICOLAIDES error [4.27], which is seen to result from an insistence to follow the initial $\overline{V}'_o$ state in ξ_{3s} along the lower dotted curve. When this energy is erroneously minimized, it keeps going lower, ending up on a poor approximation to a pR state rather than approaching the desired $\overline{V}'$ state. To avoid this difficulty one can make use of the HYLLERAAS-UNDHEIM-MACDONALD theorem (HUM) [4.75-77].

Table 4.8 for the more highly charged ions A" III, A''' IV, ... in the A I isoelectronic sequence.

Table 4.8 NCMET ψ_c wave functions for four members of the N I $2s2p^4\ ^4P$ isoelectronic sequence. The oxygen, fluorine, and silicon wave functions are lowest-root results, representing the lowest ^4P state in each case. The nitrogen result is the second ^4P root, as required by the correct variational method, which avoids variational collapse towards the lower $(2p^2 3s\ ^4P)$ pR root. Multiplication by the "renorm. factor" normalizes each vector to unity. The RHF, ψ_c and difference energies are given in a.u.[4.74]

ψ_c Configuration	Configurational Coefficients[a]			
	Nitrogen	Oxygen	Fluorine	Silicon
$2s2p^4\ ^4P$	1.00000	1.00000	1.00000	1.00000
$2s^2 2p^2 3s\ ^4P$	-0.71287	0.05402	0.03470	0.01387
$2s^2 2p^2 3d\ ^4P$	0.23569	0.13974	0.10815	0.05128
$2s2p^3 3p\ ^4P(1)$[b]	0.38523	0.00277	0.00586	0.00018
$2s2p^3 3p\ ^4P(2,3,4)$[b]	0.18957	0.08017	0.05403	0.00564
$2s2p^3 4f\ ^4P$	0.02187	0.02085	0.01875	0.01197
$2p^4 3s\ ^4P$	-0.09573	0.00152	0.00090	0.00046
$2p^4 3d\ ^4P$	0.05686	0.04129	0.03457	0.01944
Renorm. Factor	0.75358	0.98484	0.99141	0.99832
E(RHF) =	-53.9881	-73.8195	-96.9143	-261.2249
$E(\psi_c)$ =	-54.0474	-73.8626	-96.9551	-261.2578
Difference =	-0.05930	-0.04309	-0.04080	-0.03296

[a] Calculated as the square roots of the sums of squares of the coefficients of all determinants in each configuration. Phases are arbitrary, but consistent from atom to atom

[b] There are four linearly-independent ^4P terms in the $2s2p^3 3p$ configuration. In this table, the $^4P(1)$ configuration contains only single excitations from ϕ_{RHF}. For brevity, the remaining three terms are lumped together into $^4P(2,3,4)$

Consider the K lowest exact energy levels of the same symmetry of a system. According to this theorem, if K configuration functions of same symmetry are chosen, the K x K $\underset{\approx}{\mathcal{H}}$ is diagonalized and parameters varied, the K eigenvectors remain orthogonal. Thus any one of the eigenvalues can be minimized with respect to the parameters and variational collapse is prevented. However, to approximate to the mth exact level of a given symmetry, we must always look at the mth root (of that symmetry) of the K x K $\underset{\approx}{\mathcal{H}}$ diagonalization.

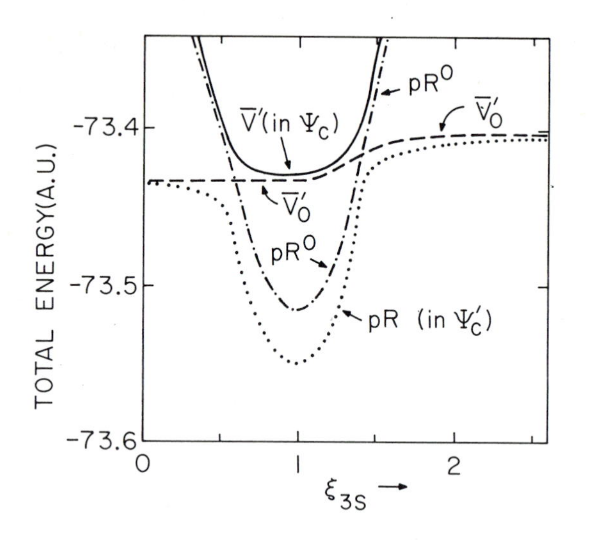

Fig.4.4 Energies of $\overline{V}'$ and pR configurations in the $2s2p^4\ {}^2P$ O II charge wave function varying with the 3s-like semi-internal orbital exponent ξ_{3s}. The dashed curve gives the energy of the $\overline{V}'_0$ configuration $2s2p^4\ {}^2P$, including all configurations of the $2s2p^4\ {}^2P$ charge wave function except $2s^22p^23s$. The pR^0 curve gives the energy of the $2s^22p^23s(\xi_{3s})\ {}^2P$ configuration. The lowest and second-lowest eigenvalues obtained from the $\mathcal{H}$-mixing of these configurations are given by the dotted and solid curves, respectively. Note the level switch and avoided crossing at $\xi_{3s} \approx 1.4$ resulting from the ψ_c $\mathcal{H}$-mixing

In practice, to have good approximations, the K configurational basis set must include all of the dominant configurations of the K exact levels.

If there is one configuration for each level experimentally and correctly assigned, then the problem is simplest [4.78]. In the $1s^22s2p^4\ {}^4P$ O II example above, the desired $\overline{V}'$ level is experimentally known to be on the second level. Thus, in the ψ_c calculation, we must, according to HUM, always look at the second root without regard to the magnitudes of the determinantal coefficients and keep following it as each parameter, e.g., ξ_{3s}, is varied. Thus in Fig.4.4, had we started with a poor ξ_{3s}, the second root would initially look like pR^0 dominant, but if we continue on the second root as we vary ξ_{3s}, we end up with the correct $\overline{V}'$ state which now has $\overline{V}'_0$ dominant.

All the SINANOĞLU-LUKEN ψ_c's (and the resulting f^{El}'s) in Table 4.8 are obtained this way, with, however, one more practical point to be observed. Better approximations to $\psi_c(\overline{V}')$ are obtained by picking the 1s, 2s, 2p from the lowest state, the other than ξ_{3s} semi-internals from the next lowest state, etc. In this way, the lower eigenvectors remain good approximations to the exact lower eigenstates, while upper state parameters are being varied. These more specialized aspects are discussed by SINANOĞLU and LUKEN [4.31].

There are a few additional non-lowest states in Table 4.3 which are not calculated by SINANOĞLU and LUKEN for Table 4.7. These are Be I $2p^2\ ^1S$; B I $2s2p^2\ ^2S$; C I $2s2p^3\ ^1D^o$, $^1P^o$; N I $2s2p^4\ ^2D$, ^2S, ^2P; O I $2s2p^5\ ^1P^o$; and F I $2s2p^6\ ^2S$, containing more than three (some an infinity of) lower states of the same symmetry. Now that it is experimentally clearer [4.30] that the Be I $1s^22p^2\ ^1S$ state lies in the continuum, one has an infinity of ^1S states below. This and similar cases require still additional considerations where rather than straightforward HUM, excited-state variational theory with subminimal bases is involved. These we leave outside the scope of the present article (see [4.31]).

Where the lowest-lying states of a symmetry consist of the interaction of valency and pre-Rydberg states to give an NCMET charge supermultiplet, we have the HUM case. The NCMET ψ_C calculations then proceed in the Type A or Type B cases (and both with our program system "ATOM") straightforwardly. This covers practically all of the low-lying spectra and lifetimes of atoms and ions and we have seen that NCMET ψ_C gives good oscillator strengths from neutral atoms to multiply-ionized ones.

In Table 4.9, our final recommended set of NCMET ψ_C oscillator strengths is given for all the $1s^22s^n2p^m \rightarrow 1s^22s^{n'}2p^{m'}$ $(\overline{V} \rightarrow \overline{V}')$ transitions of the Be I, B I, C I, N I, O I, and F I isoelectronic sequences. All the results come from the SINANOĞLU-WESTHAUS-LUKEN work with the writer's NCMET program system "ATOM". Table 4.9 is a uniform set of f^{El}'s, all coming from the NCMET ψ_C, including all the specifically non-closed-shell correlations. These values should therefore be particularly useful for comparing with experimental data and in calculating abundances in astrophysics. This is in contrast to, for example, the NBS tables and $f^{El}(1/Z)$ curves, where the points come from different theoretical and experimental sources of varying accuracy.

Note that results are given only to Z=14, even though values on the $f^{El}(1/Z \rightarrow 0)$ straight line beyond that are very easy to obtain. These high-Z values can be found in Table 5 of the SINANOĞLU BFS paper in 1973 [4.25]. However, they are non-relativistic. In this high-Z region, we have recently found unusual effects coming from the relativistic/intermediate coupling treatment [4.41]. For the highly-stripped heavy-ion regions, the reader should consult that paper [4.41] together with [4.25,Table 5].

Table 4.9 Uniform NCMET ψ_c oscillator strengths (SINANOĞLU-WESTHAUS-LUKEN) for the KL intravalency ($\overline{V} \rightarrow \overline{V}'$) transitions of the Be I, B I, C I, N I, O I, F I isoelectronic sequences. These are the recommended NCMET results to date

Isoelectronic sequence and transition	Be	B	C	N	O	F	Ne	Na	Si
Be I									
$2s^2\ ^1S - 2s2p\ ^1P^o$	1.25	1.00	0.760	0.605	0.513	0.435	0.384		0.261
$2s2p\ ^1P^o - 2s^2\ ^1D$		0.148	0.165	0.171	0.155	0.143	0.131		0.096
$-\ ^1S$			0.173	0.135	0.122	0.101	0.092		0.062
B I									
$2s^22p\ ^2P - 2s2p^2\ ^2D$		0.036	0.121	0.111	0.104	0.096			0.065
$-\ ^2S$			0.112	0.079	0.065	0.055			0.032
$-\ ^2P$		0.626	0.485	0.389	0.327	0.282			0.168
C I									
$2s^22p^2\ ^3P - 2s2p^3\ ^3D$			0.077	0.097	0.097	0.090			0.066
$-\ ^3P$			0.035	0.133	0.122	0.108			0.077
$-\ ^3S$			0.272	0.211	0.178	0.150			0.093
$^1D -\ ^1D$				0.306	0.291	0.210			0.104
$-\ ^1P$				0.158	0.208	0.180			0.106
$^1S -\ ^1P$				0.255	0.285	0.263			0.173
N I									
$2s^22p^3\ ^4S - 2s2p^4\ ^4P$				0.294	0.198	0.208	0.208		0.170
$^2D -\ ^2D$				0.052	0.141	0.131			0.101
$-\ ^2P$					0.171	0.213			0.139
$^2P -\ ^2D$					0.031	0.036			0.031
$-\ ^2S$					0.100	0.085			0.060
$-\ ^2P$					0.102	0.097			0.079
O I									
$2s^22p^4\ ^3P - 2s2p^5\ ^3P$					0.114	0.150	0.146	0.146	0.135
$^1D -\ ^1P$							0.229		0.185
$^1S -\ ^1P$							0.078		0.077
F I									
$2s^22p^5\ ^2P - 2s2p^6\ ^2S$							0.070	0.081	0.076

4.5 New Oscillator Strengths for Intershell ($KL \rightarrow KL'[M]$) Transitions to Pre-Rydberg Levels ($\overline{V} \rightarrow pR$)

In the Type B "non-lowest" state calculations of SINANOĞLU and LUKEN [4.31] above for the $\overline{V} \rightarrow \overline{V}'$ transitions, the lower roots give reasonably good approximations to the charge wave functions of the pre-Rydberg states of the same NCMET charge supermultiplets. Actually, the complete ψ_C of a KL'[M] intershell (pR) configuration would include the M-shell type new semi-internal effects. However, according to the NCMET charge supermultiplet theorem above, the main mixings into the $\overline{V}$ states are from the virtual pR states. Therefore, and, by the H-sum rule, the main mixings into the corresponding actual pR states will be from the accompanying $\overline{V}$ states. This explains why one gets reasonable ψ_C (pR) from the lower roots of the $\psi_C(\overline{V})$ calculation. Table 4.10 shows the new f^{E1} values obtained this way by SINANOĞLU and LUKEN for some $\overline{V} \rightarrow$ pR transitions in neutral and +1 ions. The agreement with experimental values is good, giving further confirmation to the NCMET/HUM results above.

Table 4.10 Neutral and +1 ion intershell $\overline{V} \rightarrow$ pR transition multiplet oscillator strengths of SINANOĞLU and LUKEN obtained from their NCMET/HUM approximate ψ_C(pR) charge wave functions for the KL'[M] states. Where two calculated values appear, the top one is from the dipole length formulation, the other from dipole velocity

Ion and Transition	λ [Å]	f^{E1}_{RHF}	f^{E1}(NCMET/HUM) from $\psi_C(\overline{V})$ and approx. ψ_C(pR)	Experiment	Reference
N II $2s^2 2p^2\ {}^1D - 2s^2 2p3s\ {}^1P^o$	747	0.077	0.168	0.060	[4.79]
		0.082	0.208	0.159	[4.80]
				0.111	[4.36]
${}^1S -$ ${}^1P^o$	858	0.091	0.026	0.020	[4.79]
		0.091	0.023	0.053	[4.80]
				0.037	[4.36]
N I $2s^2 2p^3\ {}^4S^o - 2s^2 2p^2 3s\ {}^4P$	1200	0.082	0.271	0.259	[4.66]
		0.090	0.396	0.291	[4.63]
				0.350	[4.71]
				0.278	[4.46]
				0.218	[4.81]
O II $2s^2 2p^3\ {}^2D^o - 2s^2 2p^2 3s\ {}^2P$	617	0.047	0.092	0.107	[4.39]
		0.056	0.126	0.118	[4.38]
				0.104	[4.38]
$2s^2 2p^2 3p\ {}^2P^o - 2s^2 2p^2 3s\ {}^2P$	673	0.050	0.036	0.036	[4.39]
		0.060	0.037	0.040	[4.38]
				0.046	[4.38]
O I $2s^2 2p^4\ {}^3P - 2s^2 2p^3 3s\ {}^3P^o$	878	0.032	0.074	0.13	[4.72]
		0.022	0.082	(autoionizes)	
$2s^2 2p^4\ {}^3P -$ $3d\ {}^3P^o$	816	0.0085	0.0136	(autoionizes)	
		0.0065	0.0358		[4.82]

4.6 Further Examination of Remaining Correlation Effects on Oscillator Strengths with NCMET

It is clear from the NCMET results on various properties up to now (Tables 4.1,5,etc.) that the charge wave function as predicted from the theory by SINANOĞLU gives the charge-like properties well, in particular the transition probabilities (within a few percent as compared to sizable uncertainties found, e.g., with RHF). Nevertheless, it is worth examining what other effects might influence the accuracy of the NCMET ψ_c results. This examination is possible systematically with NCMET, since NCMET does not need to stop at the ψ_c, which it delineates only as a major part of the non-closed-shell problem, but has provisions for efficiently introducing the remaining effects or for testing their magnitudes, etc. of, for example, the all-external-pair correlations. In the closed-shell parts of the theory (MET) this had been done even to the point of examining the small three- and four-particle all-external (dynamical) correlations and the one-electron polarizations beyond the closed-shell Hartree-Fock, ψ_c, etc. (see [4.5,26] for the review of these).

After NCMET ψ_c calculations of f^{E1} (or also f^{E2}, forbidden lifetimes, generalized oscillator strengths, etc.), to assess accuracy, one needs to inquire about: (i) degree of optimization of the semi-internals $f_\ell(r)$; (ii) "cancellation effects" in f^{E1}; (iii) variational collapse, etc., for non-lowest states; and (iv) any sizable all-external correlation effects needed.

We have shown in Table 4.6 above, comparing different levels of refinement on the semi-internals, that accuracy of a few percent in the f^{E1} is quickly reached in regard to (i).

"Cancellation effects" (ii), i.e., some NCMET charge supermultiplet configurations strongly stealing intensity from each other, thereby largely cancelling out, are accounted for to good accuracy in NCMET ψ_c f^{E1}. Good examples of this occur in the KLM → KLM' third row atom transitions which have very small f-values (< 0.1) due to such "cancellations" while RHF gives large ones (of the order of 0.5). Our NCMET ψ_c calculations of these difficult cases have yielded good f^{E1} values in agreement with recent BFS experiments, showing that the "cancellation effects" have been properly taken care of. Table 4.11 gives the Si I KL $3s^23p^2$ ^3P - KL $3s3p^3$ $^3D^o$ isoelectronic sequence [4.83]. These third-row results involve ψ_c's which have determinants an order of magnitude larger (of the order of 1000) as compared to 10 to 60 in the KL shell second-row problem, and yet the NCMET ψ_c calculations [4.25,83,84] were still efficiently carried out. The NCMET lowering in these f^{E1} values from orbital theories was by factors as large as 10-30, yet agreed with BFS.

Table 4.11 The NCMET ψ_c oscillator strengths in the third row (KLM → KLM') are much smaller than the RHF values due to "cancellation effects", yet they are accurately obtained with NCMET ψ_c and agree with experiment. We display results for the Si I isoelectronic sequence KL $3s^2 3p^2\ {}^3P$ - KL $3s3p^3\ {}^3D^o$. Numbers marked with an asterisk are interpolated from f(1/Z). References are indicated in square brackets

Ion	f^{E1}_{RHF}	$f^{E1}_{NCMET\ \psi_c}$	f^{E1}_{Expt}	Reference
P II	0.412	0.010	0.0076 0.013	[4.47] [4.85]
S III	0.389	0.036	0.022	[4.86]
Cℓ IV	(0.39)*	(0.045)*	0.045	[4.87]
Ca VII	0.304	0.057		
Fe XIII	0.230	0.055		

The proper minimization, maximization, variational collapse avoidance, etc., (iii) are also easily handled with NCMET and the automated programs of "ATOM" with some background on variational theory. These aspects we discussed above (see also the LUKEN-SINANOĞLU papers).

This leaves the all-external correlations, (iv), whether they have, say, an effect more than about 10% on charge-like properties.

We have now systematically covered many of the charge-like properties with NCMET ψ_c, finding good agreement with experiment (see Table 4.1 for examples). SMITH and BROWN [4.88] reached similar conclusions on the related Compton profiles. There have also been some direct calculations of the all-external part itself, confirming that, as mentioned by HIBBERT [4.30,89] and others [4.90], $\langle \chi_U | \chi_U \rangle \ll \langle \chi_c | \chi_c \rangle$. This has been shown very recently also by MCCAVERT [4.91] for the atomic electronic quadrupole moment.

Thus the NCMET charge wave function's role seems now well established, the long list of the SINANOĞLU-WESTHAUS f^{E1} values for doubly- and higher-ionized atoms and their good agreement with experiment (within a few percent) being one important piece of evidence.

Only in some special cases (see below) or where extreme accuracy is desired (as in the He atom, where charge-like correlations are not in general expected to be dominant, or where many configurations mix so strongly that there are no experimentally-definable configurational assignments, hence no definite charge supermultiplets, as in the He-like doubly-excited spectra in the far UV (treated by special group-theoretic techniques [4.92,93]) we may need to use the all-external correlations (for more than two-electron systems obtained by the $\hat{u}_{ij}$ methods of χ_U MET/NCMET [4.5]).

We note here one such special effect for even further improvement of the accuracy of the f^{El} in negative ions, neutrals, and +1 ions [4.94]. In such atoms, as shown in Table 4.7, the semi-orbitals are particularly diffuse, with large coefficients. Wherever some few-electron correlation-functions parts within the MET/NCMET N-electron χ are large, we expect their unlinked products [4.95] too to be non-negligible. Although this separation is not essential, we can split off from an all-external χ_U the unlinked products of the semi-internal functions $\hat{f}$, as, for example, in the two-electron case

$$\psi = A(12) + A(1\hat{f}_2) + A(\hat{f}_1 2) + A(\hat{f}_1\hat{f}_2) + \bar{u}_{12}^{f} \quad . \tag{4.19}$$

($\phi_{RHF} \equiv A(12)$; A the anti-symmetrizer; 1 and 2 two Hartree-Fock spin-orbitals) instead of the equivalent

$$\psi = A(12) + A(1\hat{f}_2) + A(\hat{f}_1 2) + \hat{u}_{12} \quad , \tag{4.20}$$

where the $\hat{f} \otimes \hat{f}$ unlinked product is lumped into the all-external $\hat{u}_{12}$.

Where the semi-internal products may make a non-negligible contribution to the charge density, we get a slightly-extended charge wave function by including all the configurations containing the pairs of semi-internal orbitals, $\{\hat{f} \otimes \hat{f}\}$. (In general, of course, to include these with the ψ_c, instead of leaving them in the χ_U, would be unnecessary work). Thus we define the "$\hat{f} \otimes \hat{f}$ - extended charge wave function"

$$\psi_c^e \equiv \psi_c + \{\hat{f} \otimes \hat{f}\}_{configs} \quad . \tag{4.21}$$

($\psi_{exact} = \psi_c + \chi_U = \psi_c^e + \bar{\chi}_U$). These $\{\hat{f} \otimes \hat{f}\}$ configurations look like double excitations, hence like all-external, from the ϕ_{RHF}, but they have more the $\{\hat{f} \otimes \hat{f}\}$ significance. Let us call these therefore "quasi-all-external correlations". We call the resulting finite set of configurations the "extended charge supermultiplet".

In Table 4.12 we show all the quasi-all-external configurations for the $\bar{V}'$ $1s^2 2s2p^3\ {}^1P^o$ C I, N II, O III, ... state, together with the usual charge wave function configurations.

The "extended supermultiplet" has yet another interpretation. We have seen that in neutrals and ±1 ions, whether of Type A or B states, the $\bar{V}$ and the pR components of the NCMET charge supermultiplet mix quite strongly. If the state is characterized as

Table 4.12 A charge wave function extended with quasi-all-external correlations, χ_{ff}. The list includes all the ψ_c configurations (the valency and pre-Rydberg supermultiplet) as well as the new pre-Rydberg-pre-Rydberg correlations χ_{ff}. This table is for the $1s^22s2p^3\ ^1P^o$ state of N II, O III, ...

Configuration $^1P^o$	Class	Configuration $^1P^o$	Class
$2s2p^3$	$\overline{V}$	$2s^23s3p$	$pR_s \otimes pR_p$
$2s^22p3s$	pR_s	$2p^23s3p$	$pR_s \otimes pR_p$
$2p^33s$	pR_s	$2s^23p3d$	$pR_p \otimes pR_d$
$2s2p^23p$	pR_p	$2s^23p3d$	$pR_p \otimes pR_d$
$2s^22p3d$	pR_d	$2s2p3p4f$	$pR_p \otimes pR_f$
$2p^33d$	pR_d	$2s2p3p^2$	$pR_p \otimes pR_p$
$2s2p^24f$	pR_f	$2s2p4f^2$	$pR_f \otimes pR_f$
$2s2p3s^2$	$pR_s \otimes pR_s$	$2p^23s4f$	$pR_s \otimes pR_f$
$2s2p3s3d$	$pR_s \otimes pR_d$	$2p^23d4f$	$pR_d \otimes pR_f$
$2s2p3d^2$	$pR_d \otimes pR_d$	$2p^23d4f$	$pR_d \otimes pR_f$

dominantly a $\overline{V}$ configuration-labeled state, we have its full ψ_c, all the configurations which are "internal" (χ_{INT}) and "semi-internal" (χ_F) with regard to that $\overline{V}$ - ϕ_{RHF}. Suppose, however, a pR configuration in the ψ_c also has quite a large coefficient. A "higher-order" charge-density effect may then be expected to come from <u>its</u> charge-correlation wave function, i.e., a $\psi_c(pR)$. In the limit of a 50-50 mixing of $\overline{V}$ and pR, we would expect an overall ψ_c containing both the charge correlations of $\overline{V}$ and of the pR. Such very large mixings occur in doubly-excited spectra lying in the continuum. The He-like doubly-excited spectrum is thus treated with such configuration interactions [4.93] considerably larger than just the ψ_c of the usual singly-excited spectra. Fortunately, HERRICK and SINANOĞLU [4.92,93] were able to predict the configuration interactions coefficients in these super-supermultiplets just from the properties of the O(3,1) Lie algebra, finding two new quantum numbers, K and T, which characterize the experimentally-observed "weak" and "strong" series and which predict autoionization selection rules as well.

The Be I $1s^22p^2\ ^1S$, now thought to lie in the continuum [4.30,96], is not unlike the He doubly-excited states. It is not surprising, therefore, that its ψ_c should be an extended one containing the quasi-all-external correlations, as first noted by HIBBERT [4.30]. These correlations are mainly the ones making up the extended-charge supermultiplet in the ψ_c^e and now we come back to their interpretation in still another way.

In the ψ_c^e containing dominant $\overline{V}$ and pR configurations both, what would be internal correlations with respect to the pR, looks like double excitations with regard to the accompanying $\overline{V}$, and gives the quasi-all-external correlations ψ_{ff}. For example, in Table 4.12, the $2s2p3d^2\ {}^1P^o$ looks like all-external with respect to $2s2p^3\ {}^1P^o$, but like "internal" with regard to $2s^22p3s\ {}^1P^o$. Thus the <u>extended-charge supermultiplet</u> includes all the valency plus pre-Rydberg charge correlations of the basic $\overline{V}$, but also all the (M shell) internal correlations of the accompanying pR's occurring [4.97] in that state's ψ_c.

Obviously, the process could be continued to include the semi-internals out of the M shells [4.11-16], i.e., with regard to the pR, which brings in many more configurations, the ones that occur in the ψ_c of the KLM atoms like Si I, P I, Cℓ I, etc. [4.83,84], but clearly for the excited KL atoms, these would do little and are best left in the χ_U as in the initial SINANOĞLU ψ_c of NCMET.

In concluding this section, we mention several alternate forms of the SINANOĞLU NCMET ψ_c, all of which have been noted at the inception of NCMET. These minor variants, some of which, however, would involve considerably more but unnecessary labor than our standard NCMET ψ_c (and "ATOM") are mentioned here so that the reader interested in f^{El} will recognize any future calculations that might appear under the names of C.I., MCSCF (multi-configurational, SCF), GRHF (general Hartree-Fock), etc., as essentially the same as the original NCMET ψ_c, so long as these contain the same configurations (valency plus pR) as first predicted by NCMET as a finite, complete all non-closed-shell type, and therefore charge-distribution-affecting, correlations set.

In the standard version,

$$\psi_c = \phi_{RHF} + \chi_{INT} + \chi_F \quad . \tag{4.5}$$

The 1s, 2s, 2p come from the ϕ_{RHF} calculation, and are then kept fixed while the $\underset{\approx}{\mathcal{H}}$ is varied with regard to the semi-internals. If, however, several $\overline{V}$-type configurations have very large mixings, i.e., if χ_{INT} is not small compared to ϕ_{RHF} (which happens quite routinely in molecules, and is expected also in negative ions), then we can use the GRHF version of NCMET ψ_c, i.e.

$$\psi_c^G \equiv \phi_{GRHF} + \chi'_F \quad , \tag{4.22}$$

where

$$\phi_{GRHF} \cong \frac{\phi_{RHF} + \chi_{INT}}{\sqrt{1 + \langle\chi_{INT}|\chi_{INT}\rangle}} \quad , \tag{4.23}$$

and now the 1s, 2s, 2p are obtained from a multi-configurational self-consistent field (MCSCF) calculation on the combination of the $\overline{V}$ configurations (or the main ones in it) occurring in ϕ_{RHF} and the χ_{INT}. These $\{k\}_{\overline{V}}$ orbitals are then kept fixed while semi-internals are obtained by variations as before.

In yet another version we can in principle do a super-MCSCF calculation (ϕ_{SRHF}) on the entire set of the supermultiplet configurations in ψ_C

$$\psi_C^S = \phi_{SRHF} \cong \frac{\phi_{GRHF} + \chi_F'}{1 + \langle \chi_F' | \chi_F' \rangle} \quad , \tag{4.24}$$

varying all of the orbitals at once, the {1s,2s,2p,3s,3p,3d,4f}. Thus the initial 1s, 2s, 2p obtained from the RHF or the GRHF change a little, adjust a little, to the changes in the semi-internals and to the magnitudes of the C.I. coefficients in ψ_C.

However, aside from the arbitrary normalization factors, η, all these variants are essentially the same NCMET ψ_C,

$$\psi_C \cong \eta_G \, \psi_C^G \cong \eta_S \, \psi_C^S \quad . \tag{4.25}$$

The much increased labor is not worth it in most atomic problems and should not detract from the simplicity of the original non-closed-shell many-electron theory NCMET.

4.7 Conclusion

We have seen that the Non-Closed-Shell Many-Electron Theory (NCMET) by SINANOĞLU, which started from a formal solution to the N-electron Schrödinger equation, delineating in it the various correlation effects and *a priori* singling out the physically significant ones for certain classes of experimentally-observable properties, has now been systematically applied to oscillator strengths for transitions in large numbers of atoms, ions, with good agreement with experiments. In particular, in the recent stages, this theoretical development has gone hand-in-hand with the exciting new field of beam-foil spectroscopy. This close interaction between theory and experiment has provided excellent cross checks of both and pointed out new interesting areas for development. Among the latter are the transitions of third-row atoms [4.25,83,84] (Aℓ I, Si I, P I, S I, Cℓ I, etc. isoelectronic sequences), a new theoretical delineation of special "pre-Rydberg" states in atomic (also particularly useful in molecular) spectra [4.13], new theory [4,92,93], and experiment [4.98] on multiply-excited spectra in the vacuum UV, predictions of unexpectedly large deviations from the usual straight line $(1/Z) \to 0$ regions of the $f^{El}(1/Z)$ curves for heavy ions stripped down to their $1s^2 2s^n 2p^m$ electrons [4,41], NCMET ψ_C values for forbidden transition probabilities [4.25], and finally the large differences between the electron-impact spectroscopic generalized oscillator strengths [4.14] as given by NCMET and the pre-NCMET orbital theories. We leave the detailed discussion of these last developments outside the scope of the present chapter, but refer the reader to those recent papers. Also other recent papers by a number of experts have provided opportunity for critical re-examinations and re-assessment of the theoretical f^{E1} values and of some further aspects of the theory. It appears now that we have a reliable tool in NCMET (and its automated program systems "ATOM") which can be used with ease and little computer cost by the experimentalists. These program systems which can be run by the non-specialist with hardly any prior training on their use will be made available to the interested BFS laboratories.

Acknowledgment

Support of this work by a National Science Foundation Grant is gratefully acknowledged. The author thanks Dr. W. Luken for many stimulating and helpful discussions. He also thanks his many experimental and theoretical colleagues for correspondence and discussions in the past several years.

In the various applications and testing out of hypotheses, etc., on the atomic aspects of MET/NCMET, a good many of his pre- and post-doctoral students have collaborated with this writer. I take this opportunity to thank these students, now colleagues in many institutions: V. McKoy, D. F. Tuan, H. J. Silverstone, I. Öksüz, P. Westhaus, S. LaPaglia, B. Skutnik, W. Luken, and S. Davis for their reliable, inspiring and pleasant collaboration.

References

4.1 S. Bashkin: *Beam-Foil Spectroscopy* (Gordon and Breach, New York 1968) and references therein
4.2 C. H. Liu, D. A. Church: Phys. Rev. Letters 29, 1208 (1972)
4.3 See also S. Bashkin: Progress in Optics XII, 289 (1974)
4.4 Some highly stripped ions (few electrons left) of high atomic number are now observed in BFS even without heavy-ion accelerators of high energy (see [4.3])
4.5 O. Sinanoğlu: In *Atomic Physics*, Vol. I (Plenum Press, New York 1969), First Intl. Conf. on Atomic Physics, July 1968)
4.6 O. Sinanoğlu: Advances Chem. Phys. 14, 237 (1968) (Frascati NATO Summer Institute Lectures, 1967)
4.7 O. Sinanoğlu: Advances Chem. Phys. 6, 615 (1964)
4.8 P. Westhaus, O. Sinanoğlu: Phys. Rev. 183, 56 (1969)
4.9 O. Sinanoğlu, D. R. Beck: Theoret. Chim. Acta 34, 183 (1974)

4.10 The "Many-Electron Theory" (MET) of O. Sinanoğlu developed from 1962 to 1967 [4.5]. Sinanoğlu referred to the non-closed-shell aspects of this theory as the "Non-Closed Shell Many-Electron Theory" (NCMET). Sinanoğlu and later others have applied the MET/NCMET to molecular potential energy curves [4.11], Van der Waals forces [4.12], molecular vacuum UV spectroscopy [4.13], electron impact [4.14], hyperfine structure [4.15,16], etc., as well as atomic spectra [4.5-7]

4.11 O. Sinanoğlu: J. Mol. Struct. 19A, 81 (1973) (Proceedings of the Intl. Conf. on Molecular Spectroscopy, Wrøcław, Poland)

4.12 H. Margenau, N. R. Kestner: *Theory of Intermolecular Forces*, 2d ed. (Pergamon Press, Oxford 1971)

4.13 O. Sinanoğlu: In *Molecular Spectroscopy and Photochemistry in the Vacuum Ultraviolet*, ed. by C. Sandorfy, P. J. Ausloos, M. B. Robin (D.Reidel, Dordrecht, Holland 1974)

4.14 S. L. Davis, O. Sinanoğlu: J. Chem. Phys. 62, 3664 (1975)

4.15 F. Schaefer, P. Klemm, F. Harris: Phys. Rev. 181, 137 (1969)

4.16 O. Sinanoğlu, D. R. Beck: Chem. Phys. Letters 20, 221 (1973)

4.17 D. R. Bates, A. Damgaard: Phil. Trans. Roy. Soc. (London) A242, 101 (1949)

4.18 C. C. J. Roothaan, P. S. Kelly: Phys. Rev. 131, 1177 (1963)

4.19 References to D. Layzer, Z. Horak et al. papers in: D. Layzer, R. H. Garstang: Ann. Rev. Astron. Astrophys. 6, 449 (1968)

4.20 C. D. H. Chisholm, A. Dalgarno: Proc. Roy. Soc. (London) A292, 264 (1966)

4.21 A. Yutsis: *Theory of Electronic Shells of Atoms and Molecules* (Mintis, USSR 1971) (1969 Vilnius Symposium Proceedings)

4.22 C. Froese-Fischer: Astrophys. J. 141, 1206 (1965)

4.23 A. W. Weiss: Phys. Rev. 162, 71 (1967)

4.24 H. Kelly: Phys. Rev. 136, B896 (1964)

4.25 O. Sinanoğlu: Nucl. Instr. and Meth. 110, 193 (1973) (referred to in text as Sinanoğlu BFS-1973, Third Intl. Conf. on Beam-Foil Spectroscopy, Tucson 1972)

4.26 O. Sinanoğlu, K. A. Brueckner: *Three Approaches to Electron Correlation in Atoms* (Yale Press, New Haven and London 1970)

4.27 C. A. Nicolaides: Chem. Phys. Letters 21, 242 (1973)

4.28 C. A. Nicolaides: Private communication. In the survey and listing [4.25] of all the NCMET f^{El} results available up to that time, the Nicolaides results from the Sinanoğlu Laboratory were not included as Sinanoğlu and other members of his group, in particular W. Luken, had realized that Nicolaides had used the Sinanoğlu programs and theory incorrectly, picking out the wrong roots of the ψ_C-matrix. Nicolaides was cautioned of these difficulties. Nicolaides' attempted extension of the Sinanoğlu NCMET to "non-lowest-of-their-symmetry" states reported by him in [4.29] is also incorrect, as pointed out first in print by A. Hibbert [4.30]. The conclusions of Sinanoğlu and Luken [4.31] based on their own analysis of the variational problem, are in agreement with those of Hibbert

4.29 C. A. Nicolaides, D. R. Beck: J. Phys. B 6, 535 (1973)

4.30 A. Hibbert: J. Phys. B 6, L127 (1973)

4.31 W. Luken, O. Sinanoğlu: to be published (NCMET V)

4.32 P. S. Kelly: Astrophys. J. 140, 1247 (1967)

4.33 M. Cohen, A. Dalgarno: Proc. Roy. Soc. (London) A280, 258 (1964)

4.34 H. Nussbaumer: Astrophys. Letters 4, 183 (1969)

4.35 W. L. Wiese, M. W. Smith, B. M. Glennon: NSRDS-NBS 4, Vol.I (1966)

4.36 E. J. Knystautas, M. Brochu, R. Drouin: Can. J. Spect. 18, 153 (1973)

4.37 J. P. Buchet, M. C. Poulizac, M. Carre: J. Opt. Soc. Am. 62, 623 (1972)

4.38 E. H. Pinnington, D. J. G. Irwin, A. E. Livingston, J. A. Kernahan: Can. J. Phys. 52, 1961 (1974)

4.39 C. L. Lin, D. J. G. Irwin, J. A. Kernahan, A. E. Livingston, E. H. Pinnington: Can. J. Phys. 50, 2596 (1972)

4.40 L. Heroux: In [4.1,p.205]

4.41 O. Sinanoğlu, W. Luken: Comm. At. Molec. Phys. IV/5, 139 (1973)

4.42 H. J. Silverstone, O. Sinanoğlu: *Modern Quantum Chemistry*, Vol.II (Academic Press, New York 1965) (Istanbul Lectures)

4.43 O. Sinanoğlu, I. Öksüz: Phys. Rev. Letters 21, 507 (1968)

4.44 O. Sinanoğlu, I. Öksüz: Phys. Rev. 181, 42, 54 (1969)

4.45 L. Heroux: Phys. Rev. 153, 156 (1967)

4.46 H. G. Berry, W. S. Bickel, S. Bashkin, J. Desesquelles, R. M. Schectman: J. Opt. Soc. Am. 61, 947 (1971)

4.47 B. D. Savage, G. M. Lawrence: Astrophys. J. 146, 940 (1966)

4.48 A. G. Blackman, A. Lurio: Phys. Rev. 153, 164 (1967)

4.49 O. Sinanoğlu, W. Luken: In *Computers in Chemical Research and Education*, Vol. II, ed. by D. Hadži (Elsevier, Amsterdam 1973)

4.50 In general not needed for oscillator strengths (to few persent accuracy). In special cases where non-closed shell correlations are small, or where extreme accuracy is desired, NCMET would allow χ_U to be added on too [4.5-7]

4.51 C. Froese-Fischer: J. Phys. B 7, L91 (1974)

4.52 O. Sinanoğlu, B. Skutnik: J. Chem. Phys. 61, 3670 (1974)

4.53 There are special methods and solutions developed early in MET/NCMET by Sinanoğlu for obtaining the closed-shell-like pair correlations individually using the helium-like two-electron effective variational equations of MET. Details and examples on this more specialized subject are found in [4.5-7] and in the chapters dealing with the MET/NCMET in [4.26]

4.54 H. J. Silverstone, O. Sinanoğlu: J. Chem. Phys. 44, 1899 (1966)

4.55 H. J. Silverstone, O. Sinanoğlu: J. Chem. Phys. 44, 3608 (1966)

4.56 S. K. Shrivastava, P. A. Westhaus: J. Chem. Phys. 59, 1054 (1973)

4.57 O. Sinanoğlu: Comm. At. Molec. Phys. 2d, 73 (1970)

4.58 The $\vec{r}$ and $\vec{v}$ are the only two L-shell appropriate charge-like f^{E1} operators. The others including acceleration weight the K-shell heavily. They can be compared with $\vec{r}$ and $\vec{v}$ forms accurately only after inclusion of K-shell correlations too in the ψ's. NCMET and "ATOM" have an option which yields $\psi_C(KL)$'s including K-shell correlations. We use these options on hyperfine structure, etc. for special properties sensitive to correlations in the K-shell [4.15,16]

4.59 L. Heroux: Phys. Rev. 180, 1 (1969)

4.60 P. Dumont, E. Biémont, N. Grevesse: J. Quant. Spectr. Rad. Trans. 14, 1127 (1974)

4.61 I. Martinson, H. Berry, W. Bickel, H. Oona: J. Opt. Soc. Am. 61, 519 (1971)

4.62 A few of the Sinanoğlu-Westhaus f^{E1} values [4.8] have been re-reported by C.A. Nicolaides [4.29] changing $\varepsilon_{theor.}$ to ε_{exp}, but not reporting f^{E1}, along with his incorrect neutral (A I) and +1 ion (A II) results. The $[f_r^{E1}(\varepsilon_{theor}) - f_r^{E1}(exp)]$ for the Sinanoğlu-Westhaus results [4.8] are within the $(f_r - f_v)$ spread and/or within experimental and NCMET accuracy. They are thus the same values (best are the $f_{\sqrt{rv}}$ when the spread is small as in Type A-all ions past A II) [4.8,57]. (See also Table 4.5 of this paper)

4.63 R. B. Hutchison: J. Quant. Spectr. Rad. Trans. 11, 81 (1971)

4.64 J. Bromander, R. Buchta, L. Lundin: Phys. Letters 29A, 523 (1969)

4.65 D. J. Pegg, L. W. Dotchin, E. L. Chupp: Phys. Letters 31A, 501 (1970)

4.66 G. M. Lawrence, B. D. Savage: Phys. Rev. 141, 67 (1966)

4.67 I. Martinson, W. S. Bickel, A. Ölme: J. Opt. Soc. Am. 60, 1213 (1970)

4.68 W. S. Bickel: Phys. Rev. 162, 7 (1967)

4.69 G. Boldt: Z. Naturforsch 18A, 1107 (1963)

4.70 W. H. Hayden Smith, J. Bromander, L. J. Curtis, R. Buchta: Physica Scripta 2, 211 (1970)

4.71 F. Labuhn: Z. Naturforsch 20A, 998 (1965)

4.72 G. M. Lawrence: Phys. Rev. 2, 397 (1970)

4.73 The procedure itself used by Nicolaides for wave functions, his "many-body calculations", and "correlated wave functions", unfortunately turn out incorrect as noted in [4.30,31] as he has picked the incorrect roots in the outputs of the Sinanoğlu-Öksüz-Westhaus-Luken NCMET computer programs

4.74 W. Luken, O. Sinanoğlu: J. Chem. Phys. 64, 1495 (1976)

4.75 E. A. Hylleraas, B. Undheim: Z. Phys. 65, 759 (1930)

4.76 J. K. L. MacDonald: Phys. Rev. 43, 830 (1933)

4.77 See also P. O. Löwdin: *Advances in Chemical Physics*, Vols.I,II, ed. by L. Prigogine (Interscience, New York 1959)p.266

4.78 If the level assignment is not known experimentally as in many molecular problems, one must make a thorough parameter search in the variations following the K-roots. From such calculations, the correct level scheme can then be theoretically deduced. For example, in Fig.4.3, if only the range $2 \leq \xi_{3s} \leq 3$ had been searched, one would have gotten the wrong level scheme, but going into the $\xi_{3s} \approx 1$ region where the true ground minimum is found, one has the correct one

4.79 J. E. Hesser, B. L. Lutz: J. Opt. Soc. Am. 58, 1513 (1968)

4.80 J. P. Buchet, M. C. Poulizac: In Proceedings 2nd European Conf. on Beam-Foil Spectroscopy, Lyon, France (1971)

4.81 C. L. Lin, D. A. Parkes, F. Kaufman: J. Chem. Phys. 53, 3896 (1970)

4.82 P. M. Dehmer, W. A. Chupka: J. Chem. Phys. 62, 584 (1975)

4.83 O. Sinanoğlu, D. R. Beck: Chem. Phys. Letters 24, 20 (1974)

4.84 D. R. Beck, O. Sinanoğlu: Phys. Rev. Letters 28, 945 (1972)

4.85 L. J. Curtis, I. Martinson, R. Buchta: Physica Scripta 2, 1 (1971)

4.86 H. G. Berry, R. M. Schectman, I. Martinson, W. S. Bickel, S. Bashkin: J. Opt. Soc. Am. 60, 335 (1970)

4.87 S. Bashkin, I. Martinson: J. Opt. Soc. Am. 61, 1686 (1971)

4.88 V. H. Smith, Jr., R. E. Brown: Chem. Phys. Letters 20, 424 (1973)

4.89 P. G. Burke, A. Hibbert, W. D. Robb: J. Phys. B 5, 37 (1972)

4.90 H. Tatewaki, H. Tateka, F. Sasaki: Int. J. Quan. Chem. 5, 335 (1971)

4.91 P. McCavert: Chem. Phys. Letters (to be published)

4.92 O. Sinanoğlu, D. Herrick: J. Chem. Phys. 62, 886 (1975)

4.93 D. Herrick, O. Sinanoğlu: Phys. Rev. A 11, 97 (1975)

4.94 Comparing +2 and higher ions' f^{E1} values (f_r^{E1} vs. f_v^{E1} spread) of Table 4.5 with those of the neutrals and +1 ones in Table 4.7, we see the spread to be more in the latter, even though a very large improvement from previous theoretical results has been achieved, resulting in good agreement with experiment

4.95 For example, in closed shell MET, we had found the unlinked products of pair correlations [4.5,26], e.g., $(\hat{u}_{12}\ \hat{u}_{34})$ along with the $(12\ \hat{u}_{34})$ and $(\hat{u}_{12}34)$ individual terms in χ

4.96 S. Honzeas, I. Martinson, P. Erman, R. Buchta: Physica Scripta 6, 55 (1972)

4.97 Neutral atoms and +1 ion f^{El}'s calculated for some test cases with the quasi-all-externals, i.e., the extended charge supermultiplet, will be reported later (Sinanoğlu and Luken). However, the f^{El}'s, such as those in our Table 4.7, are expected not to be affected to any extent other than possibly getting smaller $f_r^{El} - f_v^{El}$ spreads, comparable to those in our Table 4.5

4.98 See, for example, H. G. Berry, J. Desesquelles, M. Dufay: Nucl. Instr. and Meth. 110, 43 (1973); references given on p.305 of [4.3]

5. Regularities of Atomic Oscillator Strengths in Isoelectronic Sequences

Wolfgang Wiese

With 16 Figures

Spectral data for highly-ionized atoms are in strong demand in two frontier areas of research: (1) for the analysis of the very hot plasmas encountered in thermonuclear fusion research, and (2) for the interpretation of the far UV solar spectra obtained from spacecraft. Both research areas have had a tremendous growth in recent years, and thus demands for atomic data on highly-stripped atoms have grown greatly, too. Currently, the demands in the area of fusion research are particularly urgent, since minor plasma impurities in the form of very-highly-ionized atoms of heavy elements (mainly contaminants from walls of fusion machines) appear to be a major factor in the behavior of the very hot hydrogen or deuterium plasmas. The few available experimental and theoretical results [5.1-4] to date indicate that these small impurities seem to have a rather detrimental effect on the plasma energy-balance by causing strong radiation cooling, and are also responsible for certain plasma instabilities. For example, recent estimates for a fairly typical Tokamak, the ST Tokamak of the Princeton Plasma Physics Laboratory, indicate that heavy-metal impurities of only 0.2% in the plasma almost double the radiation loss and represent about 10% of the total energy loss. Furthermore, theoretical scaling-studies [5.3] predict that impurity concentrations of this order of magnitude, if present in future, larger Tokamaks, would be sufficient to prevent the plasma from ever reaching the temperature necessary for ignition of the fusion reaction. Both for determining impurity-element concentrations from the spectral radiation as well as for modeling studies, the availability of the relevant oscillator strengths (f-values) is critically important. These are the principal atomic quantities entering into the expressions for the excitation-rate coefficients (according to the normally-used SEATON-VAN REGEMORTER formula [5.5]), which in turn govern the radiative intensities of the impurity lines and thus the main part of the radiative power loss.

Observations of solar far UV spectra from spacecraft and rockets, especially the OSO and Skylab observations, have also yielded extensive spectral data, which are mainly due to the very hot coronal regions. Again the spectra of highly-stripped ions of many elements are involved, and for their quantitative interpretation, knowledge of the relevant oscillator strengths is essential.

The present data base for oscillator strengths of highly-stripped ions is very small. If one defines a highly-stripped ion as one with at least 10 electrons stripped off, then the available experimental material is essentially zero and on the theoretical side only a few principal transitions of simple atomic systems, primarily He-, Li-, Be-, B- and C-like ions, have been studied in some detail.

In view of this meager material, an additional - in fact, presently the main - source of data is interpolated f-values, obtained from utilizing systematic trends of oscillator strengths within isoelectronic sequences. Isoelectronic-sequence studies are based on the idea, made rigorous by perturbation theory, that for a fixed radiative transition (or an atomic state) along an isoelectronic sequence, the properties scale with the varying nuclear charge; we attempt to analyze this functional dependence.

Theoretical studies to investigate and utilize the nuclear-charge dependence of atomic quantities go back to 1930 to HYLLERAAS' [5.6] introduction of the nuclear charge as a variable in perturbation theory. For a long time after that, studies of isoelectronic-sequence trends were mostly concerned with atomic energy levels [5.7]. Applications to atomic oscillator strengths started only in recent years with the development of the "nuclear-charge expansion method" by DALGARNO and coworkers [5.8,9]. Independently, WEISS [5.10] performed in 1963 an oscillator-strength study on the lithium sequence, and since then, numerous other theoretical isoelectronic-sequence investigations have been undertaken.

A different approach which utilizes available data was then developed by WEISS and the present author in 1968 [5.11-13]. It is based on a general result of perturbation theory (which will be discussed later in detail) according to which the oscillator strength is expected to be a relatively simple function of the nuclear charge. Thus the available literature data on atomic f-values may be graphically assembled for a given transition within an isoelectronic sequence and a plot of f vs. nuclear charge for a few species should establish this functional relationship. Many systematic trends have indeed been established by this graphical method. The main practical importance of this approach lies in the fact that new data on very-highly-stripped ions may be obtained from interpolations based on a few available f-value data on lower stages of ionization as well as on the hydrogenic limit for an ion of infinite nuclear charge. Also, the reliability of available data may be tested by their degree of fit into established trends. It is this "empirical" approach which bears a rather direct connection to beam-foil spectroscopy, with beam-foil spectroscopy being one of the principal sources of input data which have made such studies possible.

In this chapter, then, we review the regularities of f-values along isoelectronic sequences and their utilization. The general theoretical background will be discussed, typical examples of the main types of trends will be given, and, for each case, the physical reasons for the particular oscillator-strength behavior will be analyzed. Finally, two very recent developments, namely f-value distributions along an isoelectronic sequence for an entire spectral series and relativistic effects in highly ionized species, will be described.

5.1 Theoretical Basis

As a matter of convenience in the development of the theoretical basis for our calculations, we begin with the definitions of the terms we used, and then follow with a treatment of the dependence of the f-values on nuclear charge.

5.1.1 Definitions

The (absorption) oscillator strength for an optical transition between atomic states i (lower) and k (upper) is defined as

$$f_{ik} = \frac{2}{3}(E_k - E_i)\, g_i^{-1} \left|\int \psi_i^* \vec{D}\, \psi_k\, dV\right|^2 \quad , \tag{5.1}$$

or

$$= \frac{2}{3}(E_k - E_i)\, g_i^{-1} |\langle i|\sum_p \vec{r}_p\, k\rangle|^2 = \frac{2}{3}(E_k - E_i)\, g_i^{-1}\, S \quad . \tag{5.2}$$

The dipole moment $\vec{D}$ is given by the sum of the position vectors $\sum_p \vec{r}_p$ of the p electrons of the atom; ψ_i, ψ_k are the electron wave functions and E_i, E_k are the excitation energies of lower and upper state (in atomic units); and g is the statistical weight. Any degeneracies in the states i and k may be taken into account by summing over them. The square of the matrix element in (5.2) has been introduced by CONDON and SHORTLEY [5.14] as the line strength S and is often used in the theoretical literature. For the prevailing Russell-Saunders coupling (LS-coupling) the line strength S is given by [5.14,15]

$$S = \mathcal{S}(L)\mathcal{S}(M)\, \sigma^2 \quad , \tag{5.3}$$

where $\mathcal{S}(L)$ and $\mathcal{S}(M)$ are respectively the relative strengths of lines and multiplets (also called "angular factors" and tabulated, for example, by ALLEN [5.16]) and σ is the radial transition integral.

Beam-foil experiments yield ideally the radiative lifetime τ_k of an excited atomic state k, which is related to the oscillator strength by

$$\tau_k = mcg_k(8\, e^2)^{-1} \left(\sum_i g_i f_{ik}/\lambda_{ik}^2\right)^{-1} \quad , \tag{5.4}$$

and to the atomic transition probability A_{ki} for spontaneous emission by

$$\tau_k = (\sum_i A_{ki})^{-1} \tag{5.5}$$

Here λ_{ik} is the wavelength of the transition $i \to k$, and e, m_e, c are the usual natural constants. From the above relations it is also clear that f and A are related by

$$f_{ik} = mc(8\pi e^2)^{-1} \lambda_{ik}^2 g_k g_i^{-1} A_{ki} \quad . \tag{5.6}$$

For the determination of f-values, beam-foil lifetime measurements are most valuable for transitions where the sums in (5.4) and (5.5) reduce to a single term, i.e., for those excited atomic states from which only one spontaneous emission process is allowed, for example, for the upper states of the principal resonance lines. For these special cases, (5.4) and (5.5) reduce to

$$\tau_k = A_k^{-1} = 8\pi e^2 (mc)^{-1} g_i \lambda_{ik}^{-2} g_k^{-1} f_k^{-1} \quad . \tag{5.7}$$

This relation still holds approximately for those other atomic states where a particular radiative downward transition is dominant, i.e., where

$$\tau_k = (A_1 + A_2 + A_3 + \ldots)^{-1} \approx A^{-1} \qquad (\text{if } A_1 \gg A_2, A_3, \ldots) \tag{5.8}$$

5.1.2 Nuclear Charge-Dependence of the f-Value

In the study of a fixed transition $i \to k$ along an isoelectronic sequence, the nuclear charge Z is the principal variable. Since the oscillator strength, according to (5.1), may be expressed in terms of wave functions ψ and energies E, one has to obtain the Z dependence of these latter quantities. This problem was first solved by HYLLERAAS [5.6] in his pioneering work in 1930. HYLLERAAS, using perturbation theory, treated the two-electron helium case, and we follow essentially his work, but for the generalized case of a p-electron atom:

The Hamiltonian H for an atom with p electrons and a nucleus of charge Z is given (in atomic units) by

$$H = -\sum_p \left(\frac{1}{2}\Delta_p + \Sigma r_p^{-1}\right) + \sum_{p<q} r_{pq}^{-1} \quad . \tag{5.9}$$

The nucleus is assumed to be infinitely massive, and no relativistic or spin effects are taken into account. The inter-electronic repulsion term

$$\sum_{p<q} r_{pq}^{-1} \tag{5.10}$$

may be regarded as the perturbation on the atomic system, so that, according to conventional perturbation theory, the Hamiltonian may be represented by two terms

$$H = H_0 + H_1 \quad . \tag{5.11}$$

H_0, i.e., without the mutual repulsion term, is the "unperturbed" part, and is simply a sum of one-electron hydrogen-atom Hamiltonians for which the Schrödinger equation

$$H_0\psi = E\psi \tag{5.12}$$

may be solved exactly, and

$$H_1 = \sum_{p<q} r_{pq}^{-1} \tag{5.13}$$

is the perturbation potential, which is assumed to be small.

The objective is now to make the nuclear charge Z the principal variable, i.e., introduce it into the perturbation term to obtain explicitly the Z dependence for ψ and E. According to HYLLERAAS [5.6] this is done by dividing (5.9) by Z^2, and by scaling all distances r by Z and all energies by Z^2. For convenience the quantities

$$\rho = Zr \quad \text{and} \quad \mathcal{E} = EZ^{-2} \tag{5.14}$$

are normally introduced, with which the Hamiltonian (5.9) becomes

$$H' = -\sum \left(\frac{1}{2}\Delta'_p + \rho_p^{-1}\right) + Z^{-1} \sum_{p<q} \rho_{pq}^{-1} \quad , \tag{5.15}$$

with $H' = Z^{-2}H$. In analogy to (5.9) and (5.11), one may write this Hamiltonian as

$$H' = H'_0 + Z^{-1} H'_1 \quad , \tag{5.16}$$

where Z^{-1} appears naturally as the perturbation parameter.

The Schrödinger equation for this system may be written as

$$(H'_0 + Z^{-1}H'_1)\psi = \mathcal{E}\psi \quad , \tag{5.17}$$

and is in the appropriate form for perturbation theory, with Z^{-1} as the expansion parameter and the inter-electronic repulsion as the perturbation. The general perturbation theory approach is then to develop ψ and $\mathcal{E}$ into power series in Z^{-1}, i.e.,

$$\psi = \psi_0 + Z^{-1}\psi_1 + Z^{-2}\psi_2 + \ldots = \sum_{n=0}^{\infty} Z^{-n}\psi_n \tag{5.18}$$

and

$$\mathcal{E} = \mathcal{E}_0 + Z^{-1}\mathcal{E}_1 + Z^{-2}\mathcal{E}_2 + \ldots = \sum_{n=0}^{\infty} Z^{-n}\mathcal{E}_n \quad , \tag{5.19}$$

or

$$E = \mathcal{E}_0 Z^2 + \mathcal{E}_1 Z + \mathcal{E}_2 + \mathcal{E}_3 Z^{-1} + \ldots = Z^2 \sum_{n=0}^{\infty} Z^{-n}\mathcal{E}_n \quad . \tag{5.20}$$

If these expansions are substituted into the Schrödinger equation, then the coefficients of equal power of the variable Z must vanish identically, and one obtains an infinite set of coupled linear equations. If one retains only the zeroth-order, one obtains a purely hydrogenic solution

$$(H_0' - \mathcal{E}_0)\,\psi_0 = 0 \quad , \tag{5.21}$$

since, analogous to H_0 in (5.12), H_0' is hydrogenic. Inclusion of first- and second-order terms in Z furnishes the well-known approximations of first- and second-order perturbation theory.

Substitution of (5.17) and (5.19) into the general expressions for f and S, (5.1) and (5.2), respectively, and collection of terms of the same degree in Z, yields for the Z dependence of S and f

$$S = S_0 Z^{-2} + S_1 Z^{-3} + S_2 Z^{-4} + \ldots = Z^{-2} \sum_{n=0}^{\infty} S_n Z^{-n} \tag{5.22}$$

and

$$f = f_0 + f_1 Z^{-1} + f_2 Z^{-2} + \ldots = \sum_{n=0}^{\infty} f_n Z^{-n} \quad . \tag{5.23}$$

These equations represent the general results of perturbation theory for the Z-dependence of the line strength and f-value. The lead terms are given by

$$S_0 = |\langle \psi_{0,i} | \vec{\rho} | \psi_{0,k}' \rangle|^2 \quad , \tag{5.24}$$

and

$$\rho_0 = \frac{2}{3}\,(\mathcal{E}_{0,k} - \mathcal{E}_{0,i})\, g_i^{-1} S_0 \quad , \tag{5.25}$$

and are also purely hydrogenic, since ψ_0 and $\mathcal{E}_0$ have been shown to be hydrogenic.

5.1.3 Investigation of Lim $1/Z \to 0$

For very large values of Z, or $1/Z \to 0$, $Z^2 S$ and f approach the lead terms, so that it is of interest to investigate these further. For the oscillator strength, one can distinguish between two different cases. If the principal quantum number n of the jumping electron does not change (i.e., a transition $n_i\ell_i \rightarrow n_k\ell_k$ with $n_i = n_k$), then the energy difference in the lead term f_0, which is just that of the hydrogen atom, vanishes and thus the oscillator strength for large Z asymptotically approaches zero.

If the principal quantum number changes for the transition, i.e., if $n_i \neq n_k$, the lead term f_0 remains finite. For atomic systems with one electron outside closed shells, i.e., the alkalis, this is strictly the hydrogen oscillator strength. For other atomic systems, it is either the hydrogen oscillator strength modified by an angular factor (see (5.3)), which is usually available from the literature [5.16] and is due to the different multiplet structure, or, if zeroth-order degeneracies occur, f_0 is a quantity which can be calculated on strictly hydrogenic terms [5.13]. The first zeroth-order degeneracy in atomic systems occurs for beryllium-like atoms with the well-known configuration interaction $2s^2\ ^1S + 2p^2\ ^1S$. Both 2s and 2p are degenerate for hydrogen and have identical zeroth-order energies. To remove the degeneracy, standard procedures of degenerate perturbation theory must be applied. WEISS [5.17] has, for example, carried out such calculations for some boron-sequence transitions.

5.2 Discussion of Established Trends

Systematic trends of oscillator strengths in isoelectronic sequences have been studied and empirically established for numerous prominent transitions of the lighter elements. The most comprehensive assembly of graphs depicting these systematic trends is found in a paper by SMITH and WIESE [5.18], where, in 91 figures, 98 trends are illustrated for the isoelectronic sequences from He through Mg. The Ne-sequence is, however, not represented and for the F-sequence only one trend has been found. Since this 1971 publication, many additional trends have been established, either within these same sequences, or for some others. Thus, illustrations for trends in the Aℓ sequence are given by SMITH et al. [5.19], and IRWIN et al. [5.20], and for Cu and Zn by SØRENSEN [5.21]. Systematic trends have been now definitely established for the following isoelectronic sequences: He, Li, Be, B, C, N, O, F, Na, Mg, Aℓ, Si, P, S, Cu, and Zn.

To this list, the H-sequence may be added, since the f-values of hydrogen-like ions of charge Z are identical to those for neutral hydrogen itself, which are exactly known [5.22]. For some of the sequences, only a single systematic trend is well-known, but for most other sequences, about 5-10 trends are reliably established and for He and Li numerous trends are available. Generally speaking, the well-known trends are limited to a few prominent transitions for each isoelectronic sequence and this situation is not expected to change drastically in the near future. For example, it appears unlikely that systematic trends will be soon established for really complex atomic systems, as for the iron-group elements.

From the expansion of f in powers of $1/Z$, (5.23), one should expect that graphs of f vs. $1/Z$ will be approximately straight lines for small values of $1/Z$, i.e., for the very-highly-charged ions. Higher-order terms in the expansion should become increasingly important towards the neutral end of a sequence. Indeed quite often one encounters curves which are well represented by parabolas, so that a quadratic approximation to (5.23) may be regarded as the "basic" f-value dependence on $1/Z$. But there are a number of curves which exhibit quite different shapes, usually either with a minimum or a maximum near the neutral end of a sequence. It has been found that the appearance of a maximum is in most cases related to the presence of strong configuration interaction for the neutral atom. It may however be also caused by cancellation in the one-electron transition integral for the neutral atom, diminishing cancellation effects for higher ions, i.e., stronger f-values, and then an asymptotic behavior with f tending to zero for $1/Z \to 0$. A maximum may also arise when configuration interaction effects somewhere along the sequence act in such a way as to enhance the f-value there, but not near the neutral end. If the enhancement would predominantly happen near the neutral atom, one would simply obtain the more common parabola again, with perhaps an exaggerated rise at the neutral end. (A graphical example for an enhancement effect will be given later).

The appearance of a minimum in the curves can be attributed either to strong cancellation in the transition integral or to configuration effects which are more pronounced somewhere else in the sequence than for the neutral atom. Finally, more complex effects of configuration interaction or the combined effects of this and cancellation in the transition integral may in some instances give rise to curves with anomalous shapes.

The various interference effects occur very often near the neutral end of a sequence, but in complex heavier element spectra they are still encountered for fairly high stages of ionization. For the very-highly-charged ions the effects appear to fade away and a straight line $1/Z$-dependence for the f-value is approached until relativistic effects change the functional relationship again. Thus spectral distributions of f-values may undergo several fundamental changes along a sequence.

To obtain an idea for the frequency of the occurrence of the different types of curves, one might mention that for the 98 curves presented in the comprehensive study by SMITH and WIESE [5.18], the following distribution was obtained: "basic" trend curves (approximately parabolic) - 62; curves with a maximum - 27; curves with a minimum - 7; curves with an anomalous shape - 2. Since an underlying physical reason is always found for each type of curve, it is very instructive to discuss some typical examples.

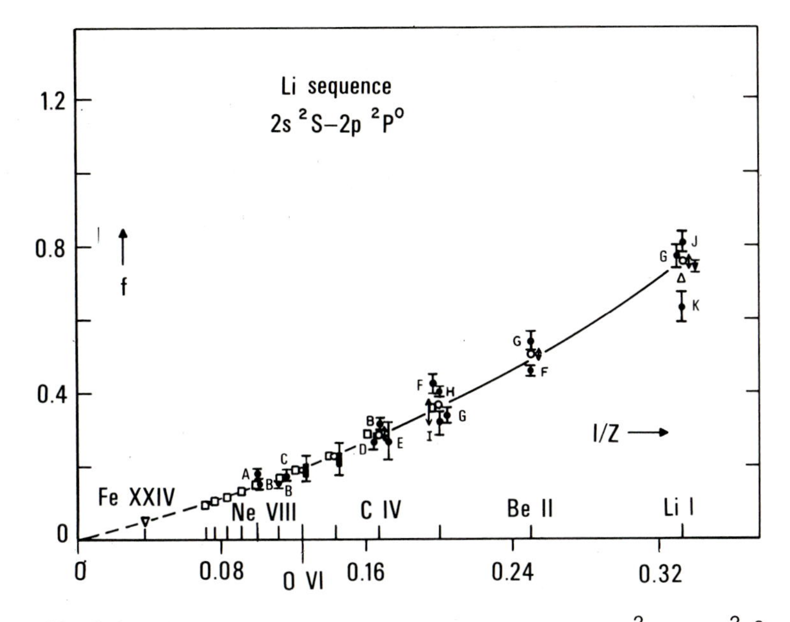

Fig.5.1 Oscillator strength vs. 1/Z for the 2s ^2S - 2p $^2P^o$ resonance doublet of the Li sequence. Presented are the multiplet f-values. The data sources are: A = [5.29]; B = [5.24]; C = [5.23]; D = [5.36]; E = [5.38]; F = [5.28]; G = [5.22]; H = [5.39]; I = [5.35]; J = [5.27]; K = [5.30]. Combined beam-foil results: For N V: [5.24-26,31-34]. For O VI: [5.24,26,37]. Other experiments: ▼ [5.40], Hanle-effect. Theory: ○ [5.10], SCF with conf. int.; △ [5.44], SCF with core polar; □ [5.21], Z-expans. method; ▽ [5.43], SCF; ↕ other theoretical methods combined (pseudo-potential method, SCF, Coulomb - Approximation) [5.41,42,45-51,53,54]. The error bars on the experimental data points represent authors' uncertainty estimates. To preserve clarity, experimental and theoretical results have been combined when they narrowly overlap

5.2.1 Basic Trends

First, as an example for a "parabolic" trend, the f vs. 1/Z plot is presented for the principal resonance transitions of the Li-isoelectronic sequence, 2s - 2p (Fig.5.1). In this case no significant configuration interaction is expected, since there are no interacting terms nearby. Indeed the dependence of f on 1/Z is very well represented by a parabola, which for the higher ions goes over into a straight line. A wealth of experimental and theoretical data [5.22-54] is available for this transition, so much in fact, that some of the closely grouped data obtained with similar methods had to be lumped together. The great importance of beam-foil spectroscopy for establishing systematic f-value trends is impressively demonstrated in this example. Of the 19 experimental data sources [5.22-40], 18 are beam-foil spectroscopy measurements [5.22-39] and these by themselves have fully established this trend.

Since the beam-foil results are obtained from separate, unrelated measurements employing ions of different elements, the formation of a systematic trend by the experimental data is an impressive experimental verification of (5.23).

The solid line drawn through the data is not just a "best" fit - it appears to be certainly very close to it - but reflects also consideration of some other constraints on the f-values which will be discussed later.

As noted earlier, well-established regularities are of great practical importance for two reasons: They may be utilized first to obtain, by graphical interpolations, additional accurate oscillator-strength values for ions not covered by existing experimental or theoretical data and, secondly, they may be utilized to evaluate the reliability of existing data from the degree of fit into an established systematic trend. Applying this to the example of Fig.5.1, it is seen that for the higher ions, i.e., beyond Z=12, many accurate f-numbers can be immediately found from the graph. But one must consider that at some point relativistic corrections become significant - roughly for Z=30 to 50 - and need to be known. Up to now, little work has been done on this subject, but fortunately some relativistic f-value data are now available for the Li-sequence and will be discussed later in a separate section. With respect to judging the reliability of experimental f-value data from the degree of fit for this Li-sequence trend, it is apparent from the error estimates provided by the authors, and indicated by the error flags, that the data of BUCHET et al.[5.30] on Li I, BROMANDER [5.28] on Be II and B III, ROBERTS and HEAD [5.39] on B III, BERKNER et al. [5.24] on C IV, and BUCHET et al. [5.29] on Ne VIII are subject to additional errors not considered by the authors.

5.2.2 Curves With a Maximum

Next, some examples will be discussed where configuration-interaction effects are known to be important and strongly influence the f-values for some ions along an isoelectronic sequence. First, we discuss the boron-sequence transition $2s^2 2p\ ^2P^o$ - $2s2p^2\ ^2D$. Figure 5.2 shows the systematic trend situation for this transition as it could be established from the available data [5.55-59] in 1968. This graph is particularly instructive insofar as it contains the results of three different theoretical approaches. It includes, first, in order of increasing sophistication, the single-configuration self-consistent field (SCF) approximation [5.55], which produces uniformly increasing and relatively large f-values (△). Secondly, it contains calculations (□) by BOLOTIN and YUTSIS [5.56] which account for the effects of ground-state configuration interaction, i.e., for the mixing of the $2s^2 2p\ ^2P^o$ and $2p^3\ ^2P^o$ states but for no other configuration mixing (two-configuration approximation). This appears to affect all ions of the sequence by about the same percentage amount, so that the f-values are reduced by 35%, but maintain the same type of Z-dependence. Finally, the

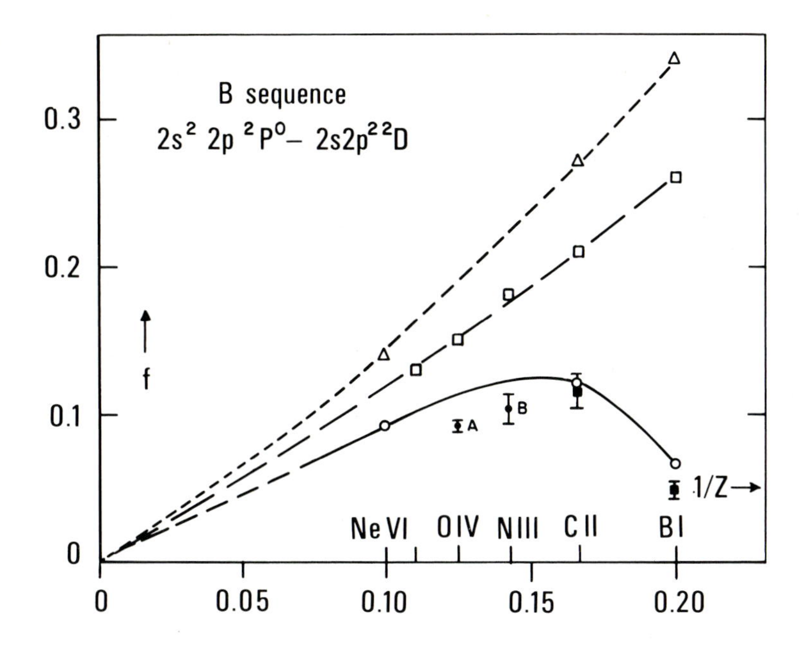

Fig.5.2 f-value vs. 1/Z for the $2s^22p\ ^2P^o$-$2s2p^2\ ^2D$ multiplet of the boron sequence (situation in 1968, from [5.11]). The symbols denote: △ [5.55], SCF; ○ [5.55], SOC; □ [5.56], two-configuration approximation; ■ [5.59], phase-shift; beam-foil data from A = [5.57]; B = [5.58]

large-scale configuration interaction treatment by WEISS [5.55] (superposition-of-configurations (SOC) approximation) shows a markedly different f-value dependence for the neutral end of the sequence, with a maximum near N III. WEISS [5.55] investigated the SOC wave functions for the ^2D state in detail and concluded that the interaction of $2s2p^2$ with configurations of the type $2s^2nd$ becomes quite important for the low stages of ionization, causing appreciable cancellation in the transition integral, and thus reducing the f-values. WEISS' comprehensive theoretical treatment of configuration interaction produces good agreement with the three experimental results [5.57-59] along the sequence. The importance of configuration interaction for this transition has been confirmed by many additional experiments [5.31,60-69] as seen in Fig.5.3, which illustrates the present situation with a definite trend established. New theoretical work [5.17,70-74] which all takes configuration interaction extensively into account, is also included and supports WEISS' earlier results.

In this case, as in many others, two different configuration-interaction effects are of importance. For the lower ions, where the (upper) $2s2p^2\ ^2D$ state is close to some $2s^2nd\ ^2D$ states, especially $2s^23d\ ^2D$, this interaction is the dominant factor. But as the nuclear charge increases, the interacting energy levels are rearranged, and the $2s2p^2$ state rapidly separates from the $2s^23d$ and higher $2s^2nd$ terms, so that this

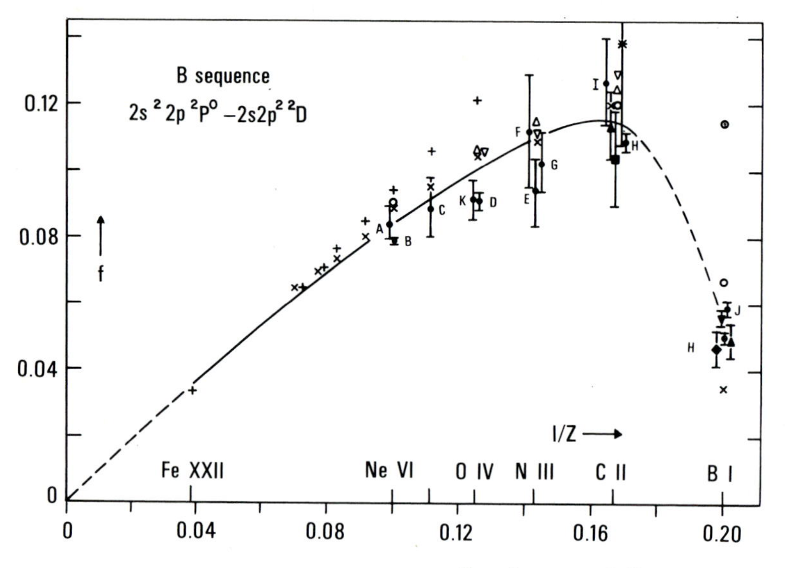

Fig.5.3 f-value vs. 1/Z for the $2s^2 2p\ ^2P^o - 2s2p^2\ ^2D$ multiplet of the boron sequence. The symbols denote: ● beam-foil data: A = [5.69]; B = [5.66]; C = [5.65]; D = [5.57]; E = [5.34]; F = [5.31]; G = [5.58]; H = [5.61]; I = [5.63]; J = [5.60]; K = [5.35]. Other experiments: ▲ [5.59], phase-shift; ■ [5.68], phase-shift; ◆ [5.62], Hanle effect; ⋇ [5.64], delayed coinc.; ▼[5.67], ion beam. Theory: △[5.74], many-electron theory (MET); ○ [5.17], SOC; ⊙ [5.72], SCF with conf. int.; + [5.71], perturbation theory; x [5.70], non-closed-shell many-electron theory (NCMET); ▽ [5.73], MET

configuration-interaction effect rapidly diminishes. What remains is the lower-state interaction between the $2s^2 2p$ and $2p^3\ ^2P^o$ states, which actually move closer together. Thus the experimental f-values as well as WEISS' sophisticated configuration-interaction calculations come gradually into better agreement with the results of the simpler 2-configuration approximation by BOLOTIN and YUTSIS [5.56].

In the next example, two complementary transitions of the Be-sequence are discussed to illustrate how configuration interaction can redistribute the oscillator strengths, enhancing the strength of one transition and reducing the other. The two transitions are $2s2p\ ^1P^o - 2p^2\ ^1S$ and $2s^2\ ^1S - 2s2p\ ^1P^o$; the configuration interaction common to both transitions involves the ^1S state, which in a configuration-interaction model would be described as (at least) a mixture of $2s^2$ and $2p^2$.

The Z-dependence graph (solid curve) for the f-value of the $2s2p\ ^1P^o - 2p^2\ ^1S$ transition (Fig.5.4) illustrates clearly that theoretical methods, which account for the above-discussed configuration mixing, produce all along the sequence enhanced f-values when compared to the single-configuration self-consistent-field approximation (dashed curve). Also, the two theoretical results [5.79,80] which are presented (these are

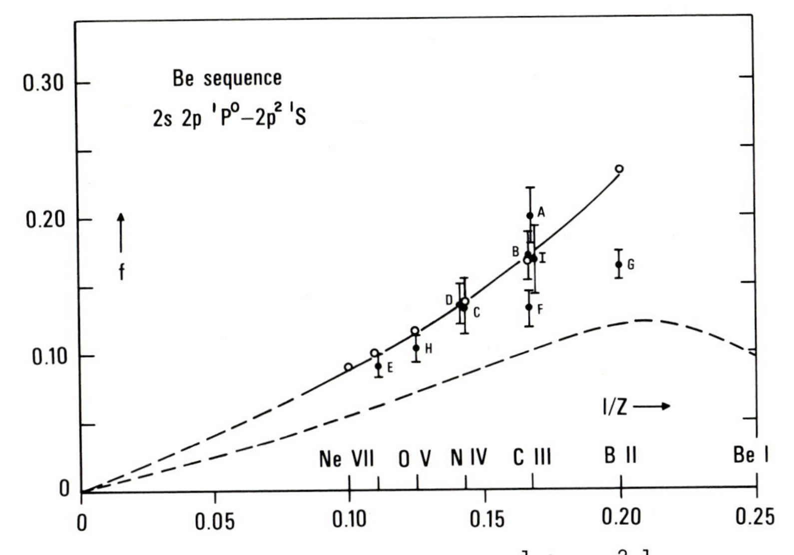

Fig.5.4 f-value vs. 1/Z for the 2s2p $^1P^o$ - $2p^2$ ^1S transition of the beryllium sequence. The symbols denote: ● beam-foil data from A = [5.75]; B = [5.63]; C = [5.31]; D = [5.34]; E = [5.65]; F = [5.36]; G = [5.77]; H = [5.35]; I = [5.76]. Theoretical data: ○ [5.78,5.79], SOC, and [5.80], SOC. These two calculations, which give almost identical results, are the most advanced theoretical treatments available and include extensive configuration mixing. To preserve clarity, none of the other similar theoretical results is presented, but to illustrate the configuration-interaction effects, the earlier single-configuration SCF-results of [5.79] are presented as the broken line

the most advanced of a score of recent calculations) are in good agreement with the experimental data, which are all from beam-foil experiments [5.31,34-36,63,65,75-77]. At the neutral end of the sequence, however, a special configuration effect couples in, which will be discussed below. For the companion transition $2s^2$ ^1S - 2s2p $^1P^o$ (Fig.5.5) the reversed effect is observed. Compared to the single-configuration SCF (dashed curve) results, an almost constant reduction in f-value is produced (solid curve) by sophisticated theoretical approximations [5.79,80,87] that include the configuration mixing of $2s^2$ and $2p^2$. This is again in good agreement with the mostly beam-foil lifetime measurements [5.28,31,34,35,37,59,60,63,65,75,81-86]. Thus a "give-and-take" situation couples these two lines all along the sequence, except for neutral Be, where another effect is present: It has been only recently realized [5.88] that the $2p^2$ ^1S state had been wrongly assigned and that it is most probably located above the ionization limit. Thus it is an autoionizing state and it is irrelevant to draw the systematic trend in Fig.5.4 beyond B II. The anomalous behavior of the Be I point in the earlier established trend (see, e.g., [Ref.5.18,Fig.17] has

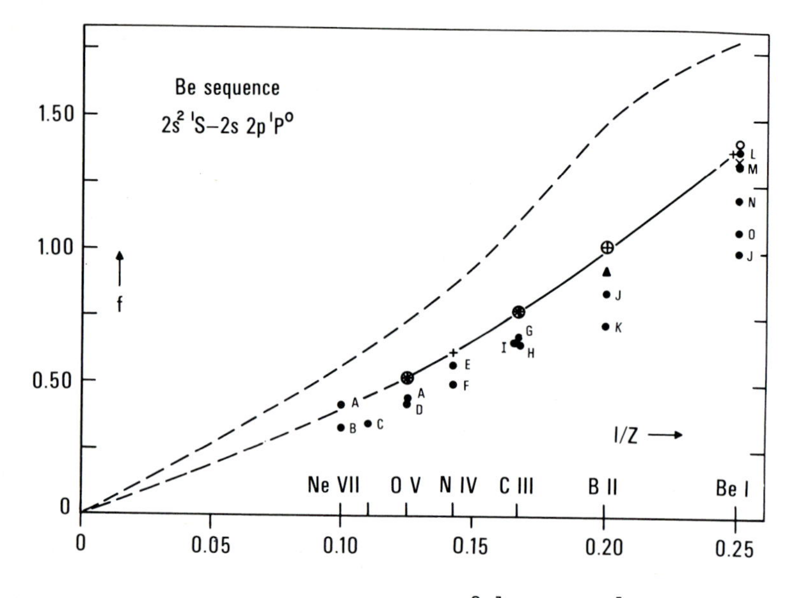

Fig. 5.5 f-value vs. 1/Z for the $2s^2\ ^1S$ - $2s2p\ ^1P^o$ transition of the beryllium sequence. The symbols denote: ● beam-foil data from A = [5.82]; B = [5.81]; C = [5.65]; D = [5.35]; E = [5.34]; F = [5.31]; G = [5.63]; H = [5.75]; I = [5.83]; J = [5.28]; K = [5.37]; L = [5.84]; M = [5.85]; N = [5.60]; O = [5.86]. Other experiments: ▲ [5.59], phase shift. Theory: ○ [5.79], SOC: x [5.87], specially-correlated SOC; + [5.80], SOC; --- single configuration SCF, [5.79]

actually been one of the main reasons for the above quoted theoretical study [5.88], and thus constitutes another example for the practical value of regularity studies.

An important recent theoretical advance has been the development of concepts, especially by WEINHOLD [5.89], to calculate rigorous upper and lower error bounds for f-values. Before this, except for a few special cases, it had not been possible to estimate reliably the accuracy of theoretical f-values, a rather unsatisfactory situation. The theoretical error bounds allow for the first time a judgment by theory on the experimental data and on the correctness of the experimental error estimates, while up to now the situation had been just the reverse, i.e., the accuracy of the theoretical results was generally checked by comparison with reliable experiments. SIMS and WHITTEN [5.87] have applied WEINHOLD's method to the above-discussed Be-sequence transition and have produced oscillator strengths with rigorous upper and lower error bounds of 7 to 10%. Thus, in Fig.5.6, which is a modified version of Fig.5.5, the region limited by the theoretical error bounds is presented as the shaded area and all experimental data (which are almost all beam-foil results) are shown with error flags, which represent the authors' uncertainty estimates.

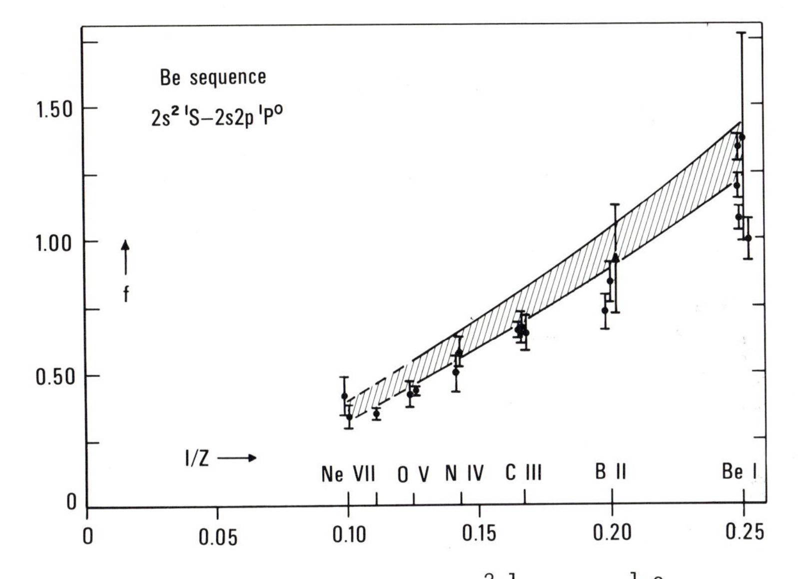

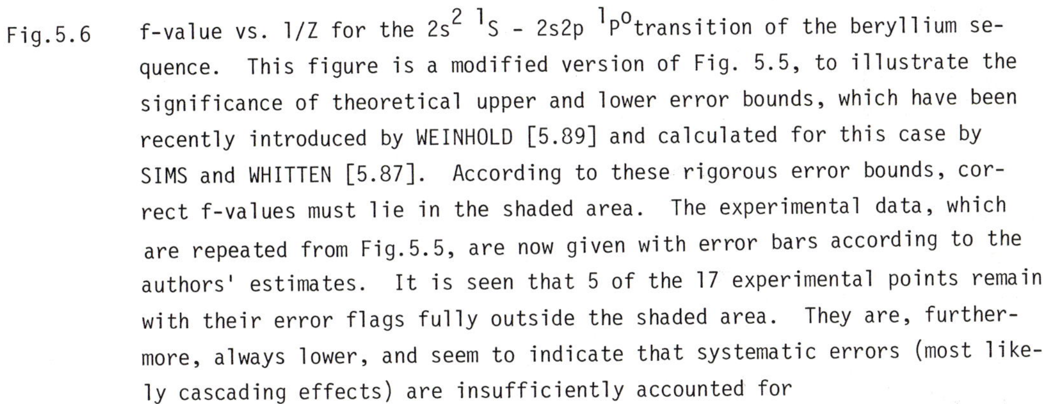
Fig.5.6 f-value vs. 1/Z for the $2s^2\ ^1S$ - $2s2p\ ^1P^o$ transition of the beryllium sequence. This figure is a modified version of Fig. 5.5, to illustrate the significance of theoretical upper and lower error bounds, which have been recently introduced by WEINHOLD [5.89] and calculated for this case by SIMS and WHITTEN [5.87]. According to these rigorous error bounds, correct f-values must lie in the shaded area. The experimental data, which are repeated from Fig.5.5, are now given with error bars according to the authors' estimates. It is seen that 5 of the 17 experimental points remain with their error flags fully outside the shaded area. They are, furthermore, always lower, and seem to indicate that systematic errors (most likely cascading effects) are insufficiently accounted for

It is seen that several of the beam-foil results lie with their estimated error ranges fully outside the shaded area, which indicates that these error estimates have apparently not included all sources of uncertainty. Indeed, several of the involved authors [5.34,35,37,60,81,84] do not give any discussion on their error estimates, thus leaving open to question if their estimates represent only statistical measurement errors or take into account uncertainties from energy losses in the foil, etc. Furthermore, it is noticed that most beam-foil results are below the theoretical f-value data range. The same observation may also be made from many other systematic-trend curves, including those presented in this chapter. The present author, among others, suggested therefore in 1970 that cascading effects due to the non-selective excitation mechanism in beam-foil spectroscopy might be responsible for this [5.90]. Thus electrons cascading into an atomic state from higher excited states would partially repopulate this state and produce a longer apparent lifetime or, according to (5.7), a smaller f-value. Since that time many beam-foil results have been subjected to elaborate analyses for cascading effects, with the overall result that the more recent beam-foil lifetime data have been in closer agreement with the advanced theoretical material than the earlier beam-foil measurements. For this particular example of Be I, however, no noticeable difference exists between the behavior of the pre-1970 beam-foil data and the later beam-foil work.

The magnitude of the cascading effects in beam-foil data is still a controversial subject. Authors of some recent theoretical investigations [5:87,91] have made detailed comparisons with beam-foil results which show that the latter still have a tendency to produce f-values that are too small, which has been interpreted to be the result of underestimating the importance of cascade repopulations. But PINNINGTON et al. [5.92] have also shown that the f-value ratio between theoretical data and their beam-foil results for 46 recently (post-1970) measured transitions is not significantly larger than one, amounting specifically to 1.11 with a standard deviation of $\pm$ 0.18. At any rate, it can be safely stated that the differences between carefully-analyzed beam-foil data and sophisticated theoretical material rarely exceed 30% and are quite often in 10% range, which is for most f-value applications more than satisfactory. Certainly the systematic-trend studies would have not been nearly as productive if the wealth of reliable data from beam-foil spectroscopy had been missing.

As a last example for a curve with a maximum, the Aℓ-sequence transition $3s^2 3p\ ^2P^o$ - $3s3p^2\ ^2D$ is presented in Fig.5.7. All available experimental data [5.20,93-98] have been plotted, and from the existing theoretical work the extensive calculations of FROESE-FISCHER [5.100] (self-consistent-field approximation with limited configuration interaction, dashed curve) have been selected and for Si II some other recent calculational results [5.78,99,102] have been added. The location of the $3s3p^2\ ^2D$ level in Aℓ I has been a mystery for a long time, and many speculations on its identification and location have been made (see, e.g., the summaries by FROESE-FISCHER [5.100] and WEISS [5.101] on this point). WEISS [5.101] finally clarified the situation in 1974 as a result of a very detailed study of the perturbation of $3s^2nd\ ^2D$ states by the $3s3p^2\ ^2D$ state. He found from a comprehensive superposition-of-configurations (SOC) treatment that very strong configuration interaction takes place between these two configurations and that the $3s3p^2$ perturber loses its identity as a discrete state, having its effect on the $3s^2nd$ series smeared out over the entire series. As a consequence, the systematic trend for the $3s^2 3p\ ^2P^o$ - $3s3p^2\ ^2D$ transition starts only at Si II, where configuration mixing is still very pronounced, causes destructive interference and thus produces a small f-value. The strength of this transition is so sensitive to the competing effects of different configurations that the theoretical data obtained from two different, but detailed configuration-interaction treatments [5.78, 99] strongly disagree with each other and with the experiments [5.97,98]. The small

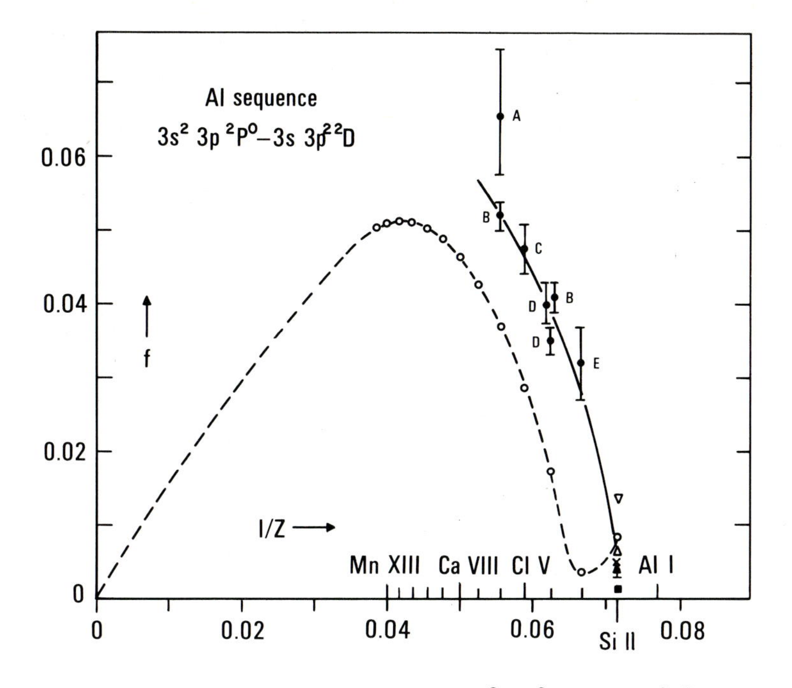

Fig.5.7 f-value vs. 1/Z for the $3s^23p\ ^2P^o$ - $3s3p^2\ ^2D$ multiplet of the Aℓ sequence. The selected data sources are: ● Beam-foil measurements by A = [5.93]; B = [5.20]; C = [5.94]; D = [5.95]; E = [5.96]. Other experiments: ■ [5.97], arc-emission; ▲ [5.98], phase shift. Theory: ▽ = [5.99], NCMET; △ [5.79], SOC; ○ [5.100], SCF with limited config. interact.; x [5.102], semi-empirical approximation

f-values for Si II are separately assembled in Table 5.1, since they cannot be too clearly shown with the rest of the sequence in Fig.5.7. For the higher ions again something interesting happens: The beam-foil results by CURTIS et al. [5.96], BERRY et al. [5.95], IRWIN et al. [5.20], and BASHKIN and MARTINSON [5.94] form one consistent systematic trend with the Si II data, while the calculations by FROESE-FISCHER [5.100] form a trend of very similar shape, but shifted to more highly ionized spectra. A likely explanation appears to be that FROESE-FISCHER's limited configuration mixing calculations do not account for some of the important configuration-interaction effects. The calculations, which include only one interacting level for the lower and two for the upper state, all within the n=3 shell, indicate that configuration interaction plays an important role all along the sequence. For example, the upper ^2D state of the transition turns out to be a mixture of about 80% $3s3p^2$ and 20% $3s^23d$, and this hardly changes along the sequence. One has thus reason to speculate that a more detailed configuration-interaction treatment would also affect the various ions by about constant amounts, as is suggested by the experimental data.

Table 5.1 Comparison of f-value data for the $3s^2 3p\ ^2P^o - 3s3p^2\ ^2D$ transition of Si II

Theory: (a) Configuration interaction calculations	
WEISS [5.78], SOC	0.0064 (dipole-length formalism) 0.0058 (dipole-velocity formalism)
SINANOǦLU [5.99], NCMET	0.0137 (dipole-length formalism) 0.0269 (dipole-velocity formalism)
FROESE-FISCHER [5.100], limited config.	0.0085
GARSTANG and SHAMEY [5.102], semi-empirical approximation	0.0049
(b) Single configuration calculation	
SINANOǦLU [5.99], restricted SCF	0.454 (dipole-length formalism) 0.504 (dipole-velocity formalism)
Experiment:	
CURTIS and SMITH [5.98], phase shift	0.004
HOFMANN [5.97], arc-emission	0.0023

5.2.3 Curves With a Minimum

Systematic-trend curves containing a minimum are usually caused by interference, or cancellation, in the transition integral σ. Figure 5.8 presents a typical example -

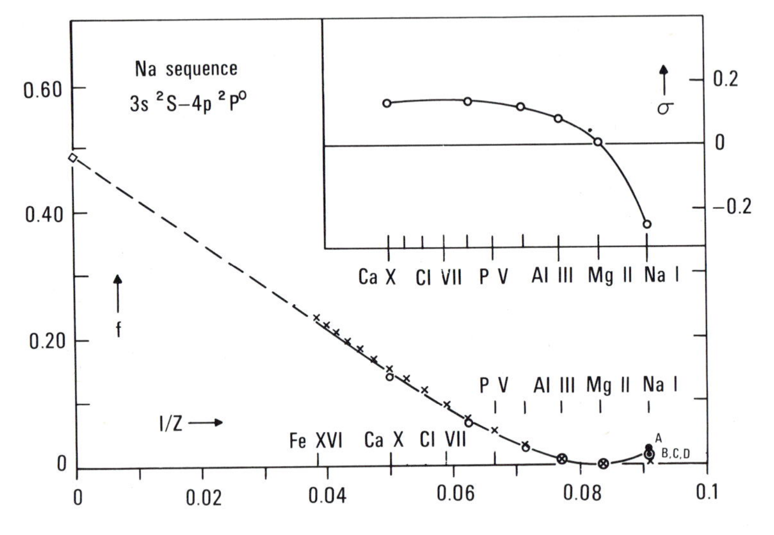

Fig.5.8 f-values and, in the insert, transition integrals, vs. 1/Z for the $3s\ ^2S$-$4p\ ^2P^o$ transition of the Na sequence. Since f is proportional to σ^2, f goes through a minimum at Mg II, where σ changes its sign. The selected data sources are: ● experimental lifetime measurements by A = [5.103], Hanle-effect; B = [5.104], delayed coincidence; C = [5.105], Hanle-effect; D = [5.106], double-resonance method. Theory: ○ Coulomb approximation [5.15] as calculated by [5.79]; x [5.107], SCF; ◇ hydrogenic value

the 3s-4p transition of the Na sequence. The graph contains, in addition to the usual f vs. 1/Z dependence, the 1/Z dependence of the transition integral. This quantity, which has been obtained here from the Coulomb approximation [5.15], changes sign at about Mg II. The f-value, which is proportional to σ^2, has therefore a minimum at Mg II, as seen in the lower curve.

The analogous transition for the Li sequence is 2s-3p, for which the f-value trend is illustrated in Fig.5.9. In this case almost complete cancellation in the transition integral occurs for neutral Li, so that the 1/Z-dependence of this f-value starts out with a minimum.

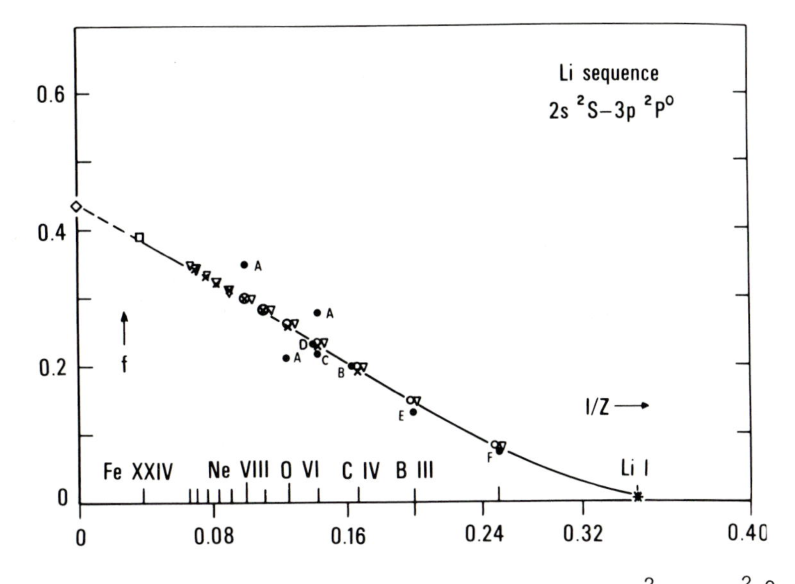

Fig.5.9 Oscillator-strength vs. 1/Z for the 2s ^{2}S - 3p ^{2}P^o transition of the lithium sequence. The data sources are: ● Beam-foil results from A = [5.29]; B = [5.75]; C = [5.32]; D = 5.58]; E = [5.37]; F = [5.84]. Theory: O [5.10], SCF; x [5.108], nucl. charge expans.; ▽ [5.50], pseudo-potential calc.; □ [5.43], SCF ◇ hydrogenic f-value; ✶ [5.109], hook method (relative value, normalized to f (2s-2p) with [5.10]). No error bars are given for the beam-foil data, since for the conversion from lifetimes to f-values the (small) contribution of the 3s-3p transition must be considered, for which another data source [5.10] is utilized

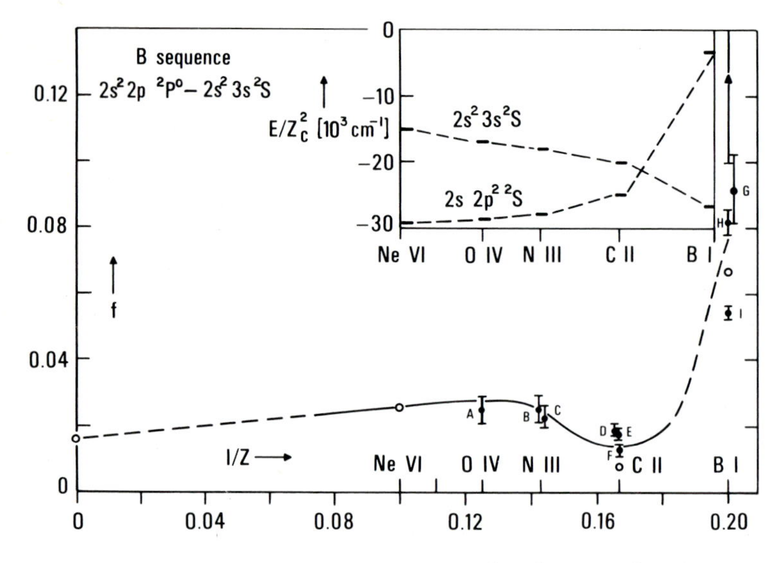

Fig.5.10 f-value vs. 1/Z for the $2s^22p\ ^2P^o$ - $2s^23s\ ^2S$ multiplet of the boron sequence. The data sources are: ● Beam-foil results from A = [5.110]; B = [5.58]; C = [5.31]; D = [5.75]; E = [5.63]; F = [5.36]; G = [5.28]; H - [5.86]; I = [5.60]. Other sources: ▲ [5.111], phase shift; ○ [5.17], SOC. The insert illustrates the relative positions of the upper $2s^23s\ ^2S$ level and the interacting $2s2p^2\ ^2S$ level for the first six spectra of the sequence. Plotted are the energies E of these levels (with reference to the ionization energies) scaled by the square of the core charge Z_c, i.e., the net charge of the core ($Z_c = Z - (N-1)$, where N is the total number of electrons)

5.2.4 Anomalous Curves

The last example, Fig.5.10, illustrates one of the few existing anomalous curves, the $2s^22p\ ^2P^o$ - $2s^23s\ ^2S$ transition of the B sequence. In this case, a gradually declining f-value trend is distorted by a minimum. The minimum at C II was first found by WEISS [5.17] from superposition-of-configuration (SOC) calculations. He suggested as the explanation a destructive interference between the interacting $2s2p^2\ ^2S$ and $2s^23s\ ^2S$ states, which happen to approach each other very closely for C II. The upper part of Fig.5.10 illustrates the positions of these energy levels along the sequence and it is seen that an inversion in the positions of these two levels takes place, with the level crossing occurring near C II. Subsequent beam-foil experiments by BUCHET et al. [5.31,110], POULIZAC and BUCHET [5.63], BUCHET-POULIZAC and BUCHET [5.75], and MARTINSON and BICKEL [5.36] have closely confirmed WEISS' theoretical prediction.

5.3 Oscillator-Strength Distributions in a Spectral Series Along an Isoelectronic Sequence

Up to now, always single transitions have been considered, and the established regularities have remained as isolated cases, which cover only a few prominent but mostly unrelated transitions in each spectrum. A further step in regularity studies is to interrelate the material for the principal transitions of a spectral series in order to study the behavior of an entire series as a function of 1/Z. The correlation of many transitions yields additional useful information and provides a strong check on the internal consistency of spectral series data. For this more ambitious goal, f-value data have to be available for all the principal transitions in a spectral series, i.e., the transitions involving the lower principal quantum numbers. Beyond this, it is desirable to have data also for the series continuum in order to be able to utilize f-sum rules. Such extensive numerical f-value data exist presently only for very simple atomic systems such as He, Li, or Na. A study of the oscillator-strength distribution in several spectral series of the Li-sequence has been recently undertaken [5.112,113] and has yielded some very instructive results on the oscillator-strength distribution in these spectral series. As an example, the f-value distribution in the principal series 2s-np along the Li sequence is presented in Fig.5.11.

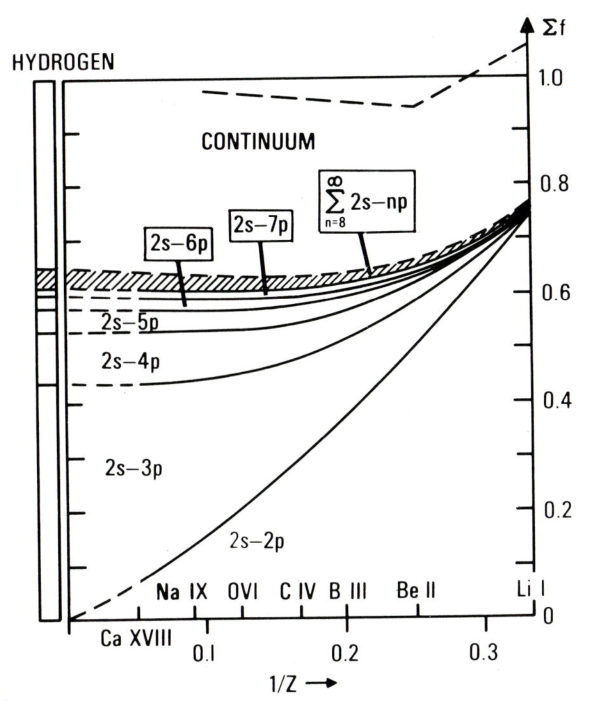

Fig.5.11 Oscillator-strength distribution for the 2s-np spectral series of the Li isoelectronic sequence and comparison with the corresponding hydrogen f-values (from [5.112,113])

The assembly and evaluation of the data have been done in 5 steps: (i) The isoelectronic sequence trends, i.e., the plots of f vs. 1/Z, were collected for the principal transitions 2s-np with n = 2,3,4,5,6,7, (e.g., the graphs for 2s-2p and 2s-3p have been presented earlier as Figs.5.1 and 5.9), and "best-fit" curves were established. (ii) With the above data, the spectral-series behavior, i.e., the dependence of f on the upper (effective) principal quantum number n, or the corresponding energy E was plotted for each ion (Li I, Be II, B III, etc.) and again curves of best fit were established. An example is shown in Fig.5.12.

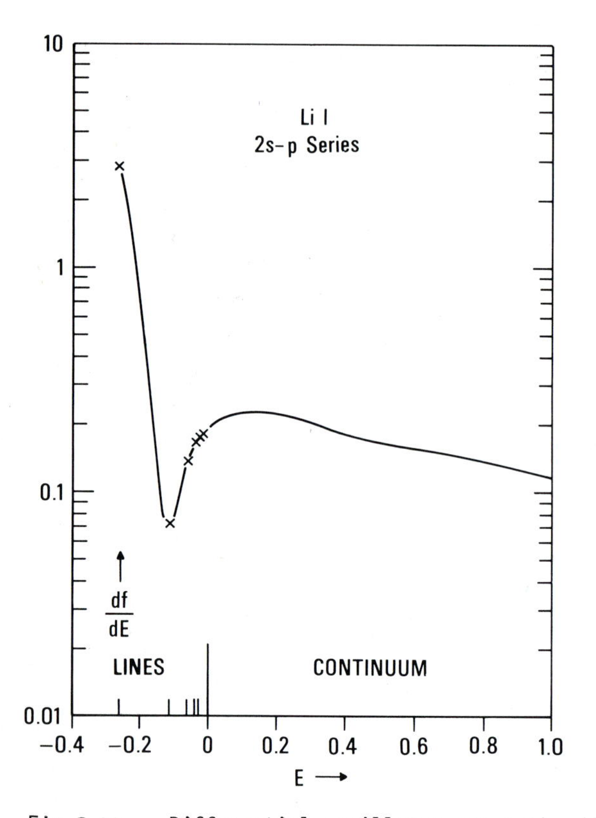

Fig.5.12 Differential oscillator strengths df/dE vs. energy E (in Rydberg units) for the 2s-np series of neutral lithium, based on the best available data. The line f-values are presented per unit energy range, using the relation $df/dE = 1/2\, n^3 f$ (see [5.112,113]), in order to compare them with the continuum data

The curves were then checked for consistency with the data obtained from (i) and, if necessary, iterations were made. (iii) The spectral-series plots were extended to include the continuum when differential f-values were available, and the condition for a smooth transition from the discrete-to-continuous spectrum was met, if necessary, by adjustments in the data. This extension to the continuum and the transition from discrete-to-continuous spectrum at the ionization limit is also shown in Fig.5.12. (iv) The WIGNER-KIRKWOOD [5.14,16] f-sum rule was applied to the spectral series for each ion along the sequence. For an s-np series, the f-sum is equal to unity, i.e., $\sum(f_{lines} + f_{cont}) = 1$. (v) Partial and total f-sums were then assembled along the whole sequence to establish the overall oscillator-strength distribution. In Fig. 5.11 the f-sums obtained directly from the existing line and continuum data are always within 5% of the f-sum rule prediction. Since the f-sum is rigorous only for a strictly one-electron spectrum, these small differences may be partly real so that no further adjustments were made. For the higher ions where no continuum data exist, the WIGNER-KIRKWOOD f-sum rule was used to obtain the total continuum f-values.

Figure 5.11 shows instructively the gradual variations in the f-value distribution along the sequence; for example, the 2s-2p transition loses rapidly f-value, which is almost all transferred to 2s-3p. Beyond O VI, the distribution of f-values for all higher series members and the continuum remains constant and is essentially the same as for the hydrogen-values, shown on the left side. A wealth of f-value information on this spectral series is thus contained, in very compact form, in a single illustration.

5.4 Relativistic Effects and Corrections

Relativistic effects, which are negligible for small values of the nuclear charge, i.e., for ions of light elements in low stages of ionization, become appreciable for highly-charged heavy ions, since under the influence of the strong core field, i.e., the net field of the nucleus and the remaining inner electrons, the radiating electrons acquire relativistic speeds.

The first relativistic effect to become important for larger Z is that of spin-orbit interaction, i.e., the transition from the usual Russell-Saunders (LS) coupling to intermediate coupling, which affects both the line strengths S and the transition energies $(E_k - E_i)$. Since $f \propto (E_k - E_i) \cdot S$, (5.2), the f-values are subject to greater changes than the line strengths. This breakdown of LS-coupling and the concomitant redistribution of f-values within multiplets away from the simple LS-coupling ratios does not directly enter into the presented systematic trends, insofar as these are always given for multiplet f-values and not for individual lines. However, when these relativistic effects become appreciable, the presented multiplet f-values (as well as the average multiplet wavelengths) become meaningless, since the line oscillator strengths cannot simply be derived from the multiplet value as in the case of LS-coupling. Another relativistic correction which becomes important only at much higher values of Z is the shrinkage of the electron orbits due to mass increase of the electrons on account of relativistic speeds. This orbital effect directly modifies the transition integral.

Until recently, no quantitative theoretical studies of relativistic effects on the f-values in highly-ionized spectra existed. Triggered by the urgent need for understanding the spectra emitted from the very hot fusion plasmas, as discussed early in this chapter, some first relativistic calculations have now been undertaken (see [5.78,114-116]), which are addressed to simple atomic systems. As we shall see below from some examples taken from this work, the relativistic regime is still essentially beyond the reach of all experimental methods, including the beam-foil technique.

The calculations with relativistic self-consistent-field wave functions performed for the Li sequence transitions 2s-2p and 2s-3p by KIM and DESCLAUX [5.116] are especially instructive, since they allow comparisons with earlier presented non-relativistic results for these transitions. See Figs.5.13 and 5.14.

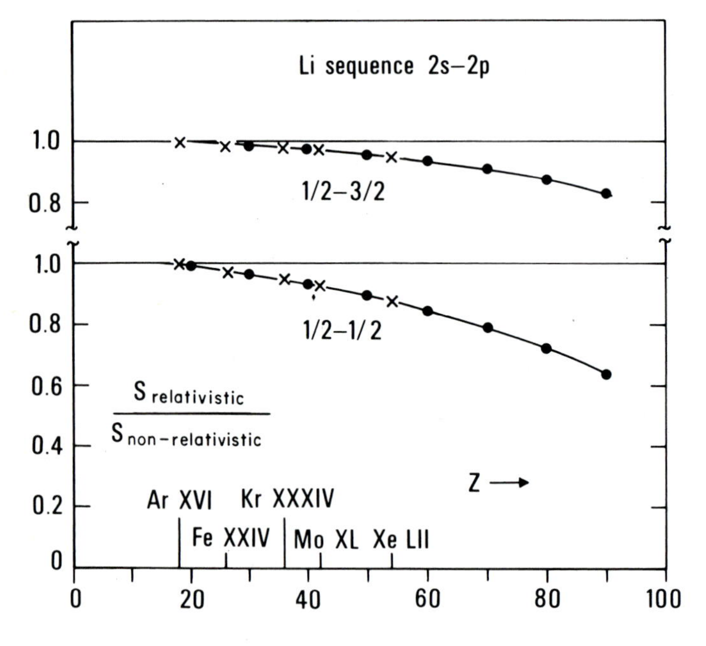

Fig.5.13 Ratio of relativistic to non-relativistic line strengths as a function of Z for the $2s_{1/2}$-$2p_{1/2}$ and $2s_{1/2}$-$2p_{3/2}$ transitions of the lithium (●) and hydrogen (x) sequences. It is seen that within the precision of this graph the hydrogen- and lithium-sequence behavior is indistinguishable. The data sources are: Li sequence, [5.116]; H sequence, [5.115]

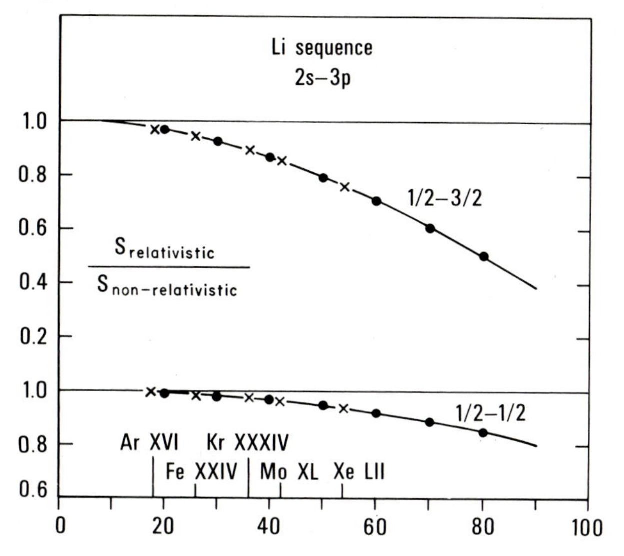

Fig.5.14 Ratio of relativistic to non-relativistic line strengths as a function of Z for the $2s_{1/2}$-$3p_{1/2}$ and $2s_{1/2}$-$3p_{3/2}$ transitions of the lithium (●) and hydrogen (x) sequences. As in Fig.5.13, the hydrogen- and lithium-sequence data are practically identical. The data sources are: Li sequence, [5.116]; H sequence, [5.115]

However, it must also be noted that the Li sequence is a special case insofar as it is a one-electron atomic system and therefore the usual intermediate coupling does not occur. In the following, we shall first discuss the effects on the line strength S rather than the f-value in order to illustrate the relativistic changes on the transition integral σ only, i.e., we exclude relativistic effects in the transition energy. In Figs.5.13 and 5.14, KIM and DESCLAUX's [5.116] results for the two Li-sequence transitions are presented as line-strength ratios between their relativistic and WEISS' [5.78] non-relativistic results. For Z=50, the difference between relativistic and non-relativistic results is still below 10%, except for the $2s_{1/2}$-$3p_{3/2}$ transition, where it is about 25%. The deviations from the normal doublet line-strength ratio of 2:1 for $(s_{1/2}\text{-}p_{3/2}):(s_{1/2}\text{-}p_{1/2})$ are even smaller, since the relativistic corrections go in the same direction for all four transitions. The lithium data are also compared in Figs.5.13 and 5.14 with the corresponding values for hydrogen-like ions which YOUNGER and WEISS' [5.115] have calculated. Perfect agreement is obtained within the precision of the graph.

We turn attention now to f-values, where the corrections will be greater due to relativistic shifts in the energies which enter into the term $(E_k - E_i)$, see (5.2). KIM and DESCLAUX [5.116] have also calculated f-values for some very highly-charged Li-like ions with relativistic self-consistent-field wave functions; their results are illustrated respectively (in Figs.5.15 and 5.16) for the $2s_{1/2}$-$2p_{1/2,3/2}$ and $2s_{1/2}$-$3p_{1/2,3/2}$ lines. These graphs are essentially a small part of the earlier-presented f vs. 1/Z curves for Li 2s-3p (Fig.5.1) and Li 2s-3p (Fig.5.9), with the regions for small 1/Z now greatly magnified, and the extensions of the earlier-presented non-relativistic data (multiplet values) given as the upper broken lines. The relativistic f-value of the $2s_{1/2}$-$2p_{1/2}$ line still tends to zero, similar to the non-relativistic case, but for the $2s_{1/2}$-$2p_{3/2}$ line the f-value increases drastically beyond Mo XL. Since Fig.5.13 showed that the relativistic line-strength S for this transition decreases slightly from the non-relativistic case, the large increase for f is all due to a very pronounced increase in the excitation energy E_k for $2p_{3/2}$, enlarging (E_k-E_i) in (5.2) (E_i=0 for this line). KIM and DESCLAUX [5.116], as well as WEISS [5.78] (using a perturbation-theory treatment) have calculated that the $2p_{3/2}$ state has for the ion Mo XL a non-relativistic energy of 5.7 Rydberg units (Ryd) and a relativistic value of 15.77 Ryd. For the $2p_{1/2}$ level, on the other hand, the relativistic energy remains with 6.56 Ryd still fairly close to the non-relativistic value of 5.7 Ryd.

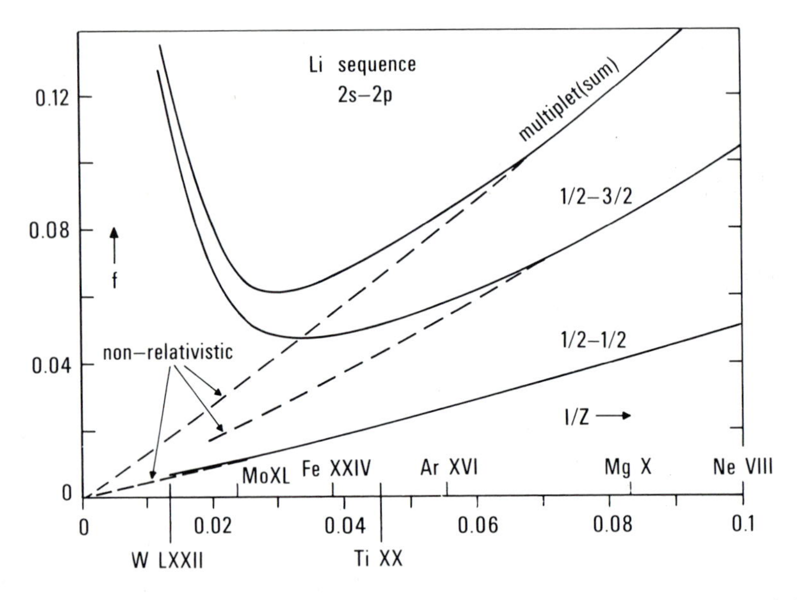

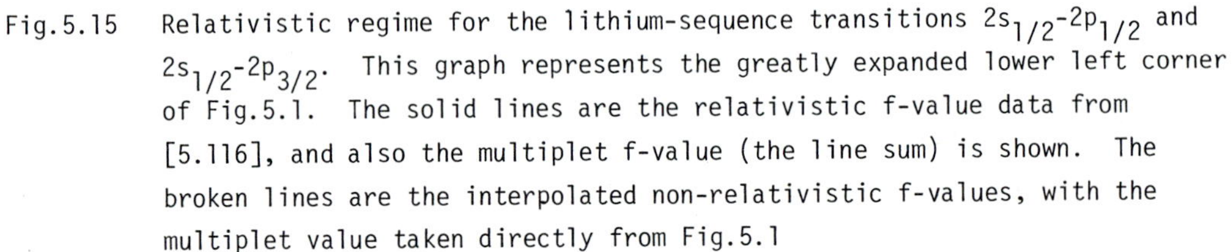
Fig.5.15 Relativistic regime for the lithium-sequence transitions $2s_{1/2}$-$2p_{1/2}$ and $2s_{1/2}$-$2p_{3/2}$. This graph represents the greatly expanded lower left corner of Fig.5.1. The solid lines are the relativistic f-value data from [5.116], and also the multiplet f-value (the line sum) is shown. The broken lines are the interpolated non-relativistic f-values, with the multiplet value taken directly from Fig.5.1

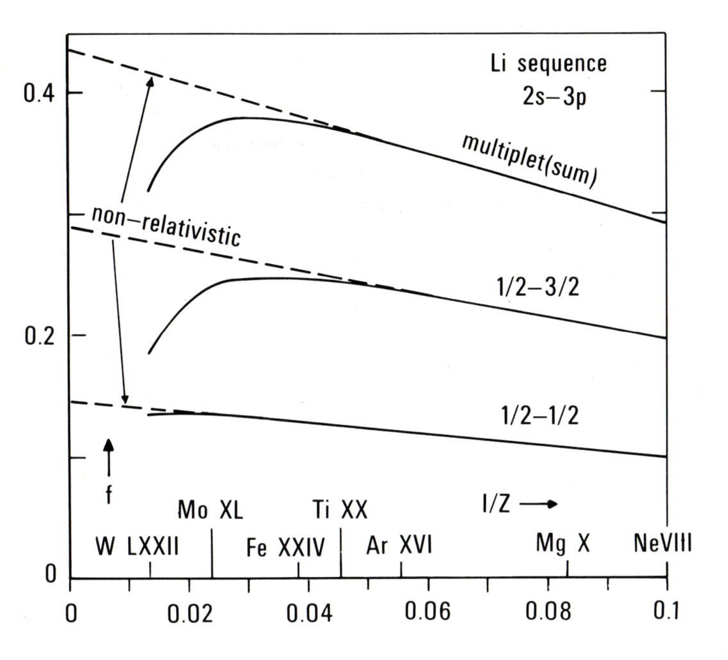

Fig.5.16 Relativistic regime for the lithium-sequence transitions $2s_{1/2}$-$3p_{1/2}$ and $2s_{1/2}$-$3p_{3/2}$. This graph represents the greatly expanded left side of Fig.5.9. The solid lines are the relativistic f-calue data from the calculations by [5.116], and also the multiplet value (the line sum) is shown. The broken lines are the interpolated non-relativistic f-values, with the multiplet value taken directly from Fig.5.9

The behavior of the 2s-3p transitions (Fig.5.16) is quite analogous but the corrections are less pronounced. While the relativistic f-value for $2s_{1/2}$-$3p_{1/2}$ still remains very close to the non-relativistic value up to the highest calculated ion, the f-value for the $2s_{1/2}$-$3p_{3/2}$ line decreases from the non-relativistic values beyond Mo XL. However, in this case the transition energy is a much less significant factor, which is not unexpected since the difference between upper and lower states $(E_k - E_i)$ is, in contrast to 2s-2p, very large. For the above-quoted ion of Mo XL, $(E_k - E_i)$ amounts to about 230 Ryd, so that calculated relativistic energy corrections of the order of 10 Ryd, quite similar in magnitude to the 2s-2p case, are not very significant yet.

From these as well as some other results of the above-quoted work, two general observations are suggested: (i) The largest relativistic effects and the first to occur along an isoelectronic sequence are spin-orbit effects, i.e., the transition to intermediate coupling as well as energy shifts, while orbital relativistic effects become important only for much higher-charged ions. (ii) f-values for transitions within the same shell (i.e., for $\Delta n=0$) are subject to the most pronounced relativistic corrections since even minor changes in the energies of upper and lower states may have drastic consequences due to the small energy differences ($E_k - E_i$) for these lines. The discussed theoretical studies give some first interesting insights into the subject of relativistic effects. Much more work, especially on more complex atomic structures, needs to be done to obtain the greatly needed f-value data for this regime. In this area lies also a new challenge to future beam-foil spectroscopy: If very highly-ionized spectra were to be excited, and the expected very short lifetimes accurately determined, some essential experimental contributions to the problem of relativistic effects could be made.

5.5 Summary

The physical principles underlying the systematic trends of atomic oscillator strengths along isoelectronic sequences have been reviewed. Several types of curves may be discerned in the more than 100 well-established trends, and the physical reasons for the different behavior are readily understood: Most common are systematic trends where the f-value uniformly decreases or increases with 1/Z and which are well approximated by a parabola. Maxima are encountered in trends where strong configuration interaction is present. This effect appears to be always most important near the neutral end of a sequence and causes there often destructive interference in the f-value, but diminishes then gradually. Thus a maximum is produced near a few-times ionized species for those transitions where the f-value becomes small or goes to zero for $1/z \to 0$. A minimum in a systematic trend is usually caused by cancellation in the transition integral, which again happens normally near the neutral end of the sequence. A few anomalous curves, containing a maximum as well as a minimum, can be also readily explained as due to special competing factors in such cases.

Studies of f-value regularities are valuable mainly for two reasons: First, the reliability and accuracy of f-value data may be judged by their degree of fit into established trends, and secondly, additional data (usually for higher ions) may be obtained simply by graphical interpolation. With respect to the first point, it was principally the systematic deviation of early beam-foil data from well-established trends in the direction towards smaller f-values which indicated strongly that cascading effects are significant and have to be taken into account. With respect to the second point one must note that the determination of new f-values by interpolation is straightforward only for not-too-highly-ionized systems, say below ionic charges of 20. Great caution must be exercised in interpolating data for more highly-ionized species because of relativistic effects. Very recent theoretical studies on light element sequences have shown that significant intermediate coupling (spin-orbit) effects as well as relativistic energy shifts occur first roughly for spectra near Z=20 and beyond, while relativistic changes in the radial transition integrals become usually significant only at much larger Z, roughly at about Z=50. The f-values for transitions within the same shell ($\Delta n=0$) are most drastically affected because the energies of upper and lower state are nearly equal, so that small changes in either one have a pronounced effect on the energy difference. By combining the isoelectronic-sequence trends of individual transitions with the systematic behavior of transitions in a spectral series, one may carry regularity studies a step further and find the overall oscillator-strength distribution in a spectral series along a sequence. By tying individual systematic trends together in this way, one may also utilize general boundary conditions such as f-sum rules to judge the overall consistency of very many f-value data.

References

5.1 E. Hinnov: Princeton Plasma Physics Lab. Reports, MATT-1022, 1024 (1974)

5.2 N. Bretz, D. Dimock, A. Greenberger, E. Hinnov, E. Meservey, W. Stodiek, S. von Goeler: In Proceedings IAEA Conf. on Plasma Physics and Contr. Nucl. Fusion Res., Tokyo (1974)

5.3 D. C. Maede: Nucl. Fusion 14, 289 (1974)

5.4 V. A. Vershkov, S. V. Mirnov: Nucl. Fusion 14, 383 (1974)

5.5 R. C. Elton: "Atomic Processes", In *Methods of Experimental Physics*, Vol.9A, ed. by H. R. Griem, R. H. Lovberg (Academic Press, New York 1970) p.135

5.6 E. Hylleraas: Z. Physik 65, 209 (1930)

5.7 B. Edlén: In *Handbuch der Physik*, Vol. XXVII, ed. by S. Flügge (Springer, Berlin, Göttingen, Heidelberg 1964)

5.8 M. Cohen, A. Dalgarno: Proc. Roy Soc. A 280, 258 (1964)

5.9 For a listing of papers on the nuclear charge expansion method, see R.J.S. Crossley: In *Advances in Atomic and Molecular Physics*, Vol. 5 (Academic Press, New York 1969) p.237

5.10 A. W. Weiss: Astrophys. J. 138, 1262 (1963)

5.11 W. L. Wiese: Appl. Opt. 7, 2361 (1968)

5.12 W. L. Wiese: In *Beam-Foil Spectroscopy*, ed. by S. Bashkin (Gordon and Breach, New York 1968) p.385

5.13 W. L. Wiese, A. W. Weiss: Phys. Rev. 175, 50 (1968)

5.14 E. U. Condon, G. H. Shortley: *The Theory of Atomic Spectra* (Cambridge Univ. Press, 1935)

5.15 D. R. Bates, A. Damgaard: Phil. Trans. Roy. Soc. A242, 101 (1949)

5.16 C. W. Allen: *Astrophysical Quantities*, 3rd ed. (The Athlone Press, London 1973)

5.17 A. W. Weiss: Phys. Rev. 188, 119 (1969)

5.18 M. W. Smith, W. L. Wiese: Ap. J. Suppl. 23, 103 (No. 196) (1971)

5.19 M. W. Smith, G. A. Martin, W. L. Wiese: Nucl. Instr. and Meth. 110, 219 (1973)

5.20 D. J. G. Irwin, A. E. Livingston, J. A. Kernahan: Nucl. Instr. and Meth. 110, 111 (1973)

5.21 G. Sørensen: Phys. Rev. A 7, 85 (1973)

5.22 T. Andersen, K. A. Jessen, G. Sørensen: Phys. Letters 29A, 384 (1969)

5.23 L. Barrette, E. J. Knystautas, B. Neveu, R. Drouin: Phys. Letters 32A, 435 (1970)

5.24 K. Berkner, W. S. Cooper III, S. N. Kaplan, R. V. Pyle: Phys. Letters 16, 35 (1965)

5.25 H. G. Berry, W. S. Bickel, S. Bashkin, J. Desesquelles, R. M. Schectman: J. Opt. Soc. Am. 61, 947 (1971)

5.26 W. S. Bickel, H. G. Berry, J. Desesquelles, S. Bashkin: J. Quant. Spectrosc. Radiat. Transfer 9, 1145 (1969)

5.27 W. S. Bickel, I. Martinson, L. Lundin, R. Buchta, J. Bromander, I. Bergström: J. Opt. Soc. Am. 59, 830 (1969)

5.28 J. Bromander: Physica Scripta 4, 61 (1971)

5.29 J. P. Buchet, M. C. Buchet-Poulizac, G. DoCao, J. Desesquelles: Nucl. Instr. and Meth. 110, 19 (1973)

5.30 J. P. Buchet, A. Denis, J. Desesquelles, M. Dufay: Comp. Rend. 265B, 471 (1967)

5.31 J. P. Buchet, M. C. Poulizac, M. Carre: J. Opt. Soc. Am. 62, 623 (1972)

5.32 J. Desesquelles: Ann. Phys. (Paris) 6, 71 (1971)

5.33 M. Dufay, A. Denis, J. Desesquelles: Nucl. Instr. and Meth. 90, 85 (1970)

5.34 P. D. Dumont: Physica 62, 104 (1972)

5.35 I. Martinson, H. G. Berry, W. S. Bickel, H. Oona: J. Opt. Soc. Am. 61, 519 (1971)

5.36 I. Martinson, W. S. Bickel: Phys. Letters 31A, 25 (1970)

5.37 I. Martinson, W. S. Bickel, A. Ölme: J. Opt. Soc. Am. 60, 1213 (1970)

5.38 M. C. Poulizac, M. Druetta, P. Ceyzeriat: J. Quant. Spectrosc. Radiat. Transfer 11, 1087 (1971)

5.39 W. A. Roberts, C. E. Head: Nucl. Instr. and Meth. 110, 99 (1973)

5.40 K. C. Brog, T. G. Eck, H. Wieder: Phys. Rev. 153, 91 (1967)

5.41 R. K. Moitra, P. K. Mukherjee: Int. J. Quantum Chem. 6, 211 (1972)

5.42 M. Cohen, P. S. Kelly: Can. J. Phys. 45, 1661 (1967)

5.43 R. D. Chapman: Astrophys. J. 156, 87 (1969)

5.44 S. Hameed, A. Herzenberg, M. G. James: J. Phys. B 1, 822 (1968)

5.45 S. Lunell: Phys. Rev. A 7, 1229 (1973)

5.46 M. G. Veselov, A. V. Shtoff: Optics and Spectroscopy (USSR) 26, 177 (1969)

5.47 T. C. Caves, A. Dalgarno: J. Quant. Spectrosc. Radiat. Transfer 12, 1539 (1972)

5.48 E. M. Leibowitz: J. Quant. Spectrosc. Radiat. Transfer 12, 299 (1972)

5.49 G. McGinn: J. Chem. Phys. 50, 1404 (1969)

5.50 B. P. Zapol', P. E. Kunin, A. V. Lyubimov, I. M. Taksar, I. I. Fabrikant: Izv. Akad. Nauk Latv. SSR, Ser. Fiz Tekh. Nauk No. 6, 14 (1971)

5.51 I. L. Beigman, L. A. Vainshtein, V. P. Shevelko: Optics and Spectroscopy (USSR) 28, 229 (1970)

5.52 J. S. Onello, L. Ford, A. Dalgarno: Phys. Rev. A 10, 9 (1974)

5.53 B. Warner: Monthly Not. Roy. Astron. Soc. 141, 273 (1968)

5.54 A. Yu. Kantseryavichyus, S. V. Zhilionite: Litov. Fiz. Sb. 7, 73 (1967)

5.55 A. W. Weiss: Phys. Rev. 162, 71 (1967)

5.56 A. B. Bolotin, A. P. Yutsis: Zh. Ekp. Teor. Fiz. 24, 537 (1953)

5.57 W. S. Bickel: Phys. Rev. 162, 7 (1967)

5.58 L. Heroux: Phys. Rev. 153, 156 (1967)

5.59 G. M. Lawrence, B. D. Savage: Phys. Rev. 141, 67 (1966)

5.60 I. Bergström, J. Bromander, R. Buchta, L. Lundin, I. Martinson: Phys. Letters 28A, 721 (1969)

5.61 J. Bromander, R. Buchta, L. Lundin: Phys. Letters 29A, 523 (1969)

5.62 A. Hese, H. P. Weise: Z. Phys. 215, 95 (1968)

5.63 M. C. Poulizac, J. P. Buchet: Physica Scripta 4, 191 (1971)

5.64 J. V. Mallow, J. Burns: J. Quant. Spectrosc. Radiat. Transfer 12, 1081 (1972)

5.65 L. Barrette, R. Drouin: Can. J. Spectrosc. 18, 50 (1973)

5.66 D. J. G. Irwin, A. E. Livingston, J. A. Kernahan: Can. J. Phys. 51, 1948 (1973)

5.67 S. A. Chin-Bing, C. E. Head: Phys. Letters A 45, 203 (1973)

5.68 R. B. Hutchinson: J. Quant. Spectrosc. Radiat. Transfer 11, 81 (1971)

5.69 J. A. Kernahan, A. Denis, R. Drouin: Physica Scripta 4, 49 (1971)

5.70 C. A. Nicolaides: Chem. Phys. Letters 21, 242 (1973)

5.71 U. I. Safranova, A. N. Ivanova, V. N. Kharitonova: Theor. Exp. Chem. (USSR) 5, 209 (1969)

5.72 Z. Sibincic: Phys. Rev. A 5, 1150 (1972)

5.73 O. Sinanoğlu: In *Theory of Electr. Shells of Atoms and Molecules*, Institute of Physics and Math., Acad. Sci. Lithuanian SSR (Publ. House Mintis, Vilnius 1971) p.31

5.74 P. Westhaus, O. Sinanoğlu: Ap. J. 157, 997 (1969)

5.75 M. C. Buchet-Poulizac, J. P. Buchet: Physica Scripta 8, 40 (1973)

5.76 M. C. Poulizac, M. Druetta: Comp. Rend. 270B, 788 (1970)

5.77 I. Martinson, W. S. Bickel: J. Opt. Soc. Am. 60, 1213 (1970)

5.78 A. W. Weiss: private communication

5.79 A. W. Weiss: Nucl. Instr. and Meth. 90, 121 (1970)

5.80 A. Hibbert: J. Phys. B 7, 1417 (1974)

5.81 G. Beauchemin, J. A. Kernahan, E. Knystautas, D. J. G. Irwin, R. Drouin: Phys. Letters A 40, 194 (1972)

5.82 D. J. G. Irwin, A. E. Livingston, J. A. Kernahan: Nucl. Instr. and Meth. 110, 105 (1973)

5.83 L. Heroux: Phys. Rev. 180, 1 (1969)

5.84 S. Hontzeas, I. Martinson, P. Erman, R. Buchta: Physica Scripta 6, 55 (1972)

5.85 I. Martinson, A. Gaupp, L. J. Curtis: J. Phys. B 7, L463 (1974)

5.86 T. Andersen, K. A. Jessen, G. Sørensen: Phys. Rev. 188, 76 (1969)

5.87 J. S. Sims, R. C. Whitten: Phys. Rev. A 8, 2220 (1973)

5.88 A. W. Weiss: Phys. Rev. A 6, 1261 (1972)

5.89 F. Weinhold: J. Chem. Phys. 54, 1874 (1971);
F. Weinhold: Phys. Rev. Letters 25, 907 (1970)

5.90 W. L. Wiese: Nucl. Instr. and Meth. 90, 25 (1970), and other papers in this volume

5.91 C. Laughlin, A. Dalgarno: Phys. Rev. A 8, 39 (1973)

5.92 E. H. Pinnington, A. E. Livingston, J. A. Kernahan: Phys. Rev. A 9, 1004 (1974)

5.93 A. E. Livingston, D. J. G. Irwin, E. H. Pinnington: J. Opt. Soc. Am. 62, 1303 (1972)

5.94 S. Bashkin, I. Martinson: J. Opt. Soc. Am. 61, 1686 (1971)

5.95 H. G. Berry, R. M. Schectman, I. Martinson, W. S. Bickel, S. Bashkin: J. Opt. Soc. Am. 60, 335 (1970)

5.96 L. J. Curtis, I. Martinson, R. Buchta: Physica Scripta 3, 197 (1971)

5.97 W. Hofmann: Z. Naturforsch. 24a, 990 (1969)

5.98 L. J. Curtis, W. H. Smith: Phys. Rev. A 9, 1537 (1974)

5.99 O. Sinanoğlu: Nucl. Instr. and Meth. 110, 193 (1973)

5.100 C. Froese-Fischer: J. Quant. Spectrosc. Radiat. Transfer 8, 755 (1968)

5.101 A. W. Weiss: Phys. Rev. A 9, 1524 (1974)

5.102 R. H. Garstang, L. J. Shamey: In *The Magnetic and Related Stars*, ed. by R. C. Cameron (Mono Book Corp., Baltimore 1967) p. 387

5.103 D. Schoenberner, D. Zimmermann: Z. Physik 216, 172 (1968)

5.104 P. Erman, J. Brzozowski, W. H. Smith: Astrophys. J. 192, 59 (1974)

5.105 R. W. Schmieder, A. Lurio, W. Happer, A. Khadjavi: Phys. Rev. A 2, 1216 (1970)

5.106 H. Krueger, K. Scheffler: J. Phys. Rad. 19, 854 (1958)

5.107 E. Biémont: J. Quant. Spectrosc. Radiat. Transfer 15, 531 (1975)

5.108 J. S. Onello: Phys. Rev. A 11, 743 (1975)

5.109 A. N. Filippov: Zh. Eksp. Teor. Fiz. 2, 34 (1932)

5.110 J. P. Buchet, M. Dufay, M. C. Poulizac: Phys. Letters A 40, 127 (1972)

5.111 W. H. Smith, H. S. Liszt: J. Opt. Soc. Am. 61, 938 (1971)

5.112 G. A. Martin, W. L. Wiese: to be published

5.113 W. L. Wiese, G. A. Martin: Proceedings VII Yugoslav Symp. on Ionized Gases, Rovinj 1974

5.114 O. Sinanoğlu, W. Luken: Chem. Phys. Letters 20, 407 (1973)

5.115 S. Younger, A. W. Weiss: to be published

5.116 Y.-K. Kim, J. P. Desclaux: Proceedings IX. Int. Conf. Phys. Electronic and Atomic Coll., Seattle 1975, and private communication

6. Applications to Astrophysics: Absorption Spectra

Ward Whaling

With 2 Figures

The relevance of beam-foil spectroscopy to astrophysics comes from the importance of radiative transition probabilities in the quantitative analysis of optical spectra. A stellar abundance measurement based on the intensity of an absorption line actually determines not the abundance alone but the product: (abundance of the element in the source) x (probability of the transition). To separate the abundance from this product one must know the transition probability, and any uncertainty in the transition probability leads to a corresponding uncertainty in the derived abundance. Abundance analyses are of basic importance to modern astrophysics, and observatories have supported continuing programs to measure the transition probabilities needed for spectral analysis. The beam-foil time-of-flight technique offers a new approach to atomic and ionic transition probabilities, an approach that is free from many of the difficulties of conventional methods. Consequently, there has been an understandable interest in the beam-foil method among astrophysical spectroscopists, and there has been a corresponding interest on the part of beam-foil spectroscopists to see the effect of their lifetime measurements on abundances.

In general, a time-consuming sequence of calculations and auxiliary measurements is needed to derive an abundance from a level lifetime and the intensity of an absorption line, and it is the purpose of this chapter to describe these steps in detail. However, in some instances it may be possible to bypass this lengthy procedure and estimate in the following simple way the impact of a new lifetime measurement on an earlier abundance calculation.

Most astrophysical abundance measurements published after 1962 have made use of the extensive tables of transition probabilities compiled by CORLISS and BOZMAN [6.1], hereafter designated CB. The CB tables are sufficiently complete to permit the computation of a level lifetime $\tau_{CB} = (\sum A_{CB})^{-1}$, where the summed transition probabilities A include all transitions that depopulate the level. By comparing τ_{CB} with the measured lifetime of the level, one finds immediately the factor by which all A_{CB} must be multiplied to put the CB transition probabilities on an absolute scale. Because the transition probabilities enter the abundance calculation only as a factor as noted above, this same correction is to be applied to any abundance calculated from the original CB transition probabilities.

If the transition probabilities of CB were an accurate set of relative transition probabilities, the procedure above would be sufficient and complete. Unfortunately, there is evidence that the A_{CB} are not a consistent set of relative transition probabilities, at least for some elements. Several experiments [6.2,3] have shown that the A_{CB} for Fe I require correction by a factor that varies with the excitation energy of the level. There is evidence [6.4] of a similar energy-dependent discrepancy in Cr I. Before the use of a single correction factor is justified, one must compare τ_{CB} and τ_{exp} for levels over a range of excitation energies. This extended comparison is not yet available for many elements so that the elements for which the A_{CB} may be useful are not yet established. A comparison of τ_{CB} and τ_{exp} will show whether an earlier abundance measurement is based on erroneous data, but is of very limited accuracy in finding the correct abundance.

In the absence of reliable relative transition probabilities which can be normalized by a single accurate lifetime, one is forced to use the more extensive procedures that are the subject of this paper. The first step is the determination of branching ratios and transition probabilities. Laboratory measurements of branching ratios are discussed in Section 6.1. The use of these transition probabilities to determine elemental abundance is discussed in Section 6.2. A brief review of work that has been done and some current problems that need attention are discussed in Section 6.3.

6.1 Branching Ratios

Before the beam-foil mean lifetime can be used to analyze astrophysical spectra, the mean lifetime or total transition probability must be divided into the probabilities for the individual transitions by which the level decays. Only the individual transition probabilities are useful for spectral analysis. If $A_{u\ell}$ is the transition probability for the emission of a photon of wavelength $\lambda_{u\ell}$ from level u of mean lifetime τ_u, the branching ratio $BR_{u\ell}$ is defined by the relation

$$BR_{u\ell} = A_{u\ell}/\sum_{\ell} A_{u\ell} \quad , \tag{6.1}$$

where the sum includes all lower levels to which level u decays. The branching ratio can be measured using the relation

$$BR_{u\ell} = I_{u\ell}/\sum_{\ell} I_{u\ell} \quad , \tag{6.2}$$

since the photon intensity $I_{u\ell}$ of the line of wavelength $\lambda_{u\ell}$ is proportional to $A_{u\ell}$. Using the relation $\tau_u = (\sum_{\ell} A_{u\ell})^{-1}$, one can express $A_{u\ell}$ in terms of measurable quantities:

$$A_{u\ell} = BR_{u\ell}\, \tau_u^{-1} \quad . \tag{6.3}$$

Because the intensities appear in the expression for BR as ratios, only relative line intensities are needed.

There are tabulations of relative line intensities in the literature, notably by MEGGERS et al. [6.5], but most are based on photographic photometry of many years ago. The accuracy is less than can be achieved by modern photoelectric techniques, and the weak lines, of particular importance as explained in section 6.2, are usually not included. In order to obtain branching ratios of accuracy comparable to their lifetime measurements, several workers have remeasured the branches from the levels they have studied. Branching-ratio measurements are a straightforward, if tedious, application of quantitative photometry. The reader will find thorough discussions of the techniques of quantitative photometry, both photographic and photoelectric, in the excellent review by LOCHTE-HOLTGREVEN and RICHTER [6.6]. In the sections below, we discuss some particular aspects of photoelectric photometry that have been used in branching ratios measured for astrophysical purposes.

6.1.1 Light Sources

An intense source is needed to measure the weak branches of primary interest for abundance analysis. The source must be free of self-absorption but local thermodynamic equilibrium is not necessary. For branching measurements in Ti and V, ROBERTS et al. [6.7] used an arc after the design of RICHTER and WULFF [6.8] in which the arc is stabilized by a stream of flowing argon. The argon flow stabilizes the intensity and position of the arc so that it can be used with photoelectric detectors. RICHTER and WULFF have analyzed the population and temperature distribution in this arc. They find that the source produces little continuum background and hence is well suited to the observation of weak lines. They were able to observe Fe I lines which have an equivalent width in the sun of only 1 mÅ and to measure transition probabilities of lines with a solar equivalent width of 3 mÅ. The equivalent width is defined in section 6.2.1.

ROBERTS et al. [6.7] have used a property of this arc to extend the scope of their branching-ratio measurements. They assume that the relative population of different levels in the same term is proportional to the statistical weight of the level. This assumption enables them to measure transition probabilities for all lines from all levels in a term if the lifetime of only one level is known. They note that this assumption is less stringent than local thermodynamic equilibrium, and they confirm its validity for terms in which the lifetimes of several levels are known.

Branching-ratio measurements on several metals of the iron group have been carried out at CalTech using a water-cooled hollow-cathode discharge with helium or argon as the carrier gas. The hollow-cathode source has excellent positional stability and freedom from rapid fluctuations in intensity, although it does display long-time drifts. MARTINEZ-GARCIA et al. [6.9] have measured transition probabilities of Fe I lines with a solar equivalent width as small as 13 mÅ with this hollow-cathode source, and SMITH and WHALING [6.10] have measured Fe II lines with a solar equivalent width of 4 mÅ.

It is essential that the source be free of self-absorption. The simplest test for self-absorption is a comparison of the relative intensity of a strong transition leading to the ground state with a weak transition from the same upper level under conditions of different source excitation. If self-absorption is present, the stronger line will be more strongly absorbed and the ratio of intensities will change. ROBERTS et al. [6.7] increased the sensitivity of this test by passing the light back and forth through the source with a mirror to enhance any absorption that might occur. (See Fig.6.1).

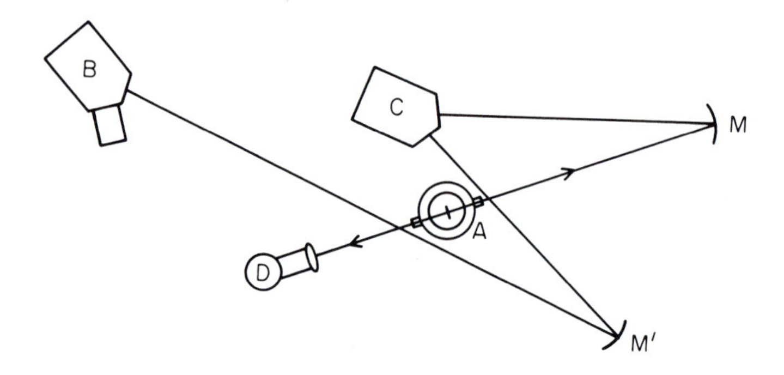

Fig.6.1 Schematic of apparatus used for branching-ratio measurements by ROBERTS et al. [6.7]. A is arc source, B and C are monochromators, D is a monitor phototube, M and M' are spherical mirrors. (Reprinted by permission of The University of Chicago Press)

6.1.2 Spectrometers

A branching-ratio measurement requires the intercomparison of relative line intensities. Variations in the output of the source can introduce error into a sequential measurement of the intensity of two lines unless some means is provided for monitoring the source excitation. Source instabilities are unavoidable when the source is driven hard to bring up the weakest lines, and all of the branching-ratio measurements in the recent literature have employed some arrangement for monitoring the source. ROBERTS et al. [6.7] employed the arrangement shown in Fig.6.1 with monitor photomultiplier D receiving light from the arc source A through a lens and interference filter. The filter should pass a strong transition from the level under study, and the intensity detected at B is divided by the monitor signal to normalize the intensity to constant source strength. WHALING et al. [6.11] and BRIDGES and KORNBLITH [6.12] placed a beam splitter ahead of the measuring monochromator to send a fraction of the light from the source into a monitor monochromator. Because the excitation of the particular level under study may vary with position in the source, care should be taken to see that the monitor receives light from the same location in the source that illuminates the entrance slit of the measuring spectrometer.

LENNARD et al. [6.13] have described a large 5-m Paschen-Runge spectrometer in which two detectors are located along the focal circle. One detector is set on the monitor line while the other detector scans over all of the lines from the level being measured. To avoid mechanical interference between the two detectors, they are located on opposite sides of the grating normal.

6.1.3 Spectrometer Calibration

Branching-ratio measurements require a spectrometer or spectrograph of known detection efficiency. Since only the ratio of line intensities appears in the expression for the branching ratio, the relative efficiency as a function of wavelength is sufficient. LOCHTE-HOLTGREVEN and RICHTER [6.6] discuss many methods for calibrating the detection efficiency of spectroscopic equipment. A few basic methods for measuring relative detection efficiency as function of wavelength are discussed in this subsection.

For wavelengths greater than 2500Å the tungsten ribbon filament radiance standard is a convenient, commercially available calibration standard. The peak output of these lamps in the infrared is so much greater than the ultraviolet output that precautions must be taken to prevent scattered IR photons from interfering with the UV calibration. ROBERTS et al. [6.7] used a predisperser monochromator C as indicated in Fig. 6.1 to exclude IR photons from the standard lamp located at position A. MARTINEZ-GARCIA et al. [6.9] used narrow-pass interference filters of measured transmission instead of a predisperser.

For wavelengths shorter than 2500Å the calibration becomes more difficult and the method to be used depends on the particular wavelengths of interest. MARTINEZ-GARCIA et al. extended the calibration of their system down to 2000Å by comparing the detection efficiency of their spectrometer with that of a photomultiplier coated with sodium salicylate, under the assumption [6.14,15] that the detection efficiency of the latter was constant over the wavelength range 2000-2600Å. Using a two-monochromator arrangement such as that in Fig.6.1, with a high pressure Xe continuum source at A, and with monochromator C set to pass a wavelength band narrower than the pass-band of monochromator B, they first inserted the coated photomultiplier in the beam just ahead of spectrometer B to measure the intensity of the beam entering B, then removed the photomultiplier and measured the response of spectrometer B.

An appealing method for calibrating spectrometers to be used in branching-ratio measurements is to measure the intensity of transitions for which the branching ratios are known. Molecular vibrational bands are useful in the range between 2500Å and the air cut-off wavelength at 1800Å. Molecular systems suitable for this type of calibration are discussed by MUMMA [6.16].

In addition to the dependence of spectrometer response on wavelength, branching-ratio measurements require attention to the linearity of photomultiplier response to illumination of different intensity. At the level of accuracy easily achieved in branching measurements, photomultipliers display nonlinearity and hysteresis behavior that varies from tube to tube. Linearity can be tested with two light sources: the response to the two sources together should equal the sum of the responses to each source alone. Filters are used to keep the illumination below the level at which nonlinear response sets in. Hysteresis and time-dependent response can be checked by turning a constant light signal on and off. Since photometry is basically a comparison of the response of the photomultiplier to an unknown flux and the response to a known flux from a calibration source, errors due to nonlinear response can be minimized by adopting experimental procedures that keep the illumination applied to the tube nearly constant. Neutral-density filters are essential accessories for such measurements.

6.1.4 Selection of Branches to be Measured

Branching-ratio measurements are time-consuming. Many branches contribute little to the analysis of spectra and can be neglected without loss. One starts by listing all of the known branches and then removes from the list those branches that, for reasons discussed below, will not repay the effort required to measure them.

Identification of Decay Branches. The Revised Multiplet Table [6.17] and the Ultraviolet Multiplet Table [6.18] include all of the decay branches classified at the time of their publication. Additional classified branches may be found in later work cited in the periodic bibliographies compiled at the National Bureau of Standards [6.19], in a critical compilation by EDLÉN [6.20] of references to the analysis of atomic spectra as of 1970, and a more extensive list prepared by ADELMAN and SNIDJERS [6.21].

LENNARD et al. [6.22] computed all possible transitions, with parity change and ΔJ = 0 or ± 1, from the level under study to lower levels in Ni I as listed by MOORE [6.23]. Branches predicted in this way were then examined in the solar spectrum to see if they were strong enough to contribute to the total transition probability, or were clean enough to justify measurement. One previously unclassified transition was probably identified in this way.

Choice of Useful Branches. The branches from any given level are likely to contain wavelengths beyond the calibrated range of the spectrometer to be used for the branching measurements. If an unmeasurable branch has a small transition probability, where "small" means that it contributes to the sum in Eq.(6.1) an amount less than the experimental uncertainties, the line may be ignored with no greater loss than the transition probability for that one line. On the other hand, a strong branch omitted from the sum in (6.1) increases by a large but unknown amount each branching ratio, and hence the transition probability, for every transition from the level. It is therefore important to determine whether unmeasurable branches contribute significantly to the sum in (6.1).

It may be possible to estimate the magnitude of the transition probability of an unmeasurable line from its intensity in the solar spectrum. This comparison is especially useful in the infrared where solar atlases extend to 25,242Å [6.24]. If one compares the (absorption) intensities of two lines from the same lower level, the line with greater reduced width in the solar spectrum (defined in Section 6.2.1) will usually have the greater transition probability. One may estimate the transition probability for an unmeasurable line from the empirical solar curve-of-growth, which is a graphical representation of the dependence of the solar equivalent width on the transition probability and is discussed in Section 6.2. LENNARD et al. [6.22] justified their omission of infrared branches in this way. Solar atlases do not extend below 3000Å, but the laboratory intensity tabulated in the Ultraviolet Multiplet Table [6.18] may give an indication of the contribution of an unmeasurable UV line relative to other lines from the same upper level.

If the purpose of the analysis is the abundance in a particular star, one may further reduce the list of branches to be measured by including only levels which have at least one branch which will contribute useful abundance information. Examination of the spectrum of the star may show that no lines are observed from some upper levels. Solar lines that have been used for abundance analysis are discussed by MULLER [6.25] and ALLER [6.26] and enumerated by MOORE et al. [6.27]. If there are many branches observed in the star, the list of lines to be measured may be further refined by requiring that the stellar lines (i) be free of interfering blends, (ii) lie in a wavelength region where the continuum is well-defined (e.g., $\lambda > 5000$Å for solar lines). By such selection the list of branches to be measured can be reduced to avoid unrewarding effort.

6.2 Curve-of-Growth Analysis

A solar curve-of-growth serves several functions in the application of BFS to astrophysical problems: It provides a test of the internal consistency of a set of relative transition probabilities and is an aid in uncovering experimental errors; it provides a method of comparing transition probabilities by different workers even when the two sets of measurements have no lines in common; and it is the simplest method of determining solar abundances. The literature on the curve-of-growth is extensive and the reader is referred to the reviews by ALLER [6.28] and by GOLDBERG and PIERCE [6.29] for thorough discussions. In the sections below we consider those aspects that are useful in deriving transition probabilities and solar abundances from BFS lifetime measurements.

The solar curve-of-growth is a graphical representation of the intensity of an absorption line of wavelength $\lambda_{u\ell}$ in the solar spectrum as a function of the product $(N_\ell\ f)$ where N_ℓ is the number density of atoms in the initial (lower) level ℓ and f is the oscillator strength for the transition. The oscillator strength is related to the transition probability by the expression: $f = 1.50 \times 10^{-16}\ \lambda^2 (g_u/g_\ell)\ A_{u\ell}$, where the statistical factor $g_i = 2J_i + 1$, and λ is in Angstrom units. The line intensity is a fairly simple function of $(N_\ell\ f)$ so long as one limits the consideration to lines of nearly the same wavelength from levels of nearly the same excitation energy. If one wishes to consider all of the lines of a particular element, the construction of the curve becomes complicated, since it must take into account the parameters of the solar atmosphere (principally the temperature, density, and composition, and their variation with solar depth) as well as the atomic parameters ($\lambda_{\ell u}$, f, excitation energy of level ℓ, ionization energy, and partition function). The perfect curve-of-growth, which precisely defines a unique intensity for every line, is a theoretical ideal which practical curves-of-growth can only approach. However, it will become clear from the following discussion that one can often extract useful information from a practical curve-of-growth.

6.2.1 Construction of a Curve-of-Growth

To construct an empirical solar curve-of-growth, the line intensity is expressed in terms of the equivalent width W, defined as the integral over wavelength of the residual intensity and equal to the width (in Å) of a completely black strip which contains an amount of radiation equivalent to that absorbed by the line. MOORE et al. [6.27, p.IX] discuss the equivalent width and how to measure it. The ordinate of the curve-of-growth is the equivalent width divided by the wavelength. The equivalent width W and the reduced width W/λ have been tabulated by MOORE et al. [6.27] for 19,182 lines in the solar spectrum in the wavelength range 3061-8770Å. It is the availability of these extensive measurements that makes the solar spectrum a convenient reference spectrum with which to compare relative intensity measurements from laboratory sources.

The complexities of the curve-of-growth are lumped into the calculation of the abscissa. In one representation used in several BFS papers [6.10,22], the abscissa is expressed as the product $g_\ell f(N_{el}/N_H)\Gamma$, where N_H and N_{el} are, respectively, the number densities of hydrogen and the element under investigation, and Γ is a function of the wavelength and lower-level excitation energy of the transition, of the ionization energy and partition function of the atom, and of the solar atmospheric model. GOLDBERG et al. [6.30] and references therein provide tables and graphs for the evaluation of the Γ parameter for Li, C, N, O, Na, Mg, Aℓ, Si, S, K, Ca, Sc, Ti, V, Cr, Mn, Fe, Co, Ni, Cu, Zn, Ga, Rb, Sr. Y, Zr, Nb, In, Hg, and Mg^+, Si^+, and Ba^+. CAYREL and JUGAKU [6.31] have tabulated values of the parameters needed to compute curves of growth for neutral and singly-ionized Na, Mg, Aℓ, Si, Ca, Ti, Cr, and Fe, and they have also provided privately the Γ parameter for other elements on request.

The factor (N_{el}/N_H) shifts the curve to right or left in the log-log plot and is not needed in some of the applications discussed below. That factor may be set equal to unity in constructing an empirical curve of growth, as shown in Fig.6.2.

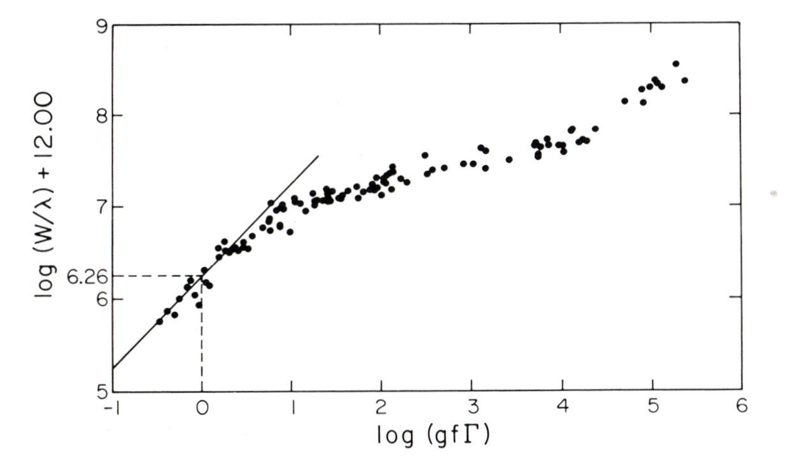

Fig.6.2 Experimental solar curve of growth for Ni I from LENNARD et al. [6.22]. Solar Ni abundance is read from the ordinate of the linear section of the curve where $\log(g_\ell f\Gamma) = 0$

6.2.2 Internal-Consistency Test

The individual points on the curve of growth will deviate from a single curve because of experimental errors in the oscillator strength f and equivalent width W used in the plot, and because of approximations in its construction and in the solar model. However, a single point which lies significantly further from the curve than the mean experimental spread casts suspicion on the f and W for that point. A point off to the right of the curve indicates an f-value that is too large and suggests a re-examination of the branching measurement on which this f-value is based for a blend or a misidentification. Shifts to the right or left cannot be recognized on the horizontal section of the curve but stand out on the rising segment where weak lines appear. If all the points representing transitions from a single upper level deviate to the same side of the curve, one should suspect the measurement of the lifetime of that level.

identification of this discrepancy is aided by using different symbols in plotting transitions from different upper levels. Points that lie above the curve usually indicate a blend in the solar spectrum. MOORE et al. [6.27] measured equivalent widths from the Ultrecht Solar Atlas [6.32]. There are now newer atlases [6.33,34] available when there is reason to question the equivalent widths given by MOORE et al.

6.2.3 Comparison of Transition Probabilities for Different Transitions

Although the BFS time-of-flight method normally yields absolute transition probabilities of high accuracy, the method is not readily applicable to many lines that are useful in astrophysical applications. An important use of BFS measurements is as a check on the systematic accuracy of other experiments which may have measured many more lines than are accessible to the BFS method. Direct comparison is not possible if the BFS measurements and the set to be checked have few, if any, lines in common. However, the curve of growth can be used to test the consistency of measurements from two experiments in just the way it can test internal consistency from a single experiment as described above. Again the weak lines provide the most meaningful comparison.

6.2.4 Solar-Abundance Determination

The solar abundance is extracted by comparing the empirical curve of growth with a theoretical curve of growth. The reader is referred to ALLER [6.28] or GOLDBERG and PIERCE [6.29] for a general discussion of this procedure that makes use of the full curve. This analysis becomes simple if one considers only the very weak, red lines that lie along the linear portion of the curve to the left of the knee. In this region the theoretical curve of growth is a straight line with 45° slope: $\log(W/\lambda) = \log(gf\Gamma) + \log(N_{el}/N_H)$. If lines weak enough to define this straight line are available, the abundance $\log(N_{el}/N_H)$ is given directly by the value of $\log(W/\lambda)$ at the point on the 45° line (extrapolated if necessary) where $\log(gf\Gamma) = 0$. It is customary to add 12 to $\log(N_{el}/N_H)$ to put all abundances on a scale where $N_H = 10^{12}$. In Fig.6.2 this "12" has been added to the ordinate. The great importance of weak lines for solar abundance analysis stems from the fact that the linear section of the curve of growth is relatively insensitive to the uncertain turbulence of the solar atmosphere.

This procedure is capable of good precision. LENNARD et al. [6.22] found the solar Ni abundance ($\log N_{Ni}/N_H + 12$) to be 6.26 ± 0.13 by this method, whereas a more exact fine analysis gave the abundance 6.28 ± 0.09. Other examples of the use of the curve of growth to extract abundances from BFS data will be found in COCKE et al. [6.4], and SMITH and WHALING [6.10].

As long as one considers only weak lines that lie on the linear section of the curve of growth, an abundance value can be derived from each line: $\log(N_{el}/N_H) = \log(W/\lambda) - \log(gf\Gamma)$. The individual abundance values are frequently plotted as a function of λ, or W, or of the excitation energy of the lower level E_ℓ, as a consistency test.

MAY et al. [6.35] have published such plots for Fe I. If the abundance calculated from individual lines varies with E_ℓ, one suspects an error in the assumed solar temperature because the abundance derived from a line of low E_ℓ depends on the temperature much less than does the abundance derived from a line of large E_ℓ. If the derived abundance varies with W, the lines used are not weak enough to lie along the linear section of the curve of growth. If the derived abundance varies with λ, one should examine the equivalent widths to see if they are influenced by the difficulty of establishing the continuum in the increasingly crowded spectrum at shorter wavelengths.

6.3 Beam-Foil-Spectroscopy Measurements Needed for Astrophysical Applications

From a broader point of view, beam-foil spectroscopy is but one method of approaching atomic transition probabilities, and BFS lifetimes should be combined with measurements by other methods to produce "best values" to be used in astrophysics and in other applications. There is, however, one area in which BFS has a unique capability to fill specific astrophysical needs which has not yet been exploited.

Multiply-charged ions are not accessible to lifetime and oscillator strength measurements based on absorption, but they present no difficulty for the BFS technique. The strongest transitions in multiply-charged ions lie in the ultraviolet beyond the atmospheric cut-off, and for this reason have been of limited interest to astronomers in the past. This restriction is fast disappearing as observations are becoming available from rockets and orbiting telescopes. One can anticipate increasing activity in ultraviolet astronomy and an increasing need for transition probabilities in the ultraviolet.

MORTON and SMITH [6.36] have compiled a list of lines useful for the analysis of interstellar absorption spectra, with the transition probabilities where they are known. They note that many of the absorption lines observed with the Copernicus Satellite spectrometer cannot be analyzed until better transition probabilities are provided, and they list specific lines in C, O, S, Mn, and Fe of greatest importance. They note that there are no values available for any of the resonance lines of F I, V III, Cr III, Co II, Co III, Ni II, and Cu II.

Other workers have listed needs met in other astrophysical applications. LAMERS [6.37] lists the second and third spectra of Fe, Cr, Cu, Mn, and Ni in the wavelength region 2000-3000Å, as the highest priority needs for the analysis of spectra of O and B-type stars. PRADERIE [6.38] lists the following critically-needed transitions in the wavelength range 1240-1370Å: In Cr II, Mult. 6; in Mn II, Mult. 41, 42, 56, and 79; in Mn III, Mult. 8; in Cu II, Mult. 2 and 3. She also notes a need for C I transition probabilities in the same wavelength region from multiplets 7, 10, 11, 15-17, 19-29, 40-47, 49-54.

Lifetime measurements in the vacuum ultraviolet present no unusual problems. The calibration of spectrometers for branching measurements will be difficult in the VUV. For the resonance lines there is frequently only one decay channel and branching measurements are not necessary.

Turning to the optical spectra of ground-based astronomy, existing and needed BFS measurements have been reviewed by SMITH [6.39] with particular reference to solar abundances. Since his review appeared, additional measurements in Pr II, Tm II, Lu II, and Ce III have been published [6.40]. Aside from these results, SMITH's review is complete and the reader is referred to his paper for the status of measurements at optical wavelengths as of January 1975.

Acknowledgment

The author is indebted to P. L. Smith, W. N. Lennard, Leon Heroux, and C. L. Cocke for their helpful comments on the material in this chapter. Preparation of this chapter was supported in part by the Office of Naval Research and the National Science Foundation.

References

6.1 C. H. Corliss, W. R. Bozman: *Experimental Transition Probabilities for Spectral Lines of Seventy Elements*, NBS Mon. 53 (Govt. Printing Office, Washington, D.C. 1962)

6.2 J. M. Bridges, W. L. Wiese: Astrophys. J. 161, L71 (1970)

6.3 M. Huber, W. H. Parkinson: Astrophys. J. 172, 229 (1972)

6.4 C. L. Cocke, A. Stark, J. C. Evans: Astrophys. J. 184, 653 (1973)

6.5 W. E. Meggers, C. H. Corliss, B. F. Schribner: *Tables of Spectral Line Intensities*, NBS Mon. 32 (Govt. Printing Office, Washington, D.C. 1961)

6.6 W. Lochte-Holtgreven, J. Richter: In *Plasma Diagnostics*, ed. by W. Lochte-Holtgreven (North-Holland Publ. Co., Amsterdam 1968) p.135

6.7 J. R. Roberts, T. Andersen, G. Sørensen: Astrophys. J. 181, 587 (1973)

6.8 J. Richter, P. Wulff: Astron. Astrophys. 9, 37 (1970)

6.9 M. Martinez-Garcia, W. Whaling, D. L. Mickey, G. M. Lawrence: Astrophys. J. 165, 213 (1971)

6.10 P. L. Smith, W. Whaling: Astrophys. J. 183, 313 (1973)

6.11 W. Whaling, R. B. King, M. Martinez-Garcia: Astrophys. J. 158, 389 (1969)

6.12 J. M. Bridges, R. L. Kornblith: Astrophys. J. 192, 793 (1974)

6.13 W. N. Lennard, W. Whaling, R. M. Sills, W. A. Zajc: Nucl. Instr. and Meth. 110, 385 (1973)

6.14 F. S. Johnson, K. Watanabe, R. Tousey: J. Opt. Soc. Am. 41, 702 (1961)

6.15 J. Hennes, J. Dunkelman: In *The Middle Ultra-Violet: Its Science and Technology*, ed. by A. E. S. Green (Wiley, New York 1966)

6.16 M. J. Mumma: J. Opt. Soc. Am. 62, 1459 (1972)

6.17 C. E. Moore: *A Multiplet Table of Astrophysical Interest*, NSRDS-NBS 40 (Govt. Printing Office, Washington, D.C. 1972)

6.18 C. E. Moore: *An Ultraviolet Multiplet Table*, NBS Circ. 488 (Govt. Printing Office, Washington, D.C. 1950)

6.19 C. E. Moore: *Bibliography on the Analyses of Optical Atomic Spectra*, NBS Spec. Publ. 306 (Govt. Printing Office, Washington, D.C. 1968)
L. Hagen, W. C. Martin: *Bibliography on Atomic Energy Levels and Spectra*, NBS Spec. Publ. 363 (Govt. Printing Office, Washington, D.C. 1972)

6.20 B. Edlén: Nucl. Instr. and Meth. 90, 1 (1970)

6.21 S. J. Adelman, M. A. J. Snijders: A Bibliography of Atomic Line Identification Lists (unpublished)

6.22 W. N. Lennard, W. Whaling, J. Scalo, L. Testerman: Astrophys.J. 197, 517 (1975)

6.23 C. E. Moore: *Atomic Energy Levels*, NBS Circ. 467, Vol. II (Govt. Printing Office, Washington, D.C. 1971)

6.24 O. C. Mohler, A. K. Pierce, R. D. McMath, L. Goldberg: *Photometric Atlas of the Near Infrared Solar Spectrum λ8465 to λ25242*, McMath-Hulbert Observatory (Univ. of Michigan Press, Ann Arbor 1950)

6.25 E. A. Muller: In *Solar Physics*, ed. by J. N. Xanthkis (Interscience Publ., New York 1967) p.45

6.26 L. H. Aller: *The Abundance of the Elements* (Interscience Publ., New York 1961) p.130

6.27 C. E. Moore, M. G. J. Minnaert, J. Houtgast: *The Solar Spectrum 2935Å-8770Å* NBS Mon. 61 (Govt. Printing Office, Washington, D.C. 1966)

6.28 L. H. Aller: In *Stellar Atmospheres*, ed. by J. Greenstein (Univ. of Chicago Press, Chicago 1960) pp.193-210

6.29 L. Goldberg, A. K. Pierce: In *Handbuch der Physik*, Vol. 52 (Springer-Verlag, Berlin 1959) p.62

6.30 L. Goldberg, E. A. Müller, L. H. Aller: Astrophys. J. Suppl. 5, 1 (1960)

6.31 R. Cayrel, J. Jugaku, Ann. d'Astrophys. 26, 495 (1963)

6.32 M. Minnaert, G. F. W. Mulders, J. Houtgast: *Photometric Atlas of the Solar Spectrum 3332Å-8771Å* (Schnobel, Amsterdam 1960)

6.33 J. Brault, L. Testerman: "Preliminary Edition of the Kitt Peak Solar Atlas", unpublished, but available on microfilm from Kitt Peak National Observatory, P.O. Box 26732, Tucson, Arizona 85726

6.34 L. Delbouille, L. Neven, G. Roland: *Photometric Atlas of the Solar Spectrum 3000-10000Å* (Institute d'Astrophysique de l'Université de Liège 1973)

6.35 M. May, J. Richter, J. Wichelmann: Astron. Astrophys. Suppl. 18, 405 (1974)

6.36 D. C. Morton, W. H. Smith: Astrophys. J. Suppl. Ser. 233 26, 333 (1973)

6.37 H. J. Lamers: private communication

6.38 F. Praderie: private communication

6.39 P. L. Smith: Nucl. Instr. and Meth. 110, 395 (1973)

6.40 T. Andersen, G. Sørensen: Solar Phys. 38, 343 (1974)

7. Applications of Beam-Foil Spectroscopy to the Solar Ultraviolet Emission Spectrum

Leon Heroux

With 5 Figures

The temperature of the sun's visible surface is about 6000°K. If temperature is measured as a function of distance outward from the surface of the sun, it decreases to about 4500°K at the top of the photosphere near 500 km and then gradually increases in the chromosphere, which extends to about 2000 km. As illustrated in Fig.7.1, the

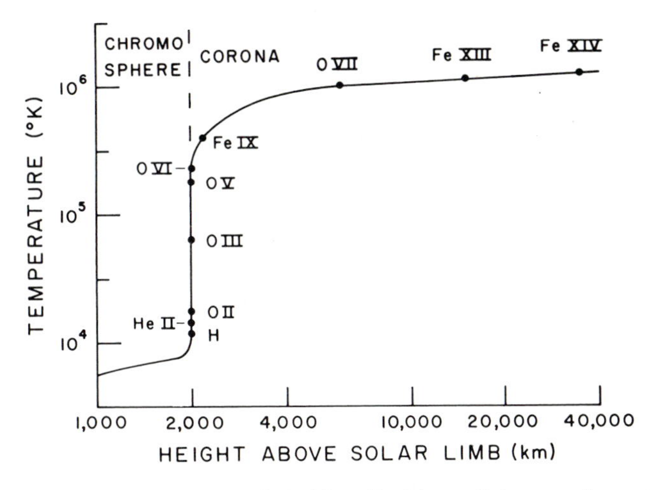

Fig.7.1 The spatial distribution of temperature in the solar chromosphere and corona. The ionization equilibrium temperatures are from JORDAN [7.3]

temperature increases abruptly from about 10^4 to 5×10^5 °K in the very narrow transition region near 2000 km, and then continues to increase gradually, finally exceeding 2×10^6 °K in the corona, which extends to large distances from the surface of the sun. The solar ultraviolet (UV) spectrum between 50 and 3000Å is produced throughout this spatial region, which encompasses a temperature range from the photospheric minimum temperature of 4500°K to the coronal temperature of 2×10^6 °K.

The solar spectrum between about 3000 and 1300Å is produced mainly in the upper photosphere and lower chromosphere. The dominant feature in this wavelength region is a continuum which becomes apparent near 1300Å and whose intensity increases toward longer wavelengths. There are numerous emission lines and absorption features superimposed on this continuum. However, with sufficiently high spectral resolution, it is possible to determine the energy distribution of this continuum as a function of wavelength and, thus, the brightness temperature of that region of the solar atmosphere emitting the radiation. The brightness temperature is established by fitting the energy distribution of the continuum to a Planck function. Because the continuum radiation originates from different temperature regions of the photosphere, the temperature deduced from the continuum intensity will depend strongly upon the wavelength interval measured.

Absorption lines are clearly evident in the solar spectrum at wavelengths longer than about 1600Å. These absorption lines originate at heights in the photosphere below that of the temperature minimum. Absorption lines do not appear in the solar spectrum at wavelengths shorter than 1600Å. Instead that part of the spectrum consists predominantly of sharp emission lines. These emission lines originate from transitions in multiply-ionized species that are produced over the extremely wide range of temperatures established in the chromosphere and corona. Figure 7.1 illustrates the temperature regions in which a few of the ion species observed in the solar spectrum are formed. The presence of the ion in the solar atmosphere is established by identifying the emission lines of the ion. This is illustrated in Fig.7.2, where multiplets in

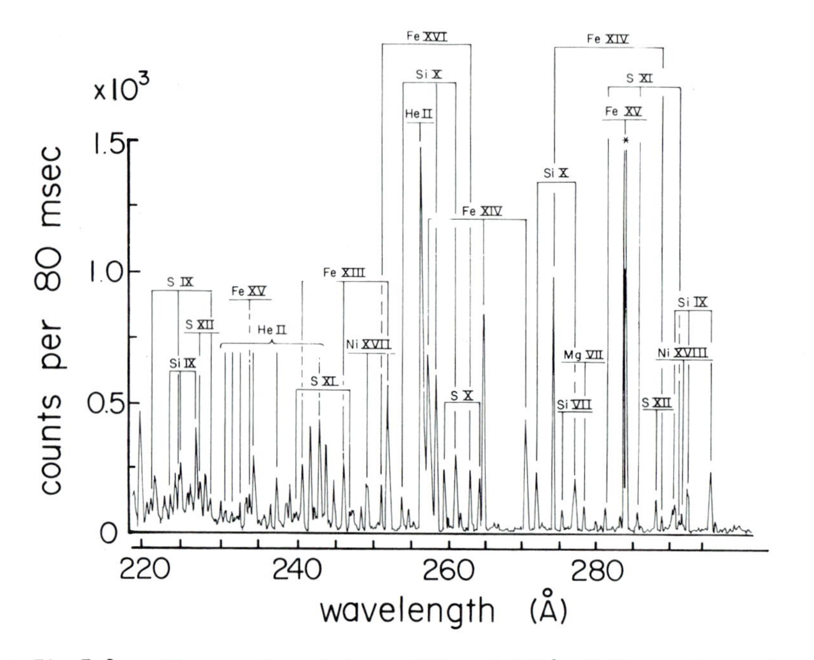

Fig.7.2 The spectrum between 220 and 290Å of the entire solar disk from MALINOVSKY and HEROUX [7.1]

the wavelength region 220 to 300Å of several ions of elements abundant in the sun are identified, according to MALINOVSKY and HEROUX [7.1]. Although this wavelength interval is small, one can see that it contains multiplets of ions produced over a wide temperature range extending from 2×10^4 to 2×10^6 °K, corresponding to the temperature region of formation of He II and Fe XVI, respectively.

A comprehensive tabulation of the wavelengths and classifications of UV spectral lines of atomic and highly-stripped atoms for the first 36 elements, from hydrogen through krypton, has been published by KELLY and PALUMBO [7.2]. This tabulation includes essentially all solar UV emission lines presently identified.

For plasmas in local thermodynamic equilibrium (LTE), the distribution of electron populations over the bound and free states of the ions and the distribution of ion species are given by the Boltzmann and Saha equations, respectively. For these equations to be valid, detailed balancing of the excitation and ionization processes must occur. Hence, processes such as collisional excitation and photoionization are balanced by the inverse processes of collisional de-excitation and radiative recombination. Neither the chromosphere nor corona is in LTE. Detailed balancing does not occur, because of the low densities in these regions of the solar atmosphere and because the plasma is optically thin. Instead, both ionization and excitation are induced only by electron impact and balanced only by radiative decay processes.

The assumption of these simple excitation and de-excitation processes allows calculation of the absolute intensities of the UV spectral lines emitted from the solar chromosphere and corona. The intensities depend upon the transition probabilities and excitation cross-sections relevant to the specific emission lines, plus the parameters of the solar plasma, such as electron temperatures, densities, and elemental abundances. If the absolute intensities of the solar UV lines are measured and the transition probabilities and cross-sections are known, one can determine these plasma parameters.

Nearly all measurements of transition probabilities for the highly ionized atoms have been made with beam-foil techniques. These measurements have strongly influenced recent advances in the determination of the physical parameters of the solar chromosphere and corona. It must be emphasized how much of our knowledge of the solar atmosphere depends on the accuracy of terrestrial laboratory experiments.

Several methods which have been used to determine the electron densities, temperatures, and elemental abundances in the solar chromosphere and corona are reviewed in this chapter. The ionization equilibrium is discussed in Section 7.1 and the excitation balance in Section 7.2. The use of experimentally-determined spectral-line intensity-ratios to determine both electron temperatures and electron densities is discussed in

Sections 7.3 and 7.4. The determination of the abundances of the elements in the solar atmosphere from a detailed analysis of the UV emission spectrum is discussed in Section 7.5. Future laboratory requirements to increase our knowledge of the solar plasma are discussed in Section 7.6.

7.1 Ionization Balance in the Chromosphere and Corona

In the solar chromosphere and corona where radiative decay processes balance ionization by electron impact, the equation for ionization balance is

$$N_{i+1}\,\alpha_{i+1} = N_i\,S_i \quad . \tag{7.1}$$

Here N_i and N_{i+1} are the ion densities for two successive stages of ionization, S_i is the electron-impact ionization coefficient for $i \to i+1$, and α_{i+1} is the radiative recombination for $i+1 \to i$. The coefficients S and α depend upon the electron temperature. The ionization balance is therefore a function of electron temperature and independent of electron density.

JORDAN [7.3] gives tables and graphs of the theoretical temperature-dependence of the ionization equilibrium of the various ionization stages of the elements abundant in the sun. The ionization equilibrium is defined by the ratio n(i)/n(E), where n(i) is the density of the ion species and n(E) the density of the element. This ratio for a particular ion peaks relatively sharply over a narrow range of temperatures. Successive stages of ionization for the same element peak at progressively higher temperatures. The temperature regions of formation of the ions that are plotted in Fig.7.1 correspond to the temperatures at which the ratios n(i)/n(E) are theoretically predicted to peak.

For the solar plasma, in which the temperature range is large, one would expect to observe emission lines from several ionization stages of a particular element. For plasmas in which the temperature is relatively uniform, one would expect to observe lines only from ions which are produced with appreciable density at the plasma temperature.

7.2 Excitation Balance in the Chromosphere and Corona

The steady-state population density of an excited level of an ion produced in a high-temperature, low-density plasma such as the solar chromosphere and corona is generally established by a balance between two processes only - electron excitation from the ground level, followed by radiative decay from the excited level. ELWERT [7.4] gives the so-called coronal-excitation equation that establishes the balance between these two processes as

$$n_p A_p = n_e n_g X(g,p) \quad . \tag{7.2}$$

In (7.2), n_p and n_g are, respectively, the population densities (cm^{-3}) of the excited level p and the ground level g, n_e is the electron density (cm^{-3}), X(g,p) the electron-excitation rate coefficient (cm^3 s^{-1}) which depends on the electron temperature T_e, and A_p is the total probability of all transitions out of level p. The left term of the equation is clearly the photon emission rate from level p, while the right term is the electron excitation rate of level p. This equation is strictly valid only if the plasma is optically thin and the ion is free of metastable levels. However, this equation can also be valid when the ion has metastable levels, provided that the excitation rate for transitions from the metastable level to level p is small in comparison to the rate from the ground level. This condition is often satisfied in the solar atmosphere because the electron density is so small. An additional assumption justifying (7.2) is that cascading transitions do not contribute significantly to the population of level p.

The excitation-rate coefficient in (7.2) is defined as

$$X(g,p) = \int_{v_{g,p}}^{\infty} Q(g,p) v\, f(v)\, dv \quad . \tag{7.3}$$

In this equation, Q(g,p) is the cross-section for excitation by electrons of velocity v, $v_{g,p}$ is the electron velocity corresponding to the threshold energy of excitation of level p, and f(v) is the Maxwellian distribution of electron velocities for temperature T_e.

The photon-emission rate ϕ (ℓ,p) in photons cm^{-3} s^{-1} for radiative transitions between the collisionally-excited level p and a lower level ℓ is obtained from (7.2) and given by

$$\phi\,(\ell,p) = n_e n_g X(g,p)\, A_{p\ell}/A_p \quad , \tag{7.4}$$

where $A_{p\ell}$ is the probability for the transition $p \rightarrow \ell$. Thus, the spectral-line intensity of an emission line of an ion in a plasma in which the coronal excitation equation is valid depends upon the parameters of the plasma and the atomic parameters of the emitting ion species. The intensity of the spectral line is a function of electron temperature, because both the ground-level density of the ion n_g and the excitation-rate coefficient X(g,p) are functions of electron temperature.

Plasma diagnostic methods that use the intensities of spectral lines to determine experimental electron temperatures and abundances are generally based on (7.4). To apply these methods, the absolute intensities of the UV lines are measured above the absorbing region of the earth's atmosphere with rocket- and satellite-borne spectrometers that are photometrically calibrated in the laboratory. The calibration of a rocket spectrometer is described in detail by HEROUX et al. [7.5]. The excitation cross-sections are calculated values in general. The calculations often rely on knowledge of the absorption oscillator strengths of the transitions between the ground levels of the ions and the upper levels of the emission lines. These oscillator strengths are directly related to the transition probabilities of the UV lines. The branching fractions $A_{p\ell}/A_p$ for the UV multiplets observed in the solar spectrum are usually identical to or close to unity. Beam-foil measurements of transition probabilities of lines of the highly-ionized atoms have contributed significantly to the appreciable amount of data now available on transition probabilities. Compilations of data on transition probabilities of many of the solar UV lines include those given by WIESE et al. [7.6,7], SMITH and WIESE [7.8], and WIESE and GLENNON [7.9].

7.3 Line-Ratio Measurements of Electron Temperature

An effective method of determining experimentally the electron temperature of a high-temperature, low-density plasma for which the coronal excitation equation is valid consists of measuring the intensity ratio of spectral lines emitted from widely-spaced energy levels in the same ion. Ions in the lithium-like isoelectronic sequence are particularly well-suited for temperature measurements because the ions have widely-spaced energy levels and are free of metastable levels. In addition, the atomic parameters of these ions are fairly accurately known.

Figure 7.3 gives the term diagram of a typical lithium-like ion and indicates some of the transitions that have been observed for several ions produced both in the sun and in laboratory plasmas. When the population distribution for the excited levels is given by (7.2), the intensity ratio for the particular transitions 2s-2p and 2s-3p, for example, is given simply by

$$I(2s\text{-}2p)/I(2s\text{-}3p) = X(2s,2p)/X(2s,3p) \quad , \tag{7.5}$$

according to (7.4). The ratios $A_{2p,2s}/A_{2p}$ and $A_{3p,2s}/A_{3p}$ do not appear in (7.5) because the former is identical to unity and the latter is very close to unity. The intensity ratio is, therefore, independent of electron and ion density and depends only on the ratio of the excitation-rate coefficients of the lines; that ratio is a function of electron temperature. It is important to form the intensity ratios from

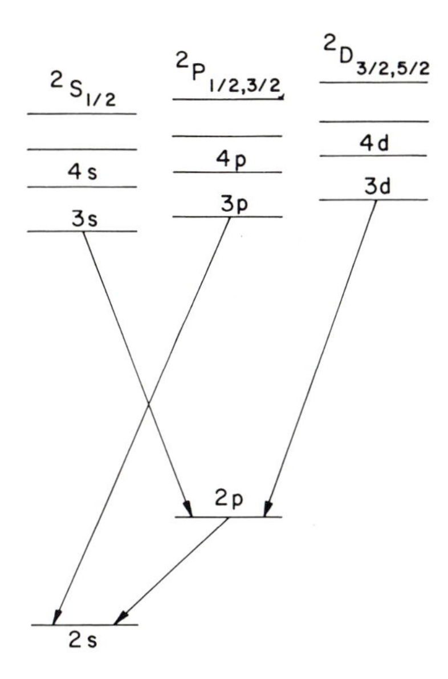

Fig.7.3 Partial term diagram for the lithium-like ions. The indicated transitions are observed in the solar spectrum

emission lines that originate from upper levels widely separated in energy to ensure that the relative populations of the levels and the corresponding relative intensity ratios are sensitive functions of electron temperature. The two other intensity ratios I(2s-2p)/I(2p-3d) and I(2s-2p)/I(2p-3s) also fulfill this requirement and are given simply by the ratios of the relevant excitation-rate coefficients.

The values of Q(g,p) that are required to evaluate X(g,p) for the collisional excitation transitions 2s → 2p, and the three transitions 2s → 3s, 3p, and 3d in the lithium-like sequence have been calculated by BURKE et al. [7.10] and by BELY [7.11,12]. With the exception of the 2s-3p transition, which may be in error by a factor of 2 near threshold, the calculated cross-sections are believed to be accurate to within 20% at threshold and to within a few percent at energies well above threshold. Since, to determine the excitation-rate coefficient, the cross-sections are averaged over a Maxwellian distribution of electron velocities, one would expect the accuracy of the rate coefficients to be somewhat better than the accuracy of the cross-sections.

Examples of the temperature dependence of the calculated excitation-rate coefficients for transitions between the 2s ground level and the 2p, 3s, 3p, and 3d levels in the lithium-like ions O VI and Mg X are given by HEROUX et al. [7.5].

The calculations of the particular intensity ratio I(2s-2p)/I(2s-3p) for several ions in the lithium-like sequence are given in Fig.7.4. All of the lithium-like ions

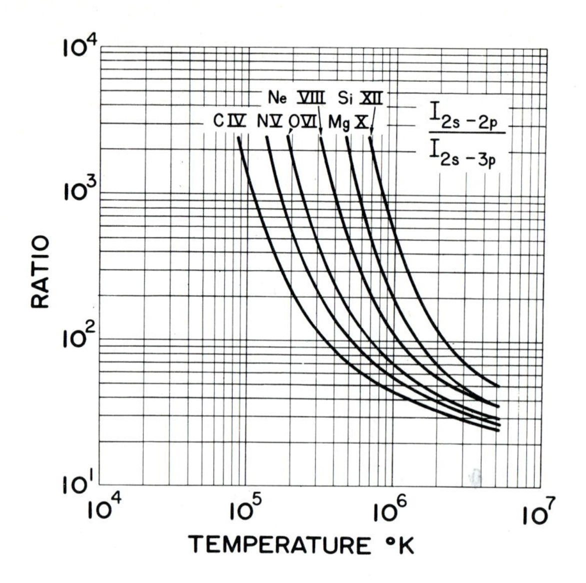

Fig.7.4 Theoretical intensity ratios for the transitions 2s-2p and 2s-3p of the lithium-like ions

given in Fig.7.4 are predicted to have an appreciable density in the temperature region where the slopes of the intensity ratios are steep. The intensity ratios of the spectral lines emitted from these regions will therefore be sensitive functions of electron temperature. As can be seen from Fig.7.4, when the emission lines of a given ion are emitted from a region of temperature where the slope is steep, a particular error in the experimental intensity ratio will be reflected as a smaller error in the experimental electron temperature.

The need for spectral lines originating from widely-spaced energy levels leads to a difficult experimental problem in the photometric calibration of the spectrometer, since such lines are widely separated in wavelength. As an example, the wavelengths of the 2s-2p and 2s-3p transitions of O VI are approximately 1030 and 150Å, respectively.

To obtain a meaningful measurement of the temperature of a plasma, both emission lines used to form the ratio must be emitted from the same temperature region of the plasma. This condition exists in many laboratory plasmas in which the electron temperature and density are relatively uniform. However, if the plasma has large gradients in density and temperature, the spectral-line intensity ratios of the lithium-like ions may not always provide a meaningful value of temperature. This difficulty occurs for particular lithium-like ions produced in the transition region of the solar atmosphere where appreciable temperature and density gradients exist. This difficulty arises because the density n(i)/n(E) of the lithium-like ion at temperatures above the peak density, although small, is in general still significant. Because of this high-temperature tail in n(i)/n(E), the intensities of the emission lines can also be appreciable at temperatures higher than the peak temperature of ion production. The high-temperature enhancement of the emission lines depends upon the excitation-rate coefficients of the lines and upon the temperature and density gradients in the region of the plasma where the tail in n(i)/n(E) is significant. For a given ion, this intensity enhancement will generally be greater for the high-excitation lines than for the low-excitation lines. Therefore, two spectral lines of the same ion may not be emitted from the same temperature region of the plasma. This effect is discussed by HEROUX et al. [7.5], where it is shown that the intensity enhancement of the 2s-3p emission line of O VI produced in the solar atmosphere is so severe that two distinct peaks in the intensity of the line appear at widely-spaced temperatures, although only one intensity peak appears for the 2s-2p line. For the spectral-line pairs of Ne VIII and Mg X, however, both the high- and low-excitation lines peak in intensity near the same temperature region. Thus, although meaningful temperature measurements cannot be made from the intensity ratio of O VI, they can be made from the intensity ratios of Ne VIII and Mg X.

The high-temperature tail in n(i)/n(E) that is a characteristic feature of ions in the lithium-like sequence is not present in many other ion sequences. The quantity for the beryllium-like ions (n(i)/n(E) vs. T_e), for example, falls off rapidly at temperatures beyond that of the peak ion density. Figure 7.5 shows a partial-term diagram

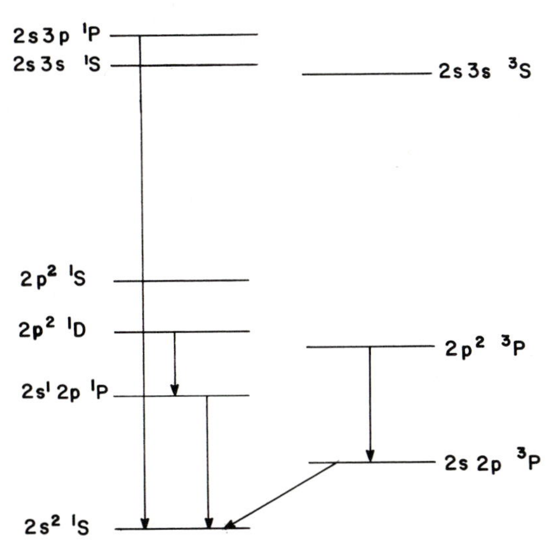

Fig.7.5 Partial term diagram for the beryllium-like ions. The indicated transitions are observed in the solar spectrum

of a typical beryllium-like ion. For ions in this sequence, the different spectral lines originating from widely-spaced levels are emitted from essentially the same temperature region of the plasma, even when the ion is produced in plasmas in which the temperature and density gradients are large.

As can be seen from this figure, the beryllium-like ions have a metastable level ($2s2p\ ^3P$). Therefore, the singlet-series levels $2s2p\ ^1P$ and $2s3p\ ^1P$ generally will be populated by collisional excitations from both the $2s^2\ ^1S$ ground level and the metastable level. However, if the plasma density is sufficiently low, the steady-state population of the metastable level will also be low in comparison to the population of the ground level, and collisional excitations from the metastable level will not contribute significantly to the populations of the singlet-series levels. For this condition, the coronal-excitation equation can be applied to determine the populations of the $2s2p\ ^1P$ and $2s3p\ ^1P$ levels. The spectral-line intensity ratio $I(2s^2\ ^1S\text{-}2s2p\ ^1P)/I(2s^2\ ^1S\text{-}2s3p\ ^1P)$, therefore, will be given by the ratio of the excitation-rate coefficients corresponding to the collisionally-induced transitions $2s^2\ ^1S \rightarrow 2s2p\ ^1P$ and $2s^2\ ^1S \rightarrow 2s3p\ ^1P$. Because the two excited levels have a large energy difference, the ratios are sensitive functions of temperature. MALINOVSKY et al. [7.13] have used this ratio for the beryllium-like ion O V to determine the electron temperature in the chromospheric-coronal transition region of the sun.

In the region of the solar atmosphere where O V is produced, the electron density is less than 10^{11} cm^{-3} and the plasma is optically thin. Under these conditions, the coronal-excitation equation is valid, and electron temperatures can be obtained from line-ratio measurements. Laboratory plasmas, however, generally have much higher electron densities, so that the presence of the metastable level in the beryllium-like ions severely limits the use of these ions for measurements of electron temperatures.

The experimental temperature determined from the emission-line intensity-ratio of a given ion will necessarily fall within the restricted temperature region in which the ion is produced with appreciable density. Line ratios of different ion species, therefore, must be used to measure a wide range of plasma temperatures. The dependence of the ion density on temperature does not have to be known to determine the plasma temperature, since the density cancels when the intensity ratio of lines from the same ion is formed. For the solar plasma, in which the temperature range is extremely large, the experimental temperature obtained from the intensity ratio of a particular ion gives the average temperature of the region in which the line pair predominates. This temperature, when compared to the temperature region of maximum emission of the spectral lines predicted by the theoretical ionization equilibrium $n(i)/n(E)$ vs. T_e, thus provides a test of the ionization-equilibrium calculations.

7.4 Line-Ratio Measurements of Electron Density

It has been mentioned in Section 7.2 that the presence of a metastable level in an ion can invalidate the use of the coronal-excitation equation for determining the population densities of the excited levels. When the coronal equation breaks down for any of the levels used to form intensity ratios, the ratios will be a function of electron density as well as electron temperature. Although a density-dependent ratio is no longer useful for determining electron temperatures, it does provide a method of measuring electron densities in the region of the plasma from which the spectral lines are emitted. The method of determining electron densities from line intensity ratios of beryllium-like ions has been developed independently by MUNRO et al. [7.14] and JORDAN [7.15]. The populations of the different levels of the ion sequence are computed as a function of electron density and temperature by solving the equations of statistical equilibrium for each level. This equation balances all processes which populate and depopulate a given level. These processes include collisional excitation and de-excitation and radiative transitions into and out of the level. The population densities of several of the levels of the beryllium-like ion shown in Fig.7.5 are given by GABRIEL and JORDAN [7.16] and MUNRO et al. [7.14]. From these calculated densities, several intensity ratios can be formed between triplet-series and singlet-series lines that are functions of electron density and temperature. The ratio of the intensity of the intercombination line $I(^1S-^3P_1)$ to that of the resonance line $I(^1S-^1P)$ is also sensitive to density and temperature. Since the ratios depend on both electron density and temperature, an independent determination of temperature must be made to determine electron density. As has been shown in Section 7.3, the electron temperature can be determined from the line ratio of singlet-series transitions in low-density plasmas. Thus, a simultaneous measurement of electron temperature and density can be made from line-ratio measurements in the same beryllium-like ion.

Transitions from metastable levels of the helium-like ions of the abundant elements between carbon and iron are also observed in the solar spectrum (JORDAN [7.17]). GABRIEL and JORDAN [7.18] have developed a theory for the intensity ratios of these ions in which the ratio of the forbidden line $1s^2\ {}^1S-1s2s\ {}^3S$ to that of the intercombination line $1s^2\ {}^1S-1s2p\ {}^3P$ is derived as a function of collisional-excitation rates, the transition probabilities, and the electron density. This ratio for the helium-like ions is essentially independent of electron temperature. Therefore, a measurement of the intensity ratio leads to a determination of electron densities in the high-temperature region of the solar corona where the helium-like ions are formed. The wavelength range for the emission lines of the helium-like ions observed in the solar spectrum extends from about 40Å for C V to about 2Å for Fe XXV. However, the wavelength spread of the line pairs for one ion is typically less than 1Å. A photometric calibration of the spectrometer used to measure the line ratios is not required, since one can assume that spectrometer efficiency will be constant for the small wavelength interval.

The first measurements of the transition probabilities of the forbidden and intercombination lines of helium-like ions were obtained with beam-foil techniques. These measurements are $A(^3P_1 \to {}^1S_0)$ for O VII, by SELLIN et al. [7.19], $A(^3P_2 \to {}^1S_0)$ for Ar XVII, by MARRUS and SCHMIEDER [7.20], and $A(^3S_1 \to {}^1S_0)$ for Ar XVII, by MARRUS and SCHMIEDER [7.21].

7.5 The Determination of Chromospheric-Coronal Abundances

The determination of the relative abundances of the elements in the sun from the analysis of the solar UV spectrum is based on the method developed by POTTASCH [7.22]. To apply this method, the absolute intensities of the spectral lines of the highly-ionized atoms emitted from the chromosphere and corona measured above the absorbing region of the earth's atmosphere must be known. The spectral lines used in the analysis are restricted to those lines that originate from levels populated by allowed (dipole) transitions from the ground level of the ion. For these allowed transitions, the excitation-rate coefficients can be expressed approximately in terms of the oscillator strengths of the transitions. If the population density of the excited level p is given by the coronal-excitation equation, the intensity $I(\ell,p)$ in photons $cm^{-2}\ s^{-1}$ emitted by the whole sun and measured at the earth above the absorbing region of the earth's atmosphere is given by

$$I(\ell,p) = \left(R_0^{\ 2}/2D^2\right)\left(A_{p\ell}/A_p\right) \int n_e n_g X(g,p)\, dh \quad . \tag{7.6}$$

In this equation, R_0 is the solar radius, D is the earth-sun distance, and the incremental height dh is a substitution for the emitting volume element dv in the solar atmosphere obtained from the relation $dh = dv/4\pi R_0^{\ 2}$. The factor 1/2 in (7.6) assumes that half of the radiation is directed outward from the sun and half directed inward.

When (7.6) is modified for the solar atmosphere, it becomes

$$I(\ell,p) = 0.4 \left(\frac{R_0}{D}\right)^2 \frac{A_{p\ell}}{A_p} \frac{n_g}{n(i)} \frac{n(E)}{n(H)} \int \frac{n(i)}{n(E)} X(g,p)\, n_e^{\ 2}\, dh \quad , \tag{7.7}$$

where the number density of the element n(E) and of hydrogen n(H) have been introduced as well as the relation $n(H) = 0.8\ n_e$. The latter relation assumes that the solar atmosphere is almost completely ionized and that hydrogen accounts for 80% of the total number density of all elements in the solar atmosphere. In (7.7), the fraction of the ion in the ground level $n_g/n(i)$ and the abundance of the element relative to hydrogen $A_E = n(E)/n(H)$ are assumed to be constant over the region where the ion is formed and, therefore, are taken outside the integral. The ionization equilibrium $n(i)/n(E)$, the rate coefficient X, and the electron density depend upon the temperature and, therefore, height in the solar atmosphere. The excitation-rate coefficient for the optically-allowed collisional transition $g \to p$ can be expressed in terms of the absorption oscillator strength f_{gp}, an effective Gaunt factor $\bar{g}$, and the excitation energy ΔE_{gp} of the transition. If the spectral-line intensity is given in ergs $cm^{-2}\ s^{-1}$, the intensity $I(\ell,p)$ is given by

$$I(\ell,p) = 2.4\times10^{-20}\ (A_{p\ell}/A_p) f_{gp}\bar{g}\ \left[n_g/n(i)\right]\ A_E \int \frac{\exp(-\Delta E_{gp}/kT)}{T^{1/2}}\ n_e^2\ dh \quad . \qquad (7.8)$$

The Pottasch method for solving (7.8) assumes that the temperature-dependent quantity

$$G(T) = \frac{\exp(-\Delta E_{gp}/kT)}{T^{1/2}}\ \frac{n(i)}{n(E)} \qquad (7.9)$$

has a maximum value $[G(T)]_{max}$ at the temperature T_{max} and that the spectral line is emitted predominantly in a narrow region R near T_{max}. An average constant value of $G(T)$ defined as $\langle G\rangle = 0.7\ [G(T)]_{max}$ is, therefore, substituted for $G(T)$ in (7.9), and (7.8) becomes

$$I(\ell,p) = 2.4\times10^{-20}\ (A_{p\ell}/A_p)\ f_{gp}\ \bar{g}\ n_g/n(i)\ \langle G\rangle\ A_E \int_R n_e^2\ dh \quad . \qquad (7.10)$$

Thus, by using the experimental line intensity and the quantities $A_{p\ell}/A_p$, f_{gp}, $\bar{g}$, $n_g/n(i)$, and $\langle G\rangle$, one can compute the quantity $A_E\int_R n_e^2\ dh$ for the spectral line. This quantity is expected to be the same for the different spectral lines of a given ion, since the different lines are emitted near the same temperature in the solar atmosphere. When the quantity $A_E\int_R n_e^2\ dh$ obtained for spectral lines of different ions of the same element is plotted as a function of the temperature region of emission of the lines, one obtains a curve of $A_E\int_R n_e^2\ dh$ vs. T for that element. Similarly constructed curves for different elements are displaced from each other along the direction of the axis of $A_E\int_R n_e^2\ dh$, and the relative displacements reflect the relative abundance of the elements.

The determination of elemental abundances from the intensity analysis of the solar UV spectrum has been carried out by several others since the original determination by POTTASCH [7.22]. In addition to the measurements of POTTASCH, abundance determinations from the intensities of solar UV lines produced in the chromosphere and corona have been made by JORDAN [7.23], DUPREE [7.24], MALINOVSKY and HEROUX [7.1], WALKER et al. [7.25,26], and ACTON et al. [7.27]. In a review article, WITHBROE [7.28] has shown that the general agreement between the abundances obtained from chromospheric-coronal radiation and those obtained from photospheric radiation is fairly good. There are, however, some significant discrepancies among the different measurements of the chromospheric-coronal abundances which are introduced primarily by errors in the measurements of the spectral-line intensities and determinations of the atomic parameters.

An important result of POTTASCH's original analysis was that the chromospheric-coronal abundance ratio A_{Fe}/A_{Si} was greater by a factor of about 10 than the generally-accepted value of that ratio obtained from photospheric radiation (GOLDBERG et al. [7.29]). A subsequent re-determination of the Fe I and Fe II oscillator strengths led to a revision of the photospheric iron abundances (GARZ et al. [7.30]) so that the photospheric and the chromospheric-coronal abundance ratio Fe/Si are now in good agreement.

7.6 Beam-Foil Measurements Needed for Diagnostic Methods

Since the first recording of the near-ultraviolet solar spectrum in October 1946 by Richard Tousey and his colleagues at the Naval Research Laboratory, the solar spectrum between 1 and 3000Å emitted from both the full solar disk and from active regions on the disk has been investigated with a variety of spectrometers flown on sounding rockets and satellites. These experiments have provided data on the solar UV spectrum with relatively high spectral resolution and data on the parameters of the solar plasma.

Future experiments in solar physics will certainly include detailed studies of the physical properties of solar active regions and solar flares. The studies will rely on the use of solar UV diagnostic techniques for the improvement of abundance determinations, detection of possible departures from ionization equilibrium, and determination of the temporal and spatial variations of electron temperatures and densities.

Abundance determinations based on the intensity analysis of the solar UV spectrum rely on accurate values of absorption oscillator strengths of the allowed transitions in highly-ionized atoms. Nearly all oscillator strengths of the higher ions of iron, nickel, silicon, and sulfur that are now being used in these abundance determinations are theoretical values. Beam-foil measurements of radiative lifetimes of the levels of these ions can provide experimental data on absorption oscillator strengths and, therefore, verify these theoretical values. Accurate measurements of transition probabilities also provide tests of the values of theoretical atomic parameters used in the calculation of excitation cross-sections required for the determination of electron temperatures and densities.

The need for accurate measurements of transition probabilities of intercombination lines originating from metastable levels of multiply-ionized atoms has been discussed by JORDAN [7.17]. These transition probabilities are of particular importance for electron-density determinations. Many of the metastable levels of the highly-stripped ions are expected to be short enough to be measured with beam-foil techniques.

Not all the solar UV emission lines have been identified. Many of the unidentified lines probably arise from transitions from metastable and doubly-excited levels of highly-stripped ions. Beam-foil measurements of the radiative lifetimes and energies of these types of transitions could contribute to the identification of many of the solar emission lines.

In conclusion, beam-foil spectroscopy has proved to be an important laboratory tool in understanding the solar atmosphere. Because of its potential use in providing additional atomic data on many of the highly-stripped ions abundant in the sun, beam-foil spectroscopy can be expected to continue its importance in the investigation of the solar atmosphere.

References

7.1 M. Malinovsky, L. Heroux: Ap. J. 181, 1009 (1973)

7.2 R. L. Kelly, L. J. Palumbo: *Atomic and Ionic Emission Lines below 2000 Angstroms - Hydrogen through Krypton* (U.S. Govt. Printing Offc, Washington, D.C. 1973)

7.3 C. Jordan: Monthly Notices Roy. Astron. Soc. 142, 501 (1969)

7.4 G. Elwert: Z. Naturforsch. 7a, 432 (1952)

7.5 L. Heroux, M. Cohen, M. Malinovsky: Solar Phys. 23, 369 (1972)

7.6 W. L. Wiese, M. W. Smith, B. M. Glennon: *Atomic Transition Probabilities*, Vol. 1 (U.S.Govt. Printing Offc, Washington, D.C. 1966)

7.7 W. L. Wiese, M. W. Smith, B. M. Miles: *Atomic Transition Probabilities*, Vol. 2 (U.S.Govt. Printing Offc., Washington, D.C. 1969)

7.8 M. W. Smith, W. L. Wiese: Ap. J. Suppl. 23, 103 (1971)

7.9 W. L. Wiese, B. M. Glennon: *American Institute of Physics Handbook*, 3rd. ed. (McGraw-Hill Book Company, Inc., New York 1972) pp. 7.200-7.263

7.10 P. G. Burke, J. H. Tait, B. A. Lewis: Proc. Phys. Soc. 87, 209 (1966)

7.11 O. Bely: Proc. Phys. Soc. 88, 587 (1966)

7.12 O. Bely: Ann. Astrophys. 29, 683 (1966)

7.13 M. Malinovsky, L. Heroux, S. Sahal-Brechot: Astron & Astrophys. 23, 291 (1973)

7.14 R. H. Munro, A. K. Dupree, G. L. Withbroe: Solar Phys. 19, 347 (1971)

7.15 C. Jordan: In *Highlights in Astronomy*, ed. by C. de Jager (D. Reidel Publishing Co., Dordrecht 1971) pp. 519-526

7.16 A. H. Gabriel, C. Jordan: In *Case Studies in Atomic Collision Physics 2*, ed. by E. W. McDaniel, M. R. C. McDowell (North-Holland Publishing Co., Amsterdam 1972) pp. 209-291

7.17 C. Jordan: In *Beam-Foil Spectroscopy*, ed. by S. Bashkin (North-Holland Publishing Co., Amsterdam 1973) pp. 373-379

7.18 A. H. Gabriel, C. Jordan: Monthly Notices Roy. Astron. Soc. 145, 241 (1969)

7.19 I. A. Sellin, M. Brown, W. W. Smith, B. Donnally: Phys. Rev. A 2, 1189 (1970)

7.20 R. Marrus, R. W. Schmieder: Phys. Rev. Letters 25, 1689 (1970)

7.21 R. Marrus, R. W. Schmieder: Phys. Rev. Letters 25, 1245 (1970)

7.22 S. R. Pottasch: Space Sci. Rev. 3, 816 (1964)

7.23 C. Jordan: Monthly Notices Roy. Astron. Soc. 132, 463 (1966)

7.24 A. K. Dupree: Ap. J. 178, 527 (1972)

7.25 A. B. C. Walker, H. R. Rugge, K. Weiss: Ap. J. 188, 423 (1974)

7.26 A. B. C. Walker, H. R. Rugge, K. Weiss: Ap. J. 192, 169 (1974)

7.27 L. W. Acton, R. C. Catura, E. G. Joki: Ap. J. 195, L93 (1975)

7.28 G. L. Withbroe: In *The Menzel Symposium on Solar Physics, Atomic Spectra, and Gaseous Nebulae*, ed. by K. B. Gebbie, NBS Spec. Publ. 353 (U.S.Govt. Printing Offc, Washington, D.C. 1971) p. 127

7.29 L. Goldberg, E. A. Muller, L. H. Aller: Ap. J. Suppl. 5, no. 45 (1960)

7.30 T. Garz, H. Holweger, M. Kock, J. Richter: Astron. & Astrophys. 2, 446 (1969)

8. Studies of Hydrogen-Like and Helium-Like Ions of High Z

Richard Marrus

With 15 Figures

When heavy-ion beams emerging from an accelerator are passed through a thin foil or gas, some fraction of them emerges in excited states of the one- and two-electron systems. For an ion of fixed Z, the fraction of the ions in the one- and two-electron charge states is determined mainly by the velocity of the incident beam. As an example of what is possible, quantities sufficient for experimentation have recently been observed of one- and two-electron krypton (Z=36) with the 8.5 MeV/amu krypton beam at the Berkeley super-HILAC. Useful beams of perhaps even higher Z should be available at the Darmstadt UNILAC when it becomes operative. Moreover, beams of one- and two-electron ions of useful intensity are almost routinely available up through about $Z \simeq 20$ at many of the lower-energy heavy-ion facilities used throughout the world.

Use of these beams has led to several interesting and important spectroscopic measurements during the last several years. In this chapter we attempt to review some of these developments, particularly: measurement of the Lamb shift in the one-electron system (Section 8.1); measurement of the Lamb shift in the two-electron system (Section 8.2); radiative decay of the metastable state of the one-electron system (Section 8.3); forbidden radiative decay in the n=2 state of the two-electron system (Section 8.4); study of doubly-excited configurations in the two-electron system (Section 8.5). All of the above problems are currently under active study by both experimental and theoretical workers.

8.1 The Lamb Shift in the One-Electron System

One of the basic predictions of the Dirac theory of an electron in a pure Coulomb field is the exact degeneracy of the $2S_{1/2}$ and $2P_{1/2}$ energy levels. Although this degeneracy is lifted by such effects as proton finite size, recoil corrections, etc., the measurement by LAMB and coworkers [8.1] of an energy separation of 1057.77(6) MHz is many orders of magnitude too large to be explained by these effects. Quantum electrodynamic (QED) effects, particularly the electron self-energy and the vacuum polarization, must be considered to explain the observed separation adequately.

Because it is one of the few direct tests of QED, and because measurements of high precision are possible, both the theory and measurement of the Lamb shift have been subjects of intense activity among physicists. The theory of the Lamb shift has been worked out [8.2] in terms of an expansion in the parameter $(Z\alpha)$, where α is the fine-structure constant. Results of the $Z\alpha$ expansion are shown in Table 8.1.

Table 8.1 Contributions to the $Z\alpha$ expansion of the Lamb Shift (Based on [8.3])

Description	Order	Magnitude [MHz]
2nd Order (Self-Energy)	$\alpha(Z\alpha)^4 m \{\log(Z\alpha),1\}$	1079.32 ± 0.02
2nd Order (Vacuum Polarization)	$\alpha(Z\alpha)^4 m$	-27.13
2nd Order (Remainder)	$\alpha(Z\alpha)^5 m$	7.14
	$\alpha(Z\alpha)^6 m \{\log^2(Z\alpha),\log(Z\alpha),1\}$	-0.38 ± 0.04
4th Order (Self-Energy)	$\alpha^2(Z\alpha)^4 m$	0.35 ± 0.07
	$\alpha^2(Z\alpha)^5 m$	± 0.02
4th Order (Vacuum Polarization)	$\alpha^2(Z\alpha)^4 m$	-0.24
Reduced Mass Corrections	$\alpha(Z\alpha)^4 \frac{m}{M} m \{\log(Z\alpha),1\}$	-1.64
Recoil	$(Z\alpha)^5 \frac{m}{M} m \{\log(Z\alpha),1\}$	0.36 ± 0.01
Proton Size	$(Z\alpha)^4 (mR_N)^2 m$	0.13
	$\Delta E(2S_{1/2} - 2P_{1/2}) =$	1057.91 ± 0.16

$\alpha^{-1} = 137.03608(26)$

In Table 8.1, m is the electron mass, M the mass of the atom, R_N the nuclear radius, and { } means terms of the order of the enclosed symbols.. More recently, ERICKSON [8.4] and MOHR [8.5,6] have calculated the electron self-energy to all orders in the parameter $(Z\alpha)$. A comparison of the n=2 Lamb shift as calculated by the various methods is shown in Table 8.2. Interestingly, there appears to be a small but identifiable discrepancy between ERICKSON's and MOHR's calculations which is presumably calculational in origin and whose magnitude in hydrogen is 0.047(14) MHz. So far, experiments are of insufficient accuracy to test the discrepancy.

Up until recently, the best values for the Lamb shift in the n=2 state of hydrogen were based on measurements made by radiofrequency or level-crossing spectroscopy on a thermal beam [8.7]. Other radiofrequency measurements [8.8] have been made in the n=2 state of Li^{2+} by LEVENTHAL and HAVEY [8.9,10]. These measurements employ ion beams excited by electron bombardment and the results are summarized in Table 8.3.

Table 8.2 Lamb shift $E(2S_{1/2})-E(2P_{1/2})$ in hydrogen-like ions [GHz]

Z	Series[a]	Erickson[b]	Mohr[c]
1	1.057900	1.057910(10)	1.057864(14)
2	14.04391	14.04464(54)	14.04188(80)
3	62.753	62.762(9)	62.732(10)
10	4869	4889(5)	4859(5)
20	5.46×10^4	$5.64(3) \times 10^4$	$5.512(5) \times 10^4$
30	2.07×10^5	$2.34(3) \times 10^5$	$2.230(5) \times 10^5$
40	4.6×10^5	$6.4(2) \times 10^5$	$6.00(3) \times 10^5$
50	6.0×10^5	$1.42(7) \times 10^6$	$1.31(2) \times 10^6$
60	-8.1×10^4	$2.7(2) \times 10^6$	$2.54(5) \times 10^6$

[a] Values obtained by evaluating the terms included in [Ref.8.2,Table 2]
[b] Ref.[8.4], and private communication
[c] Ref.[8.5,6], and private communication. These values are based on preliminary estimates of certain contributions.

Table 8.3 Measurements of the n=2 Lamb shift in hydrogenic ions [Z > 2)

Ion	Δ_L^{exp}[GHz]		Δ_L^{theory}[GHz]
He^+	14.0402(18)	[8.11]	14.0445(52)
He^+	14.0454(12)	[8.12]	
Li^{2+}	63.031(327)	[8.13]	62.762(9)
Li^{2+}	62.765(21)	[8.9,10]	
C^{5+}	780.1(8.0)	[8.14]	783.68(25)
O^{7+}	2216(8)	[8.15]	2205.2(16)
O^{7+}	2203(11)	[8.16]	

Fast ion beams have recently been employed to study the Lamb shift from two different points of view. In both, accelerator-produced heavy ions are stripped to the one-electron charge state and excited to the n=2 state by the beam-foil or beam-gas method. In one approach, the metastable state is quenched in a motional electric field produced by passing the beam through a magnetic field. These experiments have resulted in measurements of the Lamb shift through Z=8 with an accuracy of about 1%. LUNDEEN and PIPKIN [8.17] have produced a beam of fast hydrogen atoms in the n=2 state by passing 50-100 keV protons through a charge-exchange cell filled with nitrogen. In their approach, they employed the method of separated oscillatory fields [8.18] to do radiofrequency resonance, and have succeeded in producing line-widths smaller than the natural linewidth. This experiment, currently in progress, has already improved the accuracy of the earlier measurements by a factor three.

8.1.1 Quenching Measurements on Fast Ion Beams of High Z

The Lamb shift in the n=2 state of the one-electron system scales almost as fast as Z^4. Because of the rapid increase in the energy splittings among the n=2 levels, radiofrequency or level-crossing spectroscopy rapidly becomes unfeasible beyond Z=3, and so far only the inherently less-accurate quench method has been successfully used for higher Z. The quench method is based on the smallness of the Lamb shift and the widely differing transition rates for electromagnetic decay of the $2S_{1/2}$ and the $2P_{1/2}$ states. The unperturbed $2S_{1/2}$ state decays to the $1S_{1/2}$ ground state primarily by two-photon emission (see Section 8.3) with decay rate $A(2S_{1/2}) = 8.23\ Z^6 s^{-1}$, but the $2P_{1/2}$ state decays by single electric-dipole emission with rate $A(2P_{1/2}) = 6\times10^8\ Z^4 s^{-1}$. In the presence of an external electric field E, Stark mixing of the states occurs and the transition rate A(E) in the presence of the electric field is given by the Bethe-Lamb formula [8.19]

$$A(E) = A(2P_{1/2}) \frac{|\langle 2P_{1/2}|e\bar{r}\cdot\bar{E}|2S_{1/2}\rangle|^2}{\hbar^2[\Delta_L^2 + \frac{1}{4} A^2(2P_{1/2})]} + A(2S_{1/2}) \quad . \tag{8.1}$$

Hence, measurement of the transition rate in an external electric field can provide a measurement of Δ_L, the Lamb shift. Corrections to the Bethe-Lamb formula arise from the presence of other P states mixed by the electric-dipole operator.

The quench method was first employed on a heavy-ion beam by FAN et al. [8.13] to study the Lamb shift in Li^{2+}. A schematic of the apparatus used in their experiment is shown in Fig.8.1. The Li^+ beam from a Van de Graaff generator is passed through a nitrogen-filled charge-exchange cell. About 29% of the beam emerging from the cell is in the one-electron $2S_{1/2}$ state. These ions are passed into a quench region consisting of a pair of electric field plates. Photons emitted in the region between the plates are viewed by a movable and a fixed counter. The fixed counter acts as a normalization detector and a decay curve is taken by varying the separation between the two counters. From the measured decay length and the known beam velocity, the transition rate A(E) can be determined as a function of the electric field. The extension of the quench method to higher Z has been achieved by MURNICK et al. [8.14], who used the method to measure Δ_L in C^{+5}, and by LAWRENCE et al. [8.15], and LEVENTHAL et al. [8.16] in O^{+7}. The results of these experiments are indicated in Table 8.3. These

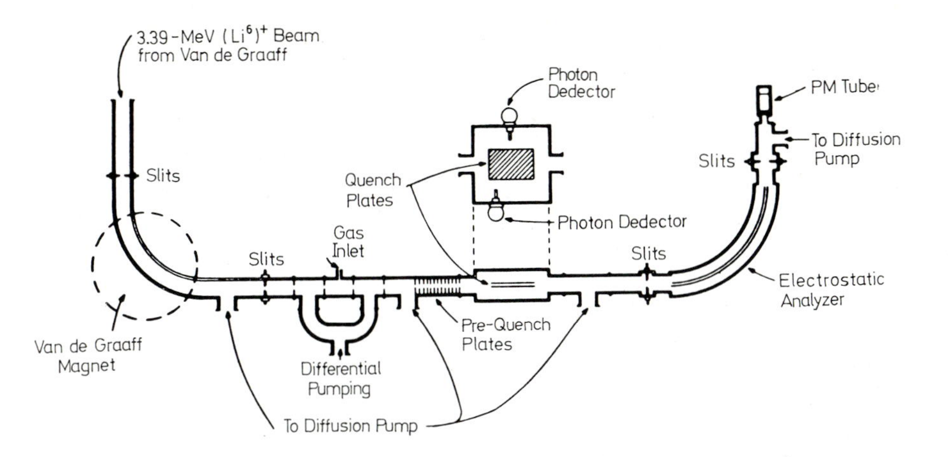

Fig.8.1 Schematic of apparatus for quench measurement of Li^{2+} Lamb shift [8.13]

later experiments have used an important modification of the original quench technique. Instead of employing an electric field, they employed a magnetic field B applied transverse to the beam, so that the atom experienced a motional electric field E in its rest frame given by

$$\overline{E} = \frac{1}{[1-(v/c)^2]^{1/2}} \frac{\overline{v}}{c} \times \overline{B} \quad . \tag{8.2}$$

There are two advantages associated with this scheme. Unlike electric fields, magnetic fields can be measured conveniently and accurately. Since the velocity of ions from a Van de Graaff accelerator is also usually known to high accuracy, the motional electric field is also accurately known. Moreover, with ion beams from an accelerator such that $v/c > 0.1$, magnetic fields are easily made corresponding to electric fields approaching 10^6 volts/cm. Stable electric fields of this magnitude are difficult to make directly.

The accuracy of the quenching experiments depends on several factors. Among the more important are (i) deflection effects on the moving beam in the magnetic field, for these alter the detection geometry, and (ii) background radiation which can arise both from two-photon counts from the metastable level and from radiative transitions from the n=2 state of any helium-like ions present in the beam. Background is also generated by nuclear reactions in the beam stop; neutrons and gamma rays may be abundant. In spite of these and other difficulties, accuracies of 1/2%-1% have been achieved.

The prospects for extending quenching measurements to higher Z seem reasonably good. Among the problems arising at higher Z are the following: (i) The electric field needed for a given fractional quenching increases rapidly with Z. This results because the Lamb shift increases rapidly with Z, and because the matrix element of the dipole operator $\bar{r}$ scales like Z^{-1}. (ii) The increased fields needed to observe the quenching give rise to increased beam deflections. Hence geometrical corrections become more significant. (iii) The lifetime of the metastable state is rapidly decreasing with Z. In spite of these difficulties, however, it appears likely that the quench method can extend knowledge of the Lamb shift to higher Z with the ultimate practical limit probably occurring at around $Z \simeq 25$.

8.1.2 Lamb Shift in Hydrogen Using Separated Oscillating Fields

The first significant improvement, by PIPKIN and his colleagues, in the accuracy of the Lamb shift in the n=2 state of hydrogen over Lamb's original result has recently been described by LUNDEEN [8.20]. A schematic diagram of their apparatus is shown in Fig.8.2.

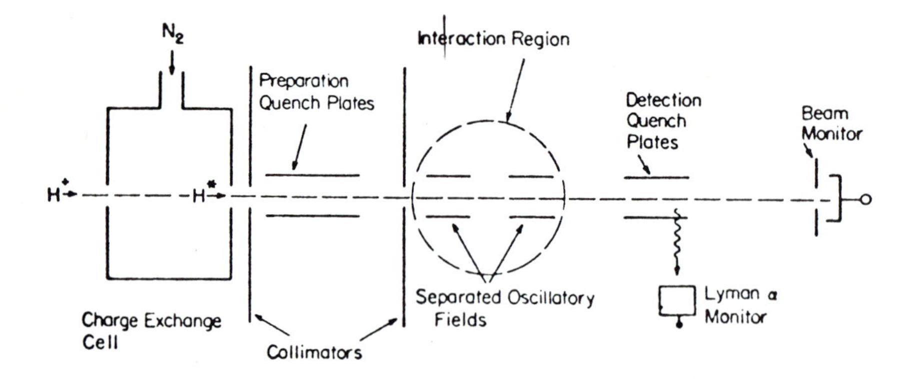

Fig.8.2 Schematic diagram of apparatus for measurement of Lamb shift in hydrogen, using separated oscillatory fields

A fast proton beam (50-100 keV) is passed through a charge-exchange cell of nitrogen gas. Some of the emerging hydrogen atoms are in the $2S_{1/2}$ state and are caused by a pair of separated oscillatory fields to undergo a radiofrequency transition to the $2P_{1/2}$ state. The transition, which is resonant for the right rf frequency, is monitored by counting Lyman-alpha quench photons emitted in the region of the detection quench plates. The improvement in the linewidth over previous experiments and over the natural width is shown in the results illustrated in Fig.8.3. Because of this improvement, a new value of the Lamb shift of 1057.893(20) MHz has been obtained [8.20]. The error of 0.02 MHz quoted here is a factor three smaller than the previous errors of 0.06 MHz quoted by TRIEBWASSER et al. [8.1], and ROBISCOE and SHYN [8.21]. This experiment is continuing and an ultimate improvement of an order of magnitude in the error is envisioned [8.20].

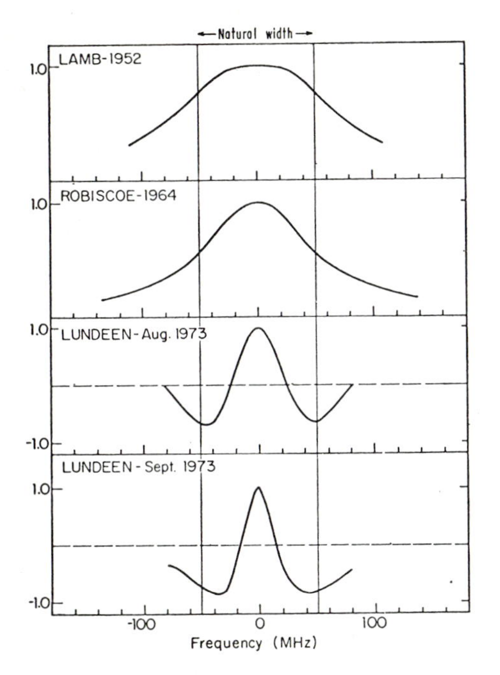

Fig.8.3 Comparison of line widths in various experiments to measure the hydrogen n=2 Lamb shift

8.2 Lamb Shift in Two-Electron Systems

The difficulty of measuring radiative corrections in two-electron systems lies in the fact that they, unlike hydrogenic systems, don't have a convenient pair of levels which are degenerate in the absence of radiative corrections. Hence, Lamb-shift values can only be inferred from very accurate measurements after subtracting theoretical values of Coulombic and relativistic effects. ACCAD et al. [8.22] in 1971 reported highly accurate calculations of energies for transitions to the $1\ ^1S_0$ ground state and to excited S and P states for the helium isoelectronic sequence up to Z=10. When compared with measurements [8.23-27], agreement could only be obtained if values for the Lamb shift were included. Because of its relative smallness, however, the value of the Lamb shifts thus obtained is accurate only to about 10%. More recently, BERRY and BACIS [8.28] made an accurate measurement of the $2\ ^3P_1 \rightarrow 2\ ^3S_1$ energy separation in Li^+ and deduced from it a Lamb-shift energy of $1.274(15)\mathrm{cm}^{-1}$. From this result and the theoretical value for the Lamb shift in the $2\ ^3S_1$ state they were able to infer an appreciable Lamb shift (about 20%) in the $2\ ^3P_1$ state. BERRY and SCHECTMAN [8.29] were the first to apply the beam-foil method to the Lamb-shift problem. They measured the energies of the transitions $3\ ^3P_{0,1,2} \rightarrow 3\ ^3S_1$ in N^{+5}. By comparing their results with calculations of ERMOLAEV [8.30,31], confirmation was obtained for a large P-state Lamb shift.

It is of some interest to note that for two-electron ions of high Z, the Lamb shift in the transition $2\ ^3P_0 \rightarrow 2\ ^3S_1$ rapidly becomes a larger part of the transition energy as Z increases. The energy associated with this transition is determined by the Coulomb energy, fine-structure effects and the Lamb shift. However, in the hydrogenic approximation, the Coulomb energy of the states is the same. The Coulomb energy splitting arises only from the electrostatic repulsion of the electrons, e^2/r_{ij}. This term scales approximately as Z for the He I isoelectronic sequence. Consider now the energy splitting arising from fine-structure effects. Note that the $2\ ^3P_0$ state can arise only from the coupling of a $2p_{1/2}$ electron with a $2s_{1/2}$ electron; i.e., there is no $2p_{3/2}$ present. Note further that the $2\ ^3S_1$ state contains a $2s_{1/2}$ electron. Since, in the hydrogenic approximation, the fine structure is identical for the $2S_{1/2}$ and $2P_{1/2}$ states, the Z^4 contribution to the fine-structure splitting of $2\ ^3S_1$ and $2\ ^3P_0$ vanishes. Rather, the leading term scales approximately with Z^3. Since the Lamb shift scales almost as Z^4, the contribution of the Lamb shift to the $2\ ^3P_1 - 2\ ^3S_1$ separation increases rapidly with Z, relative to the other terms. A lowest-order calculation of the Lamb-shift contribution to the overall energy splitting shows that in argon (Z=18), the Lamb shift is about 0.8% of the total transition energy, whereas in krypton (Z=36) it is about 4%, and in xenon (Z=54) it is about 12%.

Experimentally, studying the Lamb shift in this transition has several advantages. First, we note that the transition $2\ ^3P_0 \rightarrow 2\ ^3S_1$ lies in the vacuum ultraviolet throughout the helium isoelectronic sequence for $Z > 10$. Hence, precision spectrometers capable of measurement to parts in 10^5 are available. Second, insofar as there is no hyperfine structure, the $2\ ^3P_0$ state decays to $2\ ^3S_1$ with a lifetime of at least 1 ns. Hence the natural width will never be a significant factor affecting the accuracy. Moreover, this lifetime corresponds to a finite decay length and measurements can be made downstream of the exciting foil. Therefore, prompt decays which occur at the foil will not be observed by the spectrometer. The above considerations make it likely that the accuracy of such a measurement would be limited by the Doppler effect from the fast beam, and accuracies of about 1 part in 10^3 should be possible. Measurement of the $2\ ^3P_0 \rightarrow 2\ ^3S_1$ energy difference in Ar^{+16}, attempting to exploit these factors, is currently underway at the Berkeley super-HILAC [8.32].

8.3 Radiative Decay of the $2S_{1/2}$ Metastable State of the One-Electron System

In addition to the Lamb shift, the structure of the n=2 state of the one-electron system can be studied by means of the radiative decay of the $2S_{1/2}$ state. We consider this problem from the theoretical and experimental points of view.

8.3.1 Theory

The metastability of the $2S_{1/2}$ state of the one-electron system can be understood with reference to the energy-level diagram of Fig.8.4. The only single-photon decay

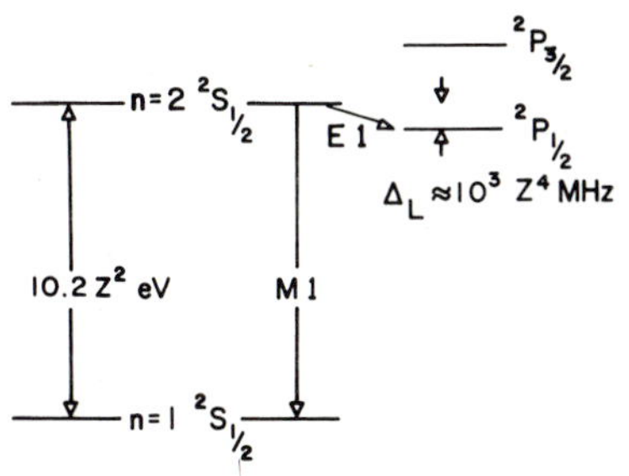

Fig. 8.4 Low-lying energy levels of the one-electron system

channel to the $1S_{1/2}$ ground state permitted by the angular momentum and parity selection rules is M1. Electric-dipole decay is permitted from the $2S_{1/2}$ state to the $2P_{1/2}$ state because of the small Lamb shift (denoted by Δ_L). A simple calculation of the rate expected for this decay can be made from the relation

$$A_{E1}(2S) = \frac{4e^2}{3\hbar c^3} \Delta_L^3 \; |\langle 2S|e\bar{r}|2P\rangle|^2 = 2 \times 10^{-10} \; Z^{10} \; [s^{-1}] \quad . \tag{8.3}$$

The numerical rate is based on an approximate value for the Lamb-shift frequency, given by

$$\Delta_L \simeq 10^3 \; Z^4 \; [\mathrm{MHz}] \quad . \tag{8.4}$$

The actual value for the Lamb shift scales somewhat more slowly than Z^4, hence $A_{E1}(2S)$ does not scale quite so fast as Z^{10}. In any case, the numerical value associated with this rate is many orders of magnitude smaller than the M1 rate to the ground state and is experimentally of no consequence.

Consider now the M1 mode to the ground state. In a non-relativistic theory, the rate $A_{M1}(2S) = 0$. This is simply deduced from the observation that the matrix element responsible for this decay is $\langle 2S|\bar{\mu}|1S\rangle$ and in non-relativistic quantum mechanics $\bar{\mu} = -\mu_0(\bar{\ell} + 2\bar{s})$. Since this expression does not involve radial operators, the matrix element vanishes identically because of the orthogonality of the radial wave functions for $2S_{1/2}$ and $1S_{1/2}$.

However, in a relativistic theory, this is no longer true. This was first noted in 1940 by BREIT and TELLER [8.33], who were motivated to make a theoretical investigation of the lifetime of this state because of astrophysical speculations [8.34] on the consequences of the metastability of this state on the Balmer absorption lines in certain super-giant stars. This can be most easily understood by noting that relativity introduces small corrections to the magnetic-moment operator; these corrections do involve radial functions. The theory of the relativistic magnetic-dipole decay has now been worked out by many authors [8.35-38] to lowest order in αZ with the result

$$A_{M1}(2S) = \frac{\pi}{35}\alpha^9 Z^{10}\, Ry = 2.496 \times 10^{-6} Z^{10}\ [s^{-1}] \quad , \tag{8.5}$$

where Ry is the Rydberg constant in frequency units. It is clear that for all values of Z, $A_{M1}(2S) \gg A_{E1}(2S)$.

BREIT and TELLER [8.33] showed that the dominant decay of the 2S state of ordinary hydrogen is a two-photon process in which two electric-dipole photons are simultaneously emitted so that the parity and angular momentum selection rules are satisfied. This is a process, first noted by GOEPPERT-MAYER [8.39] in 1931, whose properties are determined by the electric-dipole operator acting in second order of perturbation theory. Specifically,

$$A(\nu')d\nu' = \frac{2^{10}\pi^6 e^4 \nu'^3 \nu''^3}{h^2 c^6}\left(\left|\sum_n \frac{\langle f|\bar{r}\cdot\hat{\varepsilon}'|n\rangle\langle n|\bar{r}\cdot\hat{\varepsilon}''|i\rangle}{\nu_{in} + \nu''} + \frac{\langle f|\bar{r}\cdot\hat{\varepsilon}''|n\rangle\langle n|\bar{r}\cdot\hat{\varepsilon}'|i\rangle}{\nu_{in} + \nu'}\right|^2\right)_{av} d\nu' \quad . \tag{8.6}$$

Here $A(\nu')d\nu'$ is the probability per second for one of the photons to be emitted with a frequency in the interval between ν' and $\nu' + d\nu'$; ν'' is the frequency of the second photon; ν_{in} is the frequency of the transition from the initial state i to the intermediate state n; $e\bar{r}$ is the electric-dipole operator; $\hat{\varepsilon}'$ and $\hat{\varepsilon}''$ are unit wave functions and the sum is over all intermediate states $|n\rangle$.

We now consider the properties of the two-photon process as determined by (8.6)

a) *Simultaneity of emission of the two photons.* The two photons must be emitted in coincidence since the process proceeds through virtual states.

b) *Angular correlation of the two photons.* For the 2S-1S transition the only intermediate states contributing to (8.6) are the P states. Hence, there will be an angular correlation $(1 + \cos^2\theta)$ characteristic of a real S-P-S cascade.

c) *Frequency distribution of the photons.* Since the frequencies of the two photons satisfy the condition

$$\nu' + \nu'' = \frac{E_i - E_f}{h} \equiv \nu_0 \quad , \tag{8.7}$$

the frequency distribution must be symmetrical about the midpoint. The distribution function $A(\nu)$ was first evaluated by SPITZER and GREENSTEIN [8.40] and is shown as the solid curve in Fig.8.5. Note that it is a broad, flat-topped distribution.

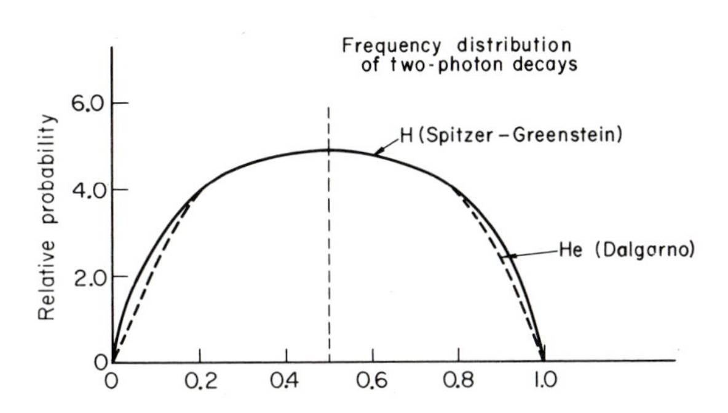

Fig.8.5 Frequency spectrum of photons emitted in two-photon decay from the $2S_{1/2}$ state of the one-electron system (solid curve) and from the $2\ ^1S_0$ state of the two-electron system (dashed curve)

d) *Two-photon transition probability.* The two-photon transition rate is given by

$$A_{2E1}(2S) = \frac{1}{2}\int_0^{\nu_0} A(\nu')d\nu' \quad . \tag{8.8}$$

The evaluation of (8.8) was first made by SPITZER and GREENSTEIN [8.40] and subsequently by SHAPIRO and BREIT [8.41], KLARSFELD [8.42], and ZON and RAPOPORT [8.43]. The most accurate value is that of KLARSFELD

$$A_{2E1}(2S) = (8.2283 \pm 0.0001)Z^6\ [s^{-1}] \quad . \tag{8.9}$$

Both KLARSFELD and ZON and RAPOPORT confirm the frequency spectrum obtained by SPITZER and GREENSTEIN.

All of the calculations cited above use the non-relativistic transition operator and non-relativistic energies and wave functions. Relativistic corrections to the transition rate and the frequency distribution were made by JOHNSON [8.44]. JOHNSON has also evaluated the M1 rate in closed analytical form and finds it to increase over the prediction of (8.5) (lowest order in αZ) by roughly a factor two as Z increases to 92. If (8.5) and (8.9) are compared, it is clear that, for low Z, the two-photon decay mode is dominant. Because of the different Z dependences, the M1 rate becomes comparable to the 2E1 rate at $Z \simeq 45$, and for higher Z the single-photon M1 rate is dominant. In this connection, we note that BOEHM and others [8.45] have reported observation of the magnetic-dipole transition $2S_{1/2} \rightarrow 1S_{1/2}$ in the X-ray spectra of heavy elements. These observations confirm the existence of a crossover point.

Further interest in the lifetime of the $2S_{1/2}$ state arises from the near degeneracy with the $2P_{1/2}$ state. The $2P_{1/2}$ state decays to the ground state via an allowed electric-dipole transition with rate [8.46]

$$A_{E1}(2P_{1/2}) = 6.25 \times 10^8 \, Z^4 \, [s^{-1}] \quad . \tag{8.10}$$

Consider now the possibility that the $2S_{1/2}$ wave function is not pure but contains a small admixture of $2P_{1/2}$ so that

$$\psi(2S_{1/2}) = \mu(2S_{1/2}) + \varepsilon\mu(2P_{1/2}) \quad , \tag{8.11}$$

where the μ's are the Schrödinger solutions for the hydrogen atoms. The actual decay rate A_{act} then becomes

$$A_{act}(2S_{1/2}) = A(2S_{1/2}) + |\varepsilon|^2 \, A(2P_{1/2}) \quad . \tag{8.12}$$

Since the Hamiltonian commutes with the parity operator, even a very small admixing parameter ε can arise only if there exists a parity-violating (P) Hamiltonian or a Hamiltonian which violates both parity and time-reversal (T) in the electron-nuclear interaction. The possibility of a P and T-violating interaction that would manifest itself in the form of a permanent electric dipole moment of the electron (EDM) was considered by SALPETER [8.47] and by FEINBERG [8.48]. Using their formulas and the current upper-limit on the EDM [8.49] we can place a limit that

$$|\mathrm{Re}\{\varepsilon\}| < 10^{-7} \quad . \tag{8.13}$$

The recent discovery of neutral currents in the weak interaction [8.50] has reawakened interest in the possibility of a P-violating interaction which does not violate T. BOUCHIAT and BOUCHIAT [8.51] have constructed a Hamiltonian describing such an interaction. Using their Hamiltonian it seems clear that the size of any expected effect on the 2S lifetime is much too small to be observable.

8.3.2 Experiments

The first experiments designed to study the radiative decay of the 2S state were performed on the ion He^{+1} by NOVICK and collaborators [8.52-54] at the Columbia Radiation Laboratory. The difficulties of working at low Z are many and formidable. The lifetime of 1.9 ms necessitates long flight paths. The endpoint energy E_0 is 40.8 eV and, hence, the photons are mainly in the vacuum UV where detectors which combine both high resolution and high efficiency do not exist. Finally, there are serious problems associated with collisional quenching and quenching due to motional fields. Despite these difficulties, an apparatus was constructed which permitted observation of the coincidences and verification of the $(1 + \cos^2\theta)$ angular correlation between the two photons. In addition, by use of broad-band photon filters, a rough verification of the predicted frequency distribution was obtained. These measurements convincingly established the two-photon nature of the decay and verified the essential details of the theory.

The first measurement of the lifetime of the 2S state was made in Ar^{+17} by a beam-foil experiment [8.55,56]. Many of the problems associated with this measurement become less serious at high Z. For example, for argon (Z=18), the theoretical lifetime of the metastable state is 3.46×10^{-9}s. The velocity of ions from the Berkeley Heavy-Ion Linear Accelerator (HILAC) in that first experiment was 4.4×10^9 cm/s, giving a mean decay length of 15.2 cm; from the point of view of experimentation, that is a very convenient length. Moreover, collisional and Stark quenching are difficult experimental problems at low Z, whereas the corresponding cross-sections for the fast HILAC beams are orders of magnitude lower.

The apparatus for studying the decay of the metastable state is shown in Fig.8.6.

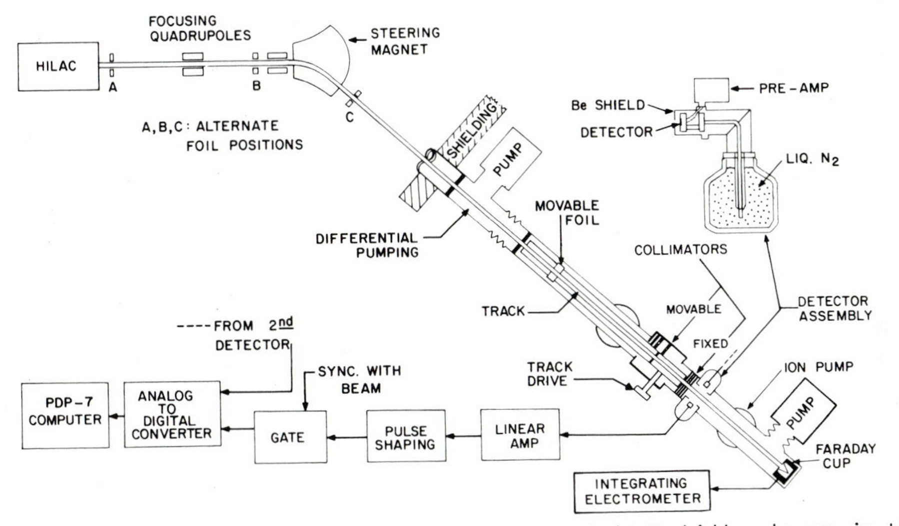

Fig.8.6 Schematic of beam-foil apparatus to study forbidden decays in the one- and two-electron systems

Argon ions from the HILAC are passed through a thick (> 100 $\mu g/cm^2$) carbon foil placed at position A. About 25% of the ions emerging from the foil are fully stripped, and these are passed into the beam pipe by a steering magnet. A thin (< 10 $\mu g/cm^2$) carbon foil is mounted on a movable platform and excites the fully-stripped beam to the 2S state of the hydrogenic argon atom. It is found that over 95% of the emerging argon atoms are in the +17 charge state. These are collimated and passed in front of a pair of Si(Li) solid-state X-ray detectors. These detectors are sensitive to X-rays in the range 600 eV (see the efficiency curve shown in Fig.8.7). Since the cut-off

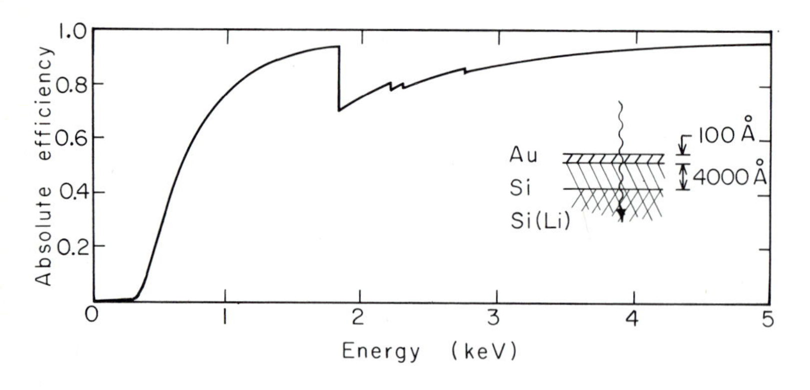

Fig.8.7 Efficiency curves for Si(Li) detectors used to study radiative decay of $2S_{1/2}$ state

frequency ν_0 = 3.24 kV, these detectors are convenient for studying the two-photon process and they have sufficient energy resolution ($\sim$ 175 eV) for establishing the frequency distribution of the emitted photons. The observation of the two-photon decay was established by a set of coincidence measurements taken with the pair of detectors. These are exhibited in Fig.8.8. The distribution of arrival times of single-photon events in the time detectors shows a pronounced correlation at zero-time difference, thus indicating real coincidences. Moreover, the sum of the energies of

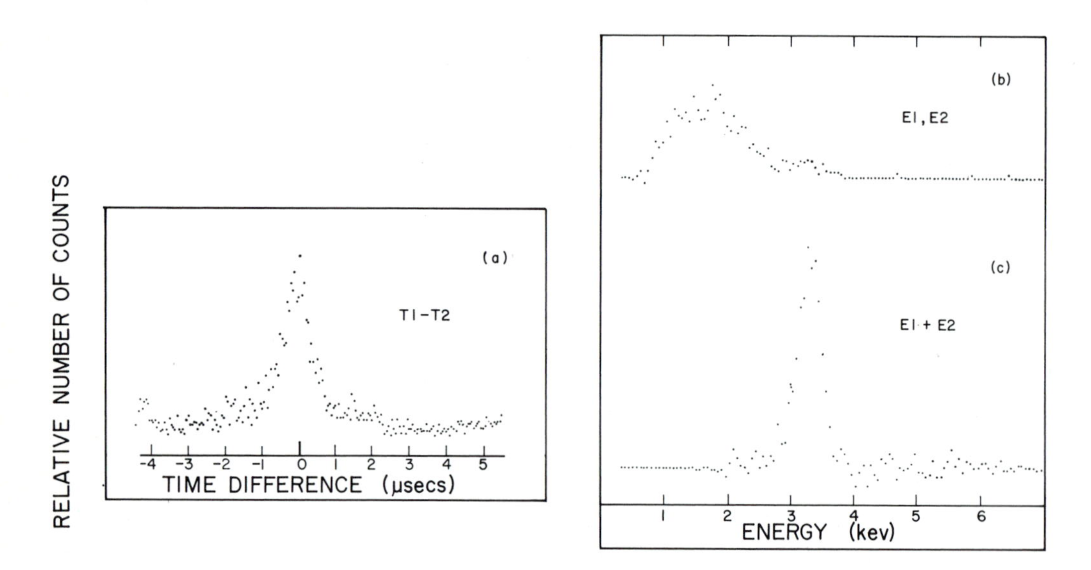

Fig.8.8 Coincidence measurements obtained for photons arising from radiative decay of $2S_{1/2}$ of Ar^{+17}

those photons arriving in coincidence shows a sharp peak at the end point energy E_0 = 3.34 kV, and finally the singles spectrum of the coincident photons agrees with the theoretical spectrum when corrected for detector efficiency. Figure 8.9 shows a

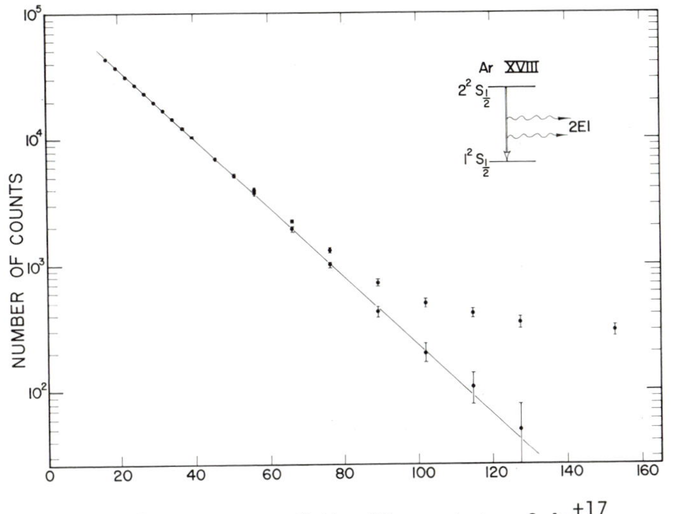

Fig.8.9 Decay curve of the $2S_{1/2}$ state of Ar^{+17}

typical decay curve taken by varying the foil detector separation. The decay curve is a single exponential over several mean lives and over several decades in the count rate. From such measurements, the metastable lifetime is found to be $\tau(2S_{1/2})$ = $3.54(25) \times 10^{-9}$s. Other beam-foil measurements by COCKE et al. [8.57] have extended the measurements to other elements. PRIOR [8.58] used an ion-trap technique to measure the lifetime in He^+, and KOCHER et al. [8.59] have made the same measurement by a time-of-flight technique. Figure 8.10 shows a comparison by JOHNSON [8.44] of the

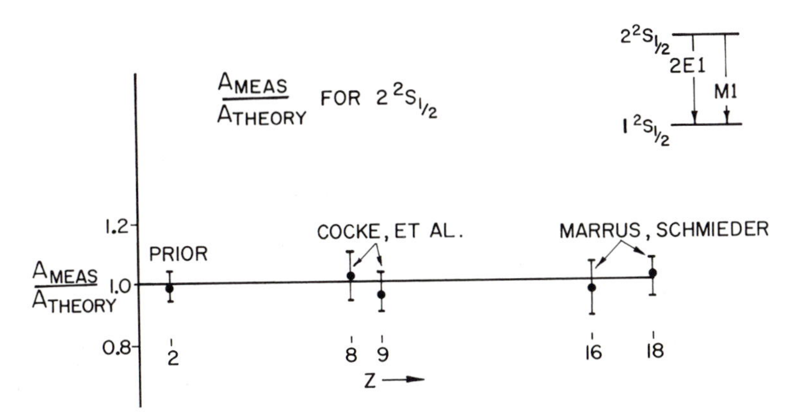

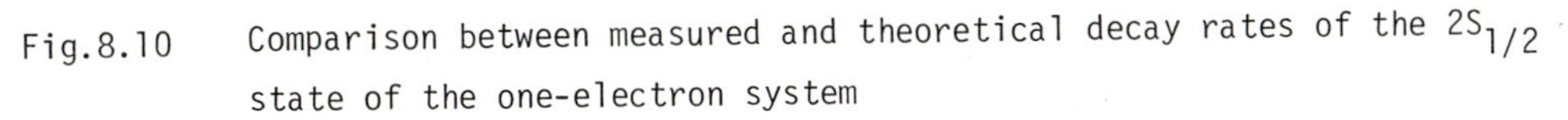
Fig.8.10 Comparison between measured and theoretical decay rates of the $2S_{1/2}$ state of the one-electron system

experimentally-measured and theoretical rates. The theory is clearly confirmed over a large range of Z.

8.4 Forbidden Radiative Decay in the n=2 State of the Two-Electron System

The beam-foil method has made possible experiments which have provided extensive information on the forbidden decay modes and the intercombination line of the two-electron atom. The origin of these decays can be understood with reference to the energy level diagram of Fig.8.11. Altogether, there are three observable forbidden (non E1)

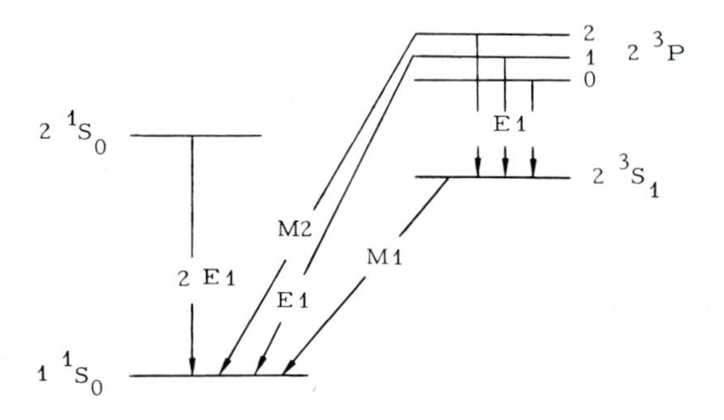

Fig.8.11 Low-lying energy levels of the two-electron system

decay modes observable in systems of high Z. We consider first the theory of these decays and then describe the experimental work.

8.4.1 Radiative Decay from $2\,^1S_0$

Radiative transitions from $2\,^1S_0$ to the $1\,^1S_0$ ground state can only occur by a two-photon process. Single-photon transitions are rigorously forbidden by the $J=0 \not\rightarrow J=0$ selection rule. The dominant two-photon process which conserves parity and angular momentum is the same (2E1) process which is dominant in the decay of the $2S_{1/2}$ state of the one-electron system and which was considered in Section 8.3. As with the $2S_{1/2}$ state, the photons are emitted in coincidence with a $(1 + \cos^2\theta)$ angular correlation. The frequency spectrum of the photons has been calculated by DALGARNO [8.60] and is shown in Fig.8.5 (dashed curve) along with the nearly identical spectrum for hydrogenic atoms. The lifetime of the $2\,^1S_0$ state has been calculated by VICTOR and DALGARNO [8.61] for a wide range of Z. In the limit of high Z, one might expect that the lifetime of $2\,^1S_0$ would approach the lifetime of the hydrogenic atom of the same Z. The 1s electron acts only as a spectator and shields the nuclear charge. In fact, the 2 $^1S_0$ rate approaches asymptotically a value of twice the hydrogenic rate and can be written in the form

$$A_{2E1}(2\ ^1S_0) \xrightarrow[\text{high } Z]{} 2(8.23)\ (Z - \sigma)^6 \quad , \qquad (8.14)$$

where σ is a shielding constant.

The theoretical lifetime of the 2 $^1S_0$ state in helium is calculated to be $\tau_{2E1}(2\ ^1S_0)$= 19.3 ms [8.62]. PEARL [8.63] and VAN DYCK et al. [8.64] have done time-of-flight experiments on a beam of metastable helium designed to measure this lifetime. In both experiments, the helium beam is excited to the metastable state by electron bombardment and the metastable atoms in the beam are detected by Auger ejection of electrons from a metal surface. The number of counts due to atoms in the 2 $^3S_1$ state is subtracted from the total metastable counts to give the number of atoms in the 2 $^1S_0$ state. By measuring the decrease in the 2 $^1S_0$ count rate as a function of distance traveled by the metastable atom, the lifetime is determined. Neither experiment observes the decay photons.

While similar in concept, the two experiments give results for the lifetime in apparent disagreement with each other. The result of PEARL [8.63] is $\tau(2\ ^1S_0)$ = 38(8) ms, while that of VAN DYCK et al. [8.64] is $\tau(2\ ^1S_0)$ = 19.7(1.0) ms. The latter result is in agreement with the theoretical value.

The beam-foil experiment of MARRUS and SCHMIEDER [8.55] established the two-photon nature of the decay and measured the lifetime in the two-electron atom Ar^{+16}. The experimental method used is the same as that described in Section 8.3, with minor changes. The helium-like beam in the 2 $^1S_0$ state was formed by bringing the Ar^{+14} ions emerging from the HILAC (see Fig. 8.6) into the apparatus pipe and through a thin (< 10 μgm/cm^2) carbon foil mounted on a movable track.

In this way a ratio of $Ar^{+16}/Ar^{+17} \geq 6:1$ could be obtained. It is important to suppress the one-electron component of the beam to avoid two-photon decays from the one-electron metastable state. As an examination of the low-lying energy levels for the one-electron and two-electron systems shows, the energy available for the two-photon decay modes is so closely the same that energy analysis is inadequate to distinguish the two different sets of photons from each other. We rely on selection of the incident charge and the thickness of the carbon foil to suppress the one-electron system to such a degree that corrections can be made (see below) for the background of two-photon decays from 2 $^2S_{1/2}$.

Results of coincidence measurements on the Ar^{+16} beam are shown in Fig.8.12. The

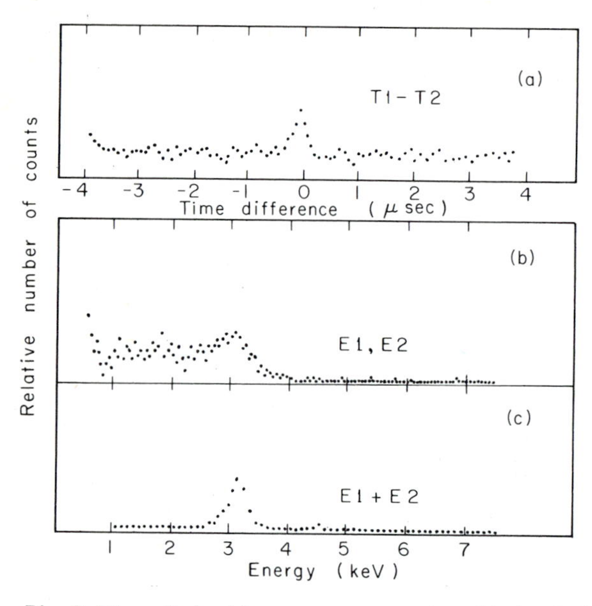

Fig.8.12 Coincidence measurements obtained for photons arising from radiative decay of $2\ ^1S_0$ of Ar^{+16}

peak at zero in the time-difference spectrum indicates the presence of true coincidences. When the energies of the two photons arriving in coincidence are summed, a peak at 3.12 keV is observed. This is in agreement with the theoretical energy [8.65] for this decay and is sufficiently well resolved to distinguish it from the one-electron, two-photon decay, which occurs at 3.34 keV.

The lifetime of the state is obtained in the usual way from a decay curve which is generated by measuring the photon intensity as a function of the separation between the exciting foil and the detectors. The accuracy of the lifetime obtained in this way is limited by two-photon counts from one-electron atoms present in the beam, as mentioned earlier in the discussion of the experimental arrangement. To take the possible contributions from the one-electron system into account, a model is constructed to estimate the hydrogen-like component in the beam; this results in a 15% correction to the uncorrected lifetime.

The value as determined from our experiments is $\tau(2\ ^1S_0) = 2.3(0.3) \times 10^{-9}$s, which is in good agreement with the non-relativistic calculations of DRAKE [8.66]. In order to summarize the results of all existing measurements on $A(2\ ^1S_0)$ and make a comparison with the theoretical two-photon rate, we plot the ratio (A_{meas}/A_{theory}) vs. Z. Figure 8.13 shows such a comparison.

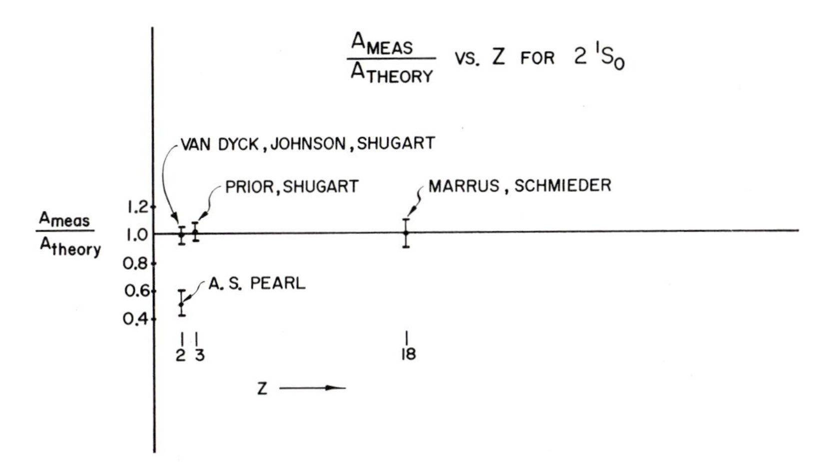

Fig.8.13 Comparison between measured and theoretical decay rates of the $2\ ^1S_0$ state of the two-electron system

8.4.2 Radiative Decay from $2\ ^3S_1$

The radiative decay of the $2\ ^3S_1$ state to the $1\ ^1S_1$ ground state can proceed in principle by a spin-orbit-induced, two-photon process. This process was first suggested by BREIT and TELLER [8.33] and can be understood with reference to (8.6). Note that if the initial and final states have different spin, the matrix elements of the dipole operator are non-vanishing only insofar as the intermediate P states are mixtures of singlet and triplet. This is accomplished by the spin-orbit and other relativistic interactions.

Accurate calculations of the rates and frequency distributions for this process have been made by DRAKE et al. [8.62], and BELY and FAUCHER [8.67]. The results show that for ordinary helium $\tau_{2E1}(2\ ^3S_1) \simeq 2 \times 10^8$s, to be compared with $\tau_{2E1}(2\ ^1S_0) = 19.5 \times 10^{-3}$s. The ratio of these lifetimes clearly illustrates the inhibiting effect resulting from the spin change in the initial and final states.

The first observation of the radiative decay $2\ ^3S_1 \rightarrow 1^1S_0$ was made by X-ray telescopes mounted in earth-orbiting satellites designed to study the X-rays emitted by highly-ionized atoms in the solar corona [8.68-70]. GABRIEL and JORDAN [8.71] pointed out that certain lines observed in these spectra corresponded to single photon M1 radiation associated with the decay $2\ ^3S_1 \rightarrow 1\ ^1S_0$. This prompted calculations [8.72-74] of the rate associated with the M1 process and led to the conclusion that the single-photon M1 rate was substantially faster than the spin-orbit-induced, two-photon rate for the entire helium isoelectronic sequence.

The single-photon, M1 process is similar in character to that described in Section 8.3 for the radiative decay of the $2S_{1/2}$ state of the one-electron system. The transition rate is finite only in the framework of a relativistic theory. A full treatment of the problem must take into account the electron-electron interaction, and has now been worked out by several authors [8.35-38]. Detailed numerical calculations of the transition rate have been made by DRAKE [8.75] and JOHNSON and LIN [8.76]. The transition rate associated with the process is $A_{M1}(2\ ^3S_1) = 1.3 \times 10^{-4}\ s^{-1}$ in ordinary helium, and scales roughly as Z^{10} at high Z.

The first laboratory observation of this decay was made in a beam-foil experiment by MARRUS and SCHMIEDER [8.77] on the helium-like ions Ar^{+16}, S^{+14} and Si^{+12}. Identification of the transitions was based mainly on the energy associated with the decay and with the observed long lifetime. Lifetime measurements based on time-of-flight were subsequently reported by MOOS and WOODWORTH [8.78] in He; GOULD et al. [8.79] for Ar^{+16} and Ti^{+20}; COCKE et al. [8.80] on $C\ell^{+15}$; GOULD et al. [8.81] on V^{+21} and Fe^{+24}.

While the measurements on $C\ell^{+15}$ and Ar^{+16} showed apparent disagreement with theory by about 25% and 20%, respectively, the higher-Z measurements on Ti^{+20}, V^{+21} and Fe^{+24} were in good agreement. This stimulated a recent remeasurement of the lifetime of $C\ell^{+15}$ by BEDNAR et al. [8.82]. They found a non-exponential character to the decay curve and that the measured lifetime approaches the theoretical value only if the time-of-flight measurements are made at more than one decay length from the exciting foil. While this criterion was satisfied by the Ti, V and Fe measurements, where the decay length is relatively short, it was not satisfied by the Cℓ and Ar measurements, where the decay length is several meters. A remeasurement of the $C\ell^{+15}$ and measurement of the S^{+14} lifetimes, while meeting the above criterion, give good agreement with theory. It should be mentioned that the reason for the non-exponential character of the decay curve is not yet understood and can't be explained by cascading effects or alignment effects.

GOULD and MARRUS [8.83] have repeated the Ar^{+17} measurements under similar conditions. The zero position of the foil was placed more than a decay length (about 5 meters) from the detectors, and the lifetime remeasured. The measurement made under these conditions is 202 ± 20 ns. This result is somewhat longer than the previous measurement [8.79] and is roughly consistent with the trend indicated by the data of BEDNAR et al. [8.82].

Figure 8.14 shows a comparison of the experimental measurements with theory.

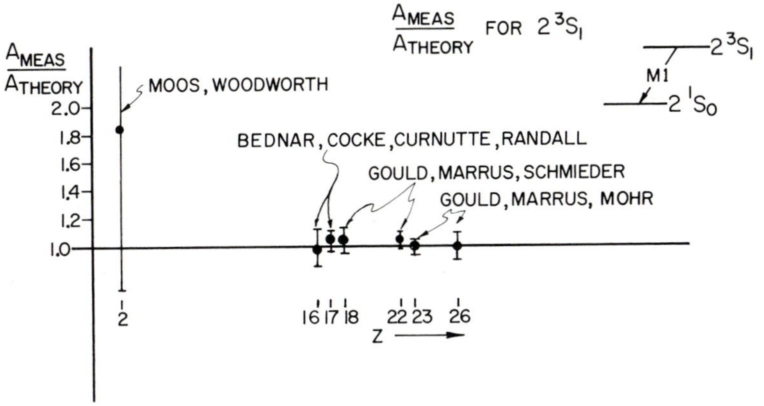

Fig.8.14 Comparison between measured and theoretical decay rates of the $2\,^3S_1$ state of the two-electron system

8.4.3 Radiative Decay from $2\,^3P_2$

The $2\,^3P_2$ level of helium decays predominantly to the $2\,^3S_1$ level by fully-allowed E1 radiation with a lifetime of 10^{-7}s [8.84]. Parity and angular-momentum selection rules permit the $2\,^3P_2$ level to decay to the $1\,^1S_0$ ground state by M2 radiation. The $2\,^3P_2$ level of the two-electron system is a special and possibly unique case where M2 and E1 decays compete with each other. If one considers the Z-dependence of these rates for the helium isoelectronic sequence, it can be seen that A_{E1} is roughly proportional to Z but that A_{M2} is roughly proportional to Z^8. Hence, for sufficiently high Z, the rates become comparable. Detailed numerical calculations of $A_{M2}(2\,^3P_2)$ have been made by GARSTANG [8.85] and DRAKE [8.86]. Their calculations show that the two rates are approximately equal at Z=18.

The first observation of the M2 radiation from the $2\,^3P_2$ state was made in Ar^{+16} by MARRUS and SCHMIEDER [8.55,56] who also made a measurement of the lifetime of the $2\,^3P_2$ state. Their result corresponds to a total transition rate from this state of $A(2\,^3P_2) = 5.9(1.0) \times 10^8\ s^{-1}$.

The calculated transition rates from this state are $A_{E1}(2\,^3P_2) = 3.55 \times 10^8\ s^{-1}$ [8.66] and $A_{M2}(2\,^3P_2) = 3.14 \times 10^8\ s^{-1}$ [8.86]. Within the limits of error, the theoretical rates agree with the experimental rate. Subsequent experiments, particularly by COCKE et al. [8.87] and GOULD et al. [8.81] have extended measurements on the $2\,^3P_2$ state to S^{+14}, $C\ell^{+15}$, Ti^{+20}, V^{+21} and Fe^{+24}.

A comparison with the theoretical rates is shown in Fig.8.15. The good agreement with the calculations over a wide range of Z confirms the prediction of M2 radiation and the calculations of the rate.

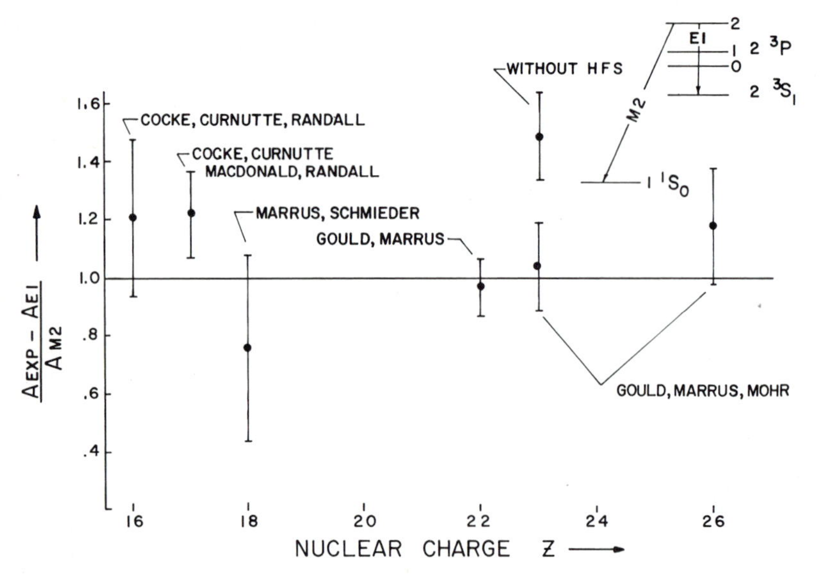

Fig.8.15 Comparison between measured and theoretical decay rates of the $2\ ^3P_2$ state of the two-electron system

An interesting aspect of the $2\ ^3P_2$ lifetime measurement in two-electron vanadium is the evidence for hyperfine quenching present in the state. Hyperfine structure in the vanadium atom causes the total angular momentum J to break down as a good quantum number. Thus, if we write $|2\ ^3P\rangle_p$ as the state vector of the perturbed $2\ ^3P_2$ state with hyperfine interaction present, then

$$|2\ ^3P\rangle_p = |2\ ^3P_2\rangle + \varepsilon|2\ ^3P_1\rangle \quad , \tag{8.15}$$

where $|2\ ^3P_2\rangle$ and $|2\ ^3P_1\rangle$ are the unperturbed eigenfunctions. The mixing parameter ε will depend on F, the total (electronic plus nuclear) angular momentum of the state. The decay rate of the state becomes

$$A(2\ ^3P)_p = A(2\ ^3P_2) + \varepsilon^2 A(2\ ^3P_1) \quad . \tag{8.16}$$

Because the rate $A(2\ ^3P_1 >> A(2\ ^3P_2)$, even a small value of ε can measurably affect the decay rate. The measurements support the presence of hyperfine quenching on two counts. First, the observed decay curve is not well fitted by a single exponential. This supports the contention that the different F states decay with different rates.

Second, the best value for the lifetime ignoring hyperfine quenching disagrees with theory, while agreement is obtained when hyperfine quenching is taken into account.

8.4.4 Radiative Decay from $2\ ^3P_1$

In ordinary helium, the $2\ ^3P_1$ state decays by E1 radiation to the $2\ ^3S_1$ state with a lifetime of 10^{-7}s. Insofar as total spin (S) is a good quantum number, direct decay to the ground state is forbidden by the $\Delta S=0$ selection rule. However, the spin-orbit and other relativistic interactions produce mixing of the $2\ ^3P_1$ and $n\ ^1P_1$ states so that the decay $2\ ^3P_1 \rightarrow 1\ ^1S_0$ can occur. The decay rate associated with this process has been calculated by DRAKE and DALGARNO [8.88]. Although the rate for helium is quite low $[A(2\ ^3P_1 - 1\ ^1S_0) = 1.8 \times 10^2\ s^{-1}]$, it scales rapidly with Z so that it can be conveniently studied at high Z by beam-foil spectroscopy.

The intercombination line was first observed in helium by LYMAN [8.89] in 1924. However, the difficulties of making quantitative measurements on this transition are illustrated by the fact that the first estimate of the branching ratio $A(2\ ^3P_1 \rightarrow 1\ ^1S_0)/A(2\ ^3P_1 \rightarrow 2\ ^3S_1) \equiv B$ was made in 1972. TANG and HAPPER [8.90] were able to set the limits $0.3 \times 10^{-5} \leq B \leq 1.8 \times 10^{-5}$.

The first beam-foil experiments on the intercombination line were reported by SELLIN et al. [8.91] who used 6.42 MeV N and O ions from a tandem Van de Graaff accelerator to measure the $2\ ^3P_1$ lifetimes. A later measurement by SELLIN et al. [8.92] used a crystal spectrometer to resolve the line in O and improve the measured lifetime. Time-of-flight measurements of this lifetime have subsequently been made by RICHARD et al. [8.93] and MONE et al. [8.94]. MOWAT et al. [8.95], RICHARD et al. [8.96], and KAUFFMAN et al. [8.97] have extended the measurement to two-electron fluorine.

In order to compare the measured results with the theoretical intercombination rates of DRAKE and DALGARNO [8.88], corrections must be made for the electric-dipole transition rate $A_{E1}(2\ ^3P_1 \rightarrow 2\ ^3S_1)$. This value is given by WIESE et al. [8.98]. The measured values thus corrected are in good agreement with the calculated rates.

Both the intercombination line and the magnetic dipole decay $2\ ^3S_1 \rightarrow 1\ ^1S_0$ have been observed for many elements present in the solar corona. GABRIEL and JORDAN [8.99] have constructed a model that makes possible the determination of the electron density in the solar corona from the relative intensities of these lines. Only atomic parameters such as the radiative lifetimes and certain collision cross-sections enter their model.

8.5 Study of Doubly-Excited Configurations in the Two-Electron System

Doubly-excited configurations are those configurations where two electrons in the ground configuration have been raised to excited levels. Obviously the two-electron system is the simplest in which such configurations exist and, therefore, provides the cleanest testing ground for calculations of these configurations. Important doubly-excited configurations [8.100] exist in the three-electron system but these will not be discussed here.

The doubly-excited configurations of the two-electron system all lie beyond the first ionization potential. In neutral helium, the lowest-lying doubly-excited state, $(2s)^2\ {}^1S_0$, is at 58.0 eV above the $(1s)^2\ {}^1S_0$ ground state and 33.4 eV above the ionization potential of He I and, hence, is connected to the ground state by vacuum ultraviolet radiation. Although the first observation of doubly-excited configurations was noted in 1928 [8.101-104], the first systematic study of their properties was made by MADDEN and CODLING [8.105,106] in the 1960's. They used the ultraviolet photons from synchrotron radiation to excite He^{**} directly from the helium ground state. Such excitations were monitored by observing the transmitted radiation with a high-resolution ultraviolet spectrometer. In this way measurements of the energies could be obtained to better than 0.1%. It was also possible to obtain lifetimes of some of the states from the observed widths associated with the transitions. This is possible because some of the He^{**} levels decay rapidly by autoionization and have large natural widths which are greater than the instrumental widths. The results obtained by MADDEN and CODLING were in good agreement with theory.

One of the limitations of photon excitation of He^{**} levels is that the electric-dipole selection rules permit excitation only of ${}^1P$ levels from the ${}^1S$ ground state. The levels populated by beam-foil excitation are not limited by this restriction. The first beam-foil excited spectra of He^{**} were made by BERRY et al. [8.107]. They observed seven lines in the vacuum ultraviolet which were ascribed to radiative decay of He^{**}. Tentative assignments of these lines were made, based mainly on agreement between the observed calculated wavelengths. They were also able to measure the lifetimes of some of the levels by observing decay curves. KNYSTAUTAS and DROUIN [8.108] also observed ultraviolet spectra associated with decay of levels from He^{**}. These decays arise mainly from radiation between doubly-excited states and levels of ordinary helium.

We note that the doubly-excited levels all lie beyond the ionization limit of He I and hence are capable of rapid de-excitation by autoionization. However, rapid de-excitation occurs by the Coulomb interaction and must satisfy the rigid selection rules, $\Delta L = \Delta S = \Delta J = 0$, with no parity change. Hence, many of the doubly-excited configurations are metastable against Coulomb autoionization. Autoionization of these levels can occur via relativistic interactions, but the transition rates associated with such processes are, in general, several orders of magnitude slower than those associated with Coulomb autoionization. It is these states which decay primarily by radiative transitions to lower-lying doubly- or singly-excited levels.

The lowest non-autoionizing term of He^{**} is $(2p)^2\ ^3P$. The energy of this term relative to the $(1s)^2\ ^1S_0$ ground state had been accurately calculated by AASHAMAR [8.109] who included mass-polarization, relativistic and radiative contributions. A very accurate measurement of this term was made by TECH and WARD [8.110], who found that $E_{exp}[(2p)^2\ ^3P - (1s)^2\ ^1S_0] = 481301.5(1.2)\ cm^{-1}$.

Transitions among the autoionizing levels of He^{**} were first observed by BERRY et al. [8.111,112] using beam-foil excitation and viewing the emitted radiation over the wavelength region 2000-6000Å with a Czerny-Turner monochromator. Twelve such lines were observed and lifetimes were measured for some of them. In subsequent work [8.113], measurements were extended to doubly-excited configurations of Li II.

The author thanks Dr. Peter J. Mohr for helpful conversations on the status of the Lamb shift and for preparing Table 8.2.

References

8.1 S. Triebwasser, E. S. Dayhoff, W. E. Lamb, Jr.: Phys. Rev. 89, 98 (1953)

8.2 A review of the theoretical situation regarding the Lamb shift through about 1969 can be found in S. J. Brodsky, S. Drell: Ann. Rev. Nucl. Sci. 20, 147 (1970)

8.3 T. Appelquist, S. J. Brodsky: Phys. Rev. Letters 24, 562 (1970)

8.4 G. W. Erickson: Phys. Rev. Letters 27, 780 (1971)

8.5 P. J. Mohr: Ann. Phys. (NY) 88, 26, 52 (1974)

8.6 P. J. Mohr: Phys. Rev. Letters (to be published)

8.7 Reviewed by R. T. Robiscoe: In *Cargese Lectures in Physics* (Gordon and Breach, New York 1968)p.3

8.8 For a review, see the article "Quantum Electrodynamics at Small Distances", published in Essays in Physics 2, 1 (1970)

8.9 M. Leventhal, P. E. Havey: Phys. Rev. Letters 32, 808 (1974)

8.10 M. Leventhal: Phys. Rev. A 11, 427 (1975)

8.11 E. Lipworth, R. Novick: Phys. Rev. 108, 1434 (1957)

8.12 M. A. Narashimham, R. L. Strombotre: Phys. Rev. A 4, 14 (1971)

8.13 C. Y. Fan. M. Garcia-Munoz, I. A. Sellin: Phys. Rev. 161, 6 (1967)

8.14 D. E. Murnick, M. Leventhal, H. W. Kugel: Phys. Rev. Letters 27, 1625 (1971)

8.15 G. P. Lawrence, C. Y. Fan, S. Bashkin: Phys. Rev. Letters 28, 1612 (1972)

8.16 M. Leventhal, D. E. Murnick, H. W. Kugel: Phys. Rev. Letters 28, 1609 (1972)

8.17 S. Lundeen, F. Pipkin: 4th Intl. Conf. Atomic Physics, Heidelberg, Germany, July 19-26, 1974

8.18 Described in W. F. Ramsey: *Molecular Beams* (Oxford University Press, Oxford 1963) Chap.5,p.124

8.19 See Eq. (7) in W. Lamb, Jr., R. Retherford: Phys. Rev. 79, 549 (1950)

8.20 S. J. Lundeen: Bull. Am. Phys. Soc. 20, 11 (1975)

8.21 R. Robiscoe, T. Shyn: Phys. Rev. Letters 24, 559 (1970)

8.22 Y. Accad, C. L. Pekeris, B. Schiff: Phys. Rev. A 4, 516 (1971)

8.23 F. Tyrén: Nova Acta Reg. Soc. Sci. Ups. (IV) 12, No. 1 (1940)

8.24 B. Edlén: Arkiv Fysik 4, 441 (1952)

8.25 G. Herzberg, H. R. Moore: Can. J. Phys. 37, 1293 (1959)

8.26 B. Edlén, B. Löfstrand: J. Phys. B 3, 1380 (1970)

8.27 L. A. Swensson: Physica Scripta 1, 246 (1970)

8.28 H. G. Berry, R. Bacis: Phys. Rev. A 8, 36 (1973)

8.29 H. G. Berry, R. M. Schectman: Phys. Rev. A 9, 2345 (1974)

8.30 A. M. Ermolaev: Phys. Rev. A 8, 1651 (1973)

8.31 A. M. Ermolaev: Phys. Rev. Letters 34, 380 (1975)

8.32 W. Davis (private communication)

8.33 G. Breit, E. Teller: Astrophys. J. 91, 215 (1940)

8.34 O. Struve, K. Wurm, L. G. Henyey: Proc. Natl. Acad. Sci. U.S. 25, 67 (1939)

8.35 G. Feinberg, J. Sucher: Phys. Rev. Letters 26, 681 (1971)

8.36 I. L. Beigman, U. I. Safronova: Zh. Eksp. Teor. Phys. 60, 2045 (1970) [Sov. Phys. - JETP 33, 1102 (1971)]

8.37 G. W. F. Drake: Phys. Rev. A 5, 1979 (1972)

8.38 S. Feneuille, E. Koenig: C. R. Acad. Sci. Ser. B 274, 46 (1972)

8.39 M. Goeppert-Mayer: Ann. Phys. 9, 273 (1931)

8.40 L. Spitzer, Jr., J. L. Greenstein: Astrophys. J. 114, 407 (1951)

8.41 J. Shapiro, G. Breit: Phys. Rev. 113, 179 (1959)

8.42 S. Klarsfeld: Phys. Letters 30A, 382 (1969)

8.43 B. A. Zon, L. P. Rapoport: JETP Letters 7, 52 (1968)

8.44 W. R. Johnson: Phys. Rev. Letters 29, 1123 (1972)

8.45 F. Boehm: Phys. Letters 33A, 417 (1970) and references therein

8.46 Evaluated by H. Bethe, E. E. Salpeter: In *Handbuch der Physik* (Springer, Berlin 1957) Vol. XXXV,p.352

8.47 E. E. Salpeter: Phys. Rev. 112, 1642 (1958)

8.48 G. Feinberg: Phys. Rev. 112, 1637 (1958)

8.49 M. C. Weisskopf, J. P. Carrico, H. Gould, E. Lipworth, T. S. Stein: Phys. Rev. Letters 21, 1645 (1968)

8.50 E. J. Hasert et al.: Phys. Letters 46B, 138 (1973)

8.51 M. A. Bouchiat, C. C. Bouchiat: Phys. Letters 48B, 111 (1974)

8.52 M. Lipeles, R. Novick, N. Tolk: Phys. Rev. Letters 15, 690 (1965)

8.53 C. J. Artura, N. Tolk, R. Novick: Astrophys. J. 157, L181 (1969)

8.54 R. Novick: In *Physics of One- and Two-Electron Atoms*, ed. by H. Bopp, F. Kleinpoppen (North-Holland, Amsterdam 1969)pp. 296-325

8.55 Richard Marrus, Robert W. Schmieder: Phys. Rev. A 5, 1160 (1972)

8.56 R. W. Schmieder, R. Marrus: Phys. Rev. Letters 25, 1245 (1970)

8.57 C. L. Cocke, B. Curnutte, J. R. Macdonald, J. A. Bednar, R. Marrus: Phys. Rev. A 9, 2242 (1974)

8.58 M. H. Prior: Phys. Rev. Letters 29, 611 (1972)

8.59 C. A. Kocher, J. E. Clendenin, R. Novick: Phys. Rev. Letters 29, 615 (1972)

8.60 A. Dalgarno: Proc. Phys. Soc. (London) 87, 371 (1966)

8.61 G. A. Victor, A. Dalgarno: Phys. Rev. Letters 25, 1105 (1967) and references therein to earlier work

8.62 G. W. F. Drake, G. A. Victor, A. Dalgarno: Phys. Rev. 180, 25 (1969)

8.63 A. S. Pearl: Phys. Rev. Letters 24, 703 (1970)

8.64 R. S. Van Dyck, Jr., C. E. Johnson, H. A. Shugart: Phys. Rev. A 4, 1327 (1971)

8.65 H. T. Doyle: *Advances in Atomic and Molecular Physics* (Academic Press, New York 1969)p.337

8.66 G. W. F. Drake (private communication cited in [8.55]

8.67 O. Bely, P. Faucher: Astron. Astrophys. 1, 37 (1969)

8.68 W. M. Neupert, M. Swartz: Astrophys. J. 160, L189 (1970)

8.69 G. A. Doschek, J. F. Meekins, R. W. Kreplin, T. A. Chubb, H. Freidman: Astrophys. J. 164, 165 (1971)

8.70 A. B. C. Walker, Jr., H. R. Rugge: Astron. Astrophys. 5, 4 (1970) and references therein

8.71 A. H. Gabriel, C. Jordan: Nature 221, 947 (1969)

8.72 C. L. Schwartz, private communication reported in M. Steinberg: Ph.D. thesis, University of California, Berkeley, 1968

8.73 H. Griem: Astrophys. J. 156, L103 (1969)

8.74 H. Griem: Astrophys. J. 161, L155 (1970)

8.75 G. W. F. Drake: Phys. Rev. A 3, 908 (1971)

8.76 W. R. Johnson, C. Lin: Phys. Rev. A 9, 1486 (1974)

8.77 R. Marrus, R. W. Schmieder: Phys. Letters 32A, 431 (1970)

8.78 H. W. Moos, J. R. Woodworth: Phys. Rev. Letters 30, 775 (1973)

8.79 H. Gould, R. Marrus, R. W. Schmieder: Phys. Rev. Letters 31, 504 (1973)

8.80 C. L. Cocke, B. Curnutte, R. Randall: Phys. Rev. Letters 31, 507 (1973)

8.81 H. Gould, R. Marrus, P. Mohr: Phys. Rev. Letters 33, 676 (1974)

8.82 J. A. Bednar, C. L. Cocke, B. Curnutte, R. Randall: Phys. Rev. A 11, 460 (1975)

8.83 H. Gould, R. Marrus: (to be published)

8.84 D. A. Landman: Bull. Am. Phys. Soc. 12, 94 (1967)

8.85 R. H. Garstang: Publ. Astron. Soc. Pac. 81, 488 (1969)

8.86 G. W. F. Drake: Astrophys. J. 158, 119 (1969)

8.87 C. L. Cocke, B. Curnutte, J. R. MacDonald, R. Randall: Phys. Rev. A 9, 57 (1974)

8.88 G. W. F. Drake, A. Dalgarno: Astrophys. J. 157, 459 (1969)

8.89 T. Lyman: Astrophys. J. 60, 1 (1924)

8.90 H. Y. S. Tang, W. Happer: Bull. Am. Phys. Soc. 17, 476 (1972)

8.91 I. A. Sellin, B. L. Donnally, C. Y. Fan: Phys. Rev. Letters 21, 717 (1968)

8.92 I. A. Sellin, M. Brown, W. W. Smith, B. Donnally: Phys. Rev. A 2, 1189 (1970)

8.93 P. Richard, R. L. Kauffman, F. Hopkins, C. W. Woods, K. A. Jamison: Phys. Rev. A 8, 2187 (1973)

8.94 C. F. Mone, W. J. Braithwaite, D. L. Matthews: Phys. Letters 44A, 199 (1973)

8.95 J. R. Mowat, I. A. Sellin, R. S. Peterson, D. J. Pegg, M. D. Brown, J. R. MacDonald: Phys. Rev. A 8, 145 (1973)

8.96 P. Richard, R. L. Kauffman, F. F. Hopkins, C. W. Woods, K. A. Jamison: Phys. Rev. Letters 30, 888 (1973)

8.97 R. L. Kauffman, C. W. Woods, F. F. Hopkins, D. D. Elliott, K. A. Jamison, P. Richard: J. Phys. B 6, 2197 (1973)

8.98 W. L. Wiese, M. W. Smith, B. M. Glennon: *Atomic Transition Probabilities* NSRDS-NBS 4 (U.S.Govt. Printing Office, Washington, D.C. 1966) Vol. 1, p.125

8.99 A. H. Gabriel, C. Jordan: Monthly Notices Roy. Astron. Soc. 145, 241 (1969)

8.100 For a review of doubly-excited levels in systems with three or more electrons see U. Fano: In *Atomic Physics*, ed. by V. W. Hughes, B. Bederson, V. W. Cohen, F. M. J. Pichanick (Plenum, New York 1969) p.209

8.101 K. T. Compton, J. C. Boyn: J. Franklin Inst. 205, 497 (1928)

8.102 P. G. Kruger: Phys. Rev. 36, 855 (1930)

8.103 R. Whiddlington, H. Priestly: Proc. Roy. Soc. (London) A145, 462 (1934)

8.104 T. Y. Wu: Phys. Rev. 66, 291 (1944)

8.105 R. P. Madden, K. Codling: Phys. Rev. Letters 10, 516 (1963)

8.106 R. P. Madden, K. Codling: Astrophys. J. 141, 364 (1965)

8.107 H. G. Berry, I. Martinson, L. J. Curtis, L. Lundin: Phys. Rev. A 3, 1934 (1971)

8.108 E. J. Knystautas, R. Drouin: Nucl. Instr. and Meth. 110, 95 (1973)

8.109 K. Aashamar: Nucl. Instr. and Meth. 90, 263 (1970)

8.110 J. L. Tech, J. F. Ward: Phys. Rev. Letters 27, 367 (1971)

8.111 H. G. Berry, J. Desesquelles, M. Dufay: Phys. Rev. A 6, 600 (1972)

8.112 H. G. Berry, J. Desesquelles, M. Dufay: Nucl. Instr. and Meth. 110, 43 (1973)

8.113 J. P. Buchet, M. C. Buchet-Poulizac, H. G. Berry, G. W. F. Drake: Phys. Rev. A 7, 922 (1973)

9. Coherence, Alignment, and Orientation Phenomena in the Beam-Foil Light Source

Joseph Macek and Donal Burns

With 10 Figures

Fluorescence is excited in the beam-foil light source by passing an energetic (50 keV to several 10's of MeV) ion beam through a thin (5 μgm/cm^2) carbon foil. Two unique properties of this source were recognized in the early 1960's and contributed to the steady growth of applications. Because the beam ions pass through the foil in times of the order of 10^{-5} nanoseconds, the time of excitation is quite well defined compared to the radiative lifetimes of the decaying levels. This enables lifetimes to be measured directly by time-of-flight techniques. Secondly, since ions make many collisions with the foil atoms and since some of the collisions are hard, the beam emerges from the foil in a highly-excited state, consisting generally of ions in various degrees of ionization and excitation. The light emitted by the beam exhibits a correspondingly rich spectrum of lines ranging from the optical through the ultraviolet and well into the X-ray region. Identification of these atomic lines, many of them previously unobserved, represents one of the principal applications of the beam-foil source.

The beam-foil source also possesses another property whose relevance was not fully appreciated until 1970, namely, the typical source geometry singles out a unique spatial axis, the beam axis, which usually coincides with the axis of the normal to the foil. Because the source has only cylindrical symmetry rather than complete spherical symmetry, the excited ions are generally in an anisotropic state. Radiation emitted in the decay exhibits the anisotropy through the angular distribution and polarization of the light.

The collision geometry loses its axis of symmetry when the foil is tilted with respect to the beam axis. Then the source has only reflection symmetry in the plane defined by the normal to the surface and the beam direction. This lower degree of symmetry for the source is mirrored, in general, by a correspondingly lower degree of symmetry in the angular and polarization distribution of the collision-induced fluorescence. Recent experiments demonstrate conclusively a pronounced variation of source anisotropy with tilt angle.

When atoms in anisotropic states are placed in external electric or magnetic fields, or when the atomic levels are split by small internal forces, the anisotropy is perturbed and varies cosinusoidally with time. Owing to the sharp definition of the time of excitation in the beam-foil source and to the high time resolution ($\sim 10^{-11}$s) of the time-of-flight technique, we have available an excellent tool for the observation of the dynamic evolution of atomic states perturbed by weak internal and external fields. The advantage of the beam-foil source for this purpose was actually realized before the anisotropy of foil-excited atomic states was discovered. This was possible because of the unique nature of the eigenstates of hydrogen-like ions, where levels with different lifetimes, e.g., the 2s and 2p levels, are mixed by relatively weak electric fields. The non-stationary state of the atom then oscillates between the two levels. When it is in the 2s eigenstate it does not decay by a dipole allowed transition, but when it is in the 2p eigenstate it does. The intensity of the decay radiation is therefore modulated, even though the unperturbed atomic state is isotropic.

Regardless of the origins of the small forces splitting the atomic levels, and thereby modulating the decay curves, we deal with coherent, non-stationary superpositions of atomic eigenstates. This article reviews the study of such states excited in collisions with thin foils, with particular emphasis on the coherence properties.

The beam-foil source is well suited to such studies for the reasons stated above, although it has some disadvantages also. Paramount among these is the absence of a reliable theory for the anisotropy. At present, each level must be examined experimentally for evidence of anisotropy, a procedure which considerably limits application of the beam-foil time-of-flight technique. Cascade population of levels represents a related, but apparently less serious problem. When an intermediate level is populated by cascades from higher levels, only a fraction of the anisotropy of the higher levels can be transferred to the intermediate level. Both radiative and Auger transitions contribute to population of levels by cascade, with Auger transitions being particularly important for the more highly-ionized states. When a level is populated mainly by cascades, its anisotropy can be quite low, especially if several successive transitions are involved.

The loss of anisotropy in cascades can be understood by considering the entire system of the excited ion and any Auger electrons it emits or the radiation field. This is an isolated system whose anisotropy cannot change with time. In a transition, the emission products, electrons or photons, carry away a portion of the anisotropy, which is manifest in the angular distribution of the electrons or photons. The lower state then has less anisotropy than the initial excited state.

In all of the work involved with the time evolution of atomic states, a knowledge of the post-foil beam velocity is required. Thus, modulation frequencies, and, hence, term splittings, can only be measured to the accuracy of the velocity determination.

A variety of different approaches have been used to obtain this velocity. In one method, the velocity is determined from the initial energy obtained by monitoring the accelerator terminal voltage and subtracting the energy decrement for passage through a foil of appropriate thickness. The post-foil beam may also be tuned to a sharply-resonant nuclear reaction of known energy. A measurement of the Doppler shift or radiation from the beam yields the beam velocity directly. The beam velocity has also been determined from the spatial modulations of light intensity from a level with previously-known fine or hyperfine structure. Finally, the post-foil energy may be measured directly by an electrostatic analyzer. The last method permits convenient and continuous monitoring of the post-foil beam while allowing for changes due to foil deterioration during the course of a run. With this approach, the energy may be determined to 0.1% [9.1], although 0.01% has already been achieved in one related experiment [9.2].

The usefulness of the foil source for measurements of fine and hyperfine structure depends critically upon further progress in overcoming these disadvantages. We do not examine such applications *per se* in this review. Rather we concentrate on the narrower question of the coherence properties of beam-foil collision-excited atomic states, reviewing the experimental manifestations of coherence within the context of perturbed angular correlations. For an excellent recent review of the status of fine structure, hyperfine structure and Lamb-shift measurements by the beam-foil technique we refer to the article by ANDRÄ [9.3].

9.1 General Theoretical Considerations

It is convenient to treat two different aspects of these phenomena. Firstly, there is the physics of the emission process, and secondly, there are symmetry considerations which are particularly relevant to the experiments in beam-foil spectroscopy. We treat these two matters in succession.

9.1.1 The Emission Process

We deal here with the interference of coherence emissions in the decay of non-stationary states. The basic question is, in what situation do we add amplitudes coherently? In the coherent case, an interesting cross-term is involved in describing intensity, while in an incoherent case, the intensities add directly with no additional term. This question was answered in the early 1930's with the advent of the quantum theory of radiation. These results are well-known and form part of the lore generally implicit in the treatment of perturbed angular correlations. We review the basic question here because the beam-foil time-of-flight technique offers the possibility for new tests of the quantum theory of radiation. Such tests are of interest, since some formulations of the semiclassical radiation theory give predictions at variance with the accepted quantum theory.

Consider the two alternative decay schemes illustrated in Fig.9.1.

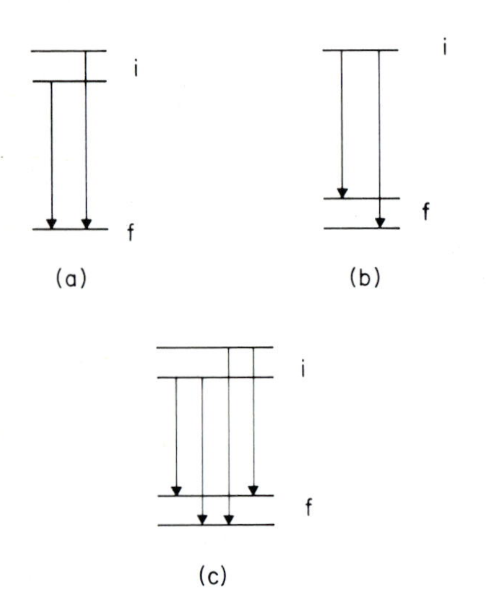

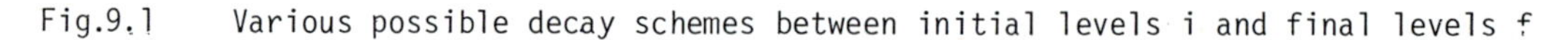
Fig.9.1 Various possible decay schemes between initial levels i and final levels f

In case (a), the upper levels have an energy splitting small compared to the energy separation of the upper and lower levels, while in case (b), the lower levels have a small splitting. In case (a), the final state of the atom and radiation-field system consists of an atom in level f and one photon in the external field. The two downward arrows represent different paths to the same final state of atom-plus-field. According to the basic principles of quantum theory, the probability for finding the system in its final state is obtained by first adding the amplitudes for different paths and then squaring the sum, i.e., we add the amplitudes coherently. In case (b), the two arrows represent different paths to different eigenstates of the atom plus radiation field. Accordingly, the two contributions are combined incoherently, that is, we add probabilities. In general, we have both possibilities simultaneously, as in case (c).

Formally, one demonstrates these arguments as outlined by BREIT [9.4]. The final eigenstate of the atom plus radiation field is

$$|\psi\rangle = \int d\underset{\sim}{k}' \sum_{f} a_f(\hat{\varepsilon},\underset{\sim}{k}',t)|\phi_f\rangle|1_{k',\hat{\varepsilon}}\rangle \quad , \tag{9.1}$$

where $|\phi_f\rangle$ denotes an eigenstate of the atom in its lower level and $|1_{\underset{\sim}{k}',\hat{\varepsilon}}\rangle$ is an eigenstate of the radiation field corresponding to one elementary excitation (photon) with wave vector $\underset{\sim}{k}'$ and polarization vector $\hat{\varepsilon}$. The amplitudes $a_f(\hat{\varepsilon},\underset{\sim}{k}',t)$ represent sums of amplitudes for all paths leading to the same lower atomic eigenstate $|\phi_f\rangle$ and photon eigenstate $|1_{\underset{\sim}{k}',\hat{\varepsilon}}\rangle$.

$$a_f(\hat{\varepsilon},\underset{\sim}{k}',t) = \sum_p a_f^{(p)}(\hat{\varepsilon},\underset{\sim}{k}',t) \quad , \tag{9.2}$$

where p denotes alternative paths. The amplitudes $a_f^{(p)}(\hat{\varepsilon},\underset{\sim}{k}',t)$ are calculated, for example, in the Wigner-Weisskopf approximation [9.4].

The probability amplitude that a photon of wave-vector $\underset{\sim}{k}$ exists in the external field equals the projection of $|\psi\rangle$ onto $|1_{\underset{\sim}{k},\hat{\varepsilon}}\rangle$. Using $\langle 1_{\underset{\sim}{k},\hat{\varepsilon}}|1_{\underset{\sim}{k}',\hat{\varepsilon}}\rangle = \delta(\underset{\sim}{k} - \underset{\sim}{k}')$, this projection becomes

$$\langle 1_{\underset{\sim}{k},\hat{\varepsilon}}|\psi\rangle = \sum_f a_f(\underset{\sim}{k},\hat{\varepsilon},t)|\phi_f\rangle \quad . \tag{9.3}$$

Equation (9.3) corresponds to a non-stationary atomic wave function

$$\psi(\xi) = \sum_f a_f(\underset{\sim}{k},\hat{\varepsilon},t)\Phi_f(\xi) \quad , \tag{9.4}$$

where ξ represents the totality of atomic coordinates. The probability for finding a photon of polarization $\hat{\varepsilon}$ and wave-vector $\underset{\sim}{k}$ in the external field when the atomic coordinates have values between ξ and $\xi + d\xi$ is just $|\psi(\xi)|^2\, d\xi$. The probability for finding a photon in the external field independent of the value of the atomic coordinate equals the integral of $|\psi(\xi)|^2$ over all atomic coordinates. By the orthogonality of atomic wave functions we have

$$\int |\psi(\xi)|^2\, d\xi = \sum_f |a_f(\underset{\sim}{k},\hat{\varepsilon},t)|^2 \quad . \tag{9.5}$$

The probability $\rho(t)$ that the photon detector register a count equals the integral of (9.5) over the detector response function $F(\underset{\sim}{k})$

$$\rho(t) = F(\underset{\sim}{k}) \sum_f |a_f(\underset{\sim}{k},\hat{\varepsilon},t)|^2\, d\underset{\sim}{k} \quad . \tag{9.6}$$

This result expresses the conclusions stated above, namely, in (9.6) and (9.2), we sum incoherently over <u>alternative</u> lower eigenstates, but coherently over alternate paths to the <u>same</u> final eigenstate. For our purposes here, one important consequence of this is that oscillations in $\rho(t)$ (or $\dot{\rho}(t)$) occur only at frequencies corresponding to the splittings of the upper levels. In contrast, some formulations of the semi-classical theory predict oscillations at frequencies corresponding to splittings of the lower levels as well. For a detailed discussion emphasizing the differences and similarities of the quantum and semi-classical theories, the reader is referred to the article by KORNBLITH and EBERLY [9.5].

Once the basic coherence rules are established it is unnecessary for us to develop (9.6) further, using approximate expressions for the amplitudes $a_f(\underset{\sim}{k},\hat{\varepsilon},t)$. Rather, we simply take over the results of the quantum theory of radiation as it applies to observation with high time resolution and correspondingly broad spectral resolution, basing our expressions for radiation from non-stationary states on plausibility arguments and the "coherence rules".

The intensity I measured by an ideal detector sensitive to light with polarization vector $\hat{\varepsilon}$ is proportional to $\sum_{m_f} \langle |(f|\hat{\varepsilon}^*\cdot\underset{\sim}{r}|i)|^2\rangle$, where $\underset{\sim}{r}$ is the transition dipole operator, $\sum_{m_f}$ indicates summation over all values of the final magnetic quantum numbers m_f, and $\langle\ \rangle$ indicates an averaging over the initial i, according to

$$\langle (i'|O|i)\rangle = \sum_{m_i m_i'} \sigma_{m_i m_i'} (i'm_i'|O|im_i) \quad , \tag{9.7}$$

where $\sigma_{m_i m_i'}$ is the density matrix of the upper levels populated in the beam-foil collision, and O is any operator. The indices m_i and m_i' refer, in general, to the set of magnetic quantum numbers specifying the z-projection of electronic and nuclear angular momenta. The intensity I is written in the form of (9.7) as

$$I = C \sum_f \langle (i'|\exp(iHt/\hbar)\hat{\varepsilon}\cdot\underset{\sim}{r}'|f)(f|\hat{\varepsilon}^*\cdot\underset{\sim}{r}\ \exp(-iHt/\hbar)|i)\rangle\ \exp(-t/\tau) \quad , \tag{9.8}$$

where C is a constant incorporating various geometrical factors, fundamental constants, and the population density of the initial levels. The time dependence is introduced via the two factors exp $(\pm iHt/\hbar)$ and $\exp(-t/\tau)$. The damping factor merely expresses the exponential damping of an initial level according to its mean life τ. Here we have also used the result that levels split by a small perturbation decay with the same mean life. The factors $\exp(\pm iHt/\hbar)$, where H is the Hamiltonian for the small interactions splitting the initial levels, operate on the eigenstates $|i)$ of the initial levels. The absence of similar factors operating on the final eigenstates follows from (9.6), where the lower levels are summed over incoherently.

Equation (9.8) expresses the intensity I in terms of a large number of initial-state density-matrix elements $\sigma_{ii'}$. The equations of perturbed angular correlation theory take a more transparent form if we use as our basic source parameters an alternative set of quantities, namely the alignment tensor $\underset{\approx}{A}$ and the orientation vector $\underset{\sim}{O}$ expressed as mean values of angular momentum operators $\langle (i|T_q^{[k]}|i')\rangle$ as in the review by FANO and MACEK [9.6]. Specifically, $\underset{\sim}{O}$ equals $\langle (i'|\underset{\sim}{J}|i)\rangle / \langle (i'|J^2|i)\rangle$, while the alignment tensor consists of elements such as $\langle (i|3J_z^2 - J^2|i)\rangle / \langle (i'|J^2|i)\rangle$. The time dependence is introduced as in (9.8) by inserting the factors exp $(\pm iHt/\hbar)$ operating on $|i)$ and $|i')$: thus, $\langle T_q^{[k]}\rangle \to \langle (i'|\exp(iHt/\hbar)\ T_q^{[k]} \exp(-iHt/\hbar)|i)\rangle$. Here k=1 refers to orientation and k=2 to alignment.

The choice of angular momentum operators requires some care. We could choose $\underset{\sim}{J}$, or $\underset{\sim}{I}$, or $\underset{\sim}{F} = \underset{\sim}{J}+\underset{\sim}{I}$. Our choice is dictated by the following consideration. The nuclear spin plays only a very minor role in the initial excitation process. Accordingly, the parameters describing the collision-excited initial state depend only trivially on the nuclear spin I and would be essentially unchanged if I were equal to zero. The appropriate parameters to use relate only to the electronic state. Thus we use $\underset{\sim}{J}$ rather than $\underset{\sim}{F}$ to construct $T_q^{[k]}$. Similarly, if the spin-orbit interaction is weak and plays no role in the initial excitation we construct $T_q^{[k]}$ from components of $\underset{\sim}{L}$ instead of $\underset{\sim}{J}$.

9.1.2 Symmetry Considerations

The most common geometry in use in beam-foil collisions has cylindrical symmetry, as mentioned earlier. For such symmetry, all of the components of O vanish, and all components of A vanish except $A_0^{col} = \langle (i'|3J_z^2 - J^2|i)\rangle / j_i(j_i+1)$. The superscript "col" means that the coordinate system is defined by the collision geometry with the z-axis along the incident beam direction. For collision-excited atoms, A_0^{col} is given, in terms of the cross-sections σ_{m_j} for exciting the magnetic substates m_j, by

$$A_0^{col} = \textstyle\sum_{m_j} [3m_j^2 - j(j+1)]\sigma_{m_j} / [j(j+1) \sum_{m_j} \sigma_{m_j}] \quad .$$

The light intensity in this case is given by [Ref.9.6, Eqs.(10),(17)]

$$I = \frac{1}{3}\,CS\{1 - \frac{1}{2}\,h^{(2)}(j_i,j_f)A_0^{col}[\tfrac{1}{2}(3\cos^2\theta-1) - \tfrac{3}{2}\sin^2\theta\,\cos 2\psi\,\cos 2\beta]\}\exp(-t/\tau) \tag{9.9}$$

The factor $h^{(k)}(j_i,j_f)$ equals

$$(-1)^{j_i-j_f}\begin{Bmatrix} j_i & j_i & k \\ 1 & 1 & j_f \end{Bmatrix}\begin{Bmatrix} j_i & j_i & k \\ 1 & 1 & j_i \end{Bmatrix}^{-1} \quad ,$$

and S equals the Condon-Shortley "line-strength" parameter to within a factor $e^2(2j_i+1)$. Notice that I is independent of the sign of β, indicating that the radiation from a source with cylindrical symmetry exhibits only linear polarization and no circular polarization. Consequently all applications of (9.9) will suppose β=0.

β specifies the elliptical polarization selected by the detector. For a given experiment β is fixed; β=0 represents selection of linear polarization and $\beta=\pi/4$ represents selection of circular polarization. The angle θ in Fig.9.2 is the angle between the beam axis and the photon detector axis, ϕ is the azimuth of the plane containing the beam axis and detector axis, and ψ specifies the angle between the polarization analyzer's axis of maximum transmission of linearly polarized light and a line perpendicular to the detector axis drawn to intersect the beam axis.

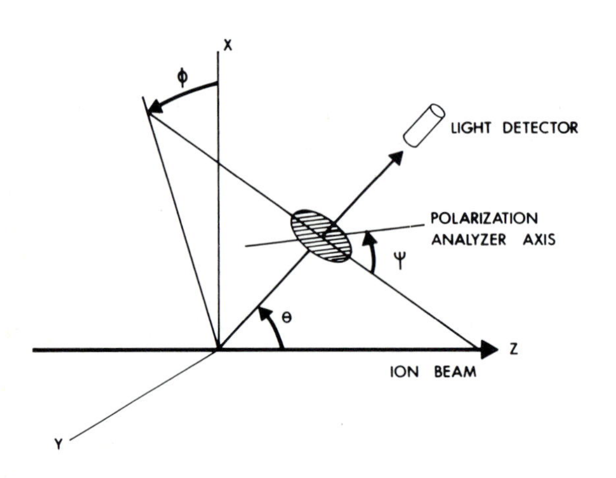

Fig.9.2 Coordinate system in which θ and ϕ represent the polar coordinates of the photon detector in the collision frame with the z-axis along the beam axis. ψ is the angle between the axis of a linear polarization analyzer and a normal to the detector axis drawn to intersect the z-axis

Equation (9.9) with β=0 forms the basis for our discussion of sources with cylindrical symmetry (and reflection symmetry in any plane through the symmetry axis). Cylindrical symmetry is destroyed if the foil itself has a crystalline axis of symmetry, or if the emitted light is detected in coincidence with an atom or ion scattered into a final direction not collinear with the initial direction, or if the foil is tilted with respect to the incident beam direction. Some recent measurements have employed the tilted-foil geometry. In this instance, the light has a more complicated polarization and angular distribution since now one component of the orientation vector does not vanish and three components of the alignment tensor do not vanish.. Defining the collision reference frames as in Fig.9.2 with the z-axis along the beam axis, the x-axis in the plane of the normal to the foil and the beam axis, and the y-axis perpendicular to the xz plane, we have

$$I = \frac{1}{3}CS\Big\{1 - \frac{h^{(2)}}{2}(j_i,j_f)[A_0^{col}(\frac{3\cos^2\theta-1}{2} + A_{1+}^{col}\frac{3\sin2\theta\cos\phi}{2} + A_{2+}^{col}\frac{3\sin^2\theta\cos2\phi}{2}]$$

$$+ \frac{3h^{(2)}}{2}(j_i,j_f)[A_0^{col}\frac{\sin^2\theta\cos2\psi}{2}$$

$$+ A_{1+}^{col}\{\sin\theta\sin\phi\sin2\psi + \sin\theta\cos\theta\cos\phi\cos2\psi\}$$

$$+ A_{2+}^{col}\{(\frac{1+\cos^2\theta}{2})\cos2\phi\cos2\psi - \cos\theta\sin2\phi\sin2\psi\}]\cos2\beta$$

$$+ \frac{3h^{(1)}}{2}(j_ij_f)O_{1-}^{col}\sin\theta\sin\phi\sin2\beta\Big\} \quad (9.10)$$

where

$$O_{1-}^{col} = \langle(i'|J_y|i)\rangle/j_i(j_i+1)$$

$$A_{1+}^{col} = \langle(i'|J_xJ_z + J_zJ_x|i)\rangle/j_i(j_i+1)$$

$$A_{2+}^{col} = \langle(i'|J_x^2 - J_y^2|i)\rangle/j_i(j_i+1) \quad . \quad (9.11)$$

Other components of $\underset{\sim}{O}$ and $\underset{\approx}{A}$ change sign under reflection in the xz plane and therefore vanish.

Equations (9.9) and (9.10) apply only to coherent superpositions of eigenstates of the same angular momentum. In the case of atomic hydrogen, eigenstates of different parity and orbital angular momentum have their energy levels sufficiently close that effects due to interference of eigenstates with different orbital angular momenta become relevant. We will discuss such coherence using the density matrix language (9.7), incorporating the time-dependence into the theory via the replacement $\sigma_{ii'} = \sigma_{ii'}(0)\exp(i\omega_{ii'}t)$, where $\omega_{ii'} = (E_i - E_{i'})/\hbar$. At present, no comprehensive expression for the intensity including all of the possible interference terms has been given (see, however, [9.7]). We can state, however, that for cylindrical symmetry it must be of the form

$$I = \sum_i C_i\exp(-t/\tau_i) + \sum_{ii'}\bar{C}_{ii'}(t)\exp[-\frac{1}{2}(\tau_i^{-1} + \tau_{i'}^{-1})t]$$

$$\times [-\frac{1}{4}(3\cos^2\theta - 1) + \frac{3}{4}\sin^2\theta\cos2\psi] \quad (9.12)$$

where $\bar{C}_{ii'}(t)$ are oscillatory terms proportional to $|\sigma_{ii'}(0)|\cos(\omega_{ii'}t + \phi_{ii'})$, $\phi_{ii'}$ being the phase of the $\sigma_{ii'}(0)$ density matrix element. Equation (9.12) is considerably more complicated than (9.9), but the dependence upon θ and ψ is the same.

In our applications of (9.9) and (9.10), the extraction of the time-dependence from the initial alignment and orientation parameters represents an essential task. Normally this requires a dynamical calculation of the energies and eigenstates of the Hamiltonian H. Our discussion will emphasize those instances where the time-dependence can be given a particularly simple form.

9.2 Alignment and Linear Polarization

By alignment, we mean that magnetic substates of opposite sign are equally populated, but that the population is a non-uniform function of $|m_j|$. Alignment is frequently observed in the beam-foil source and gives rise to linear polarization of the emitted light as observed in a particular direction. Observable manifestations of alignment appear either in the absence or presence of external fields, and we examine several possibilities.

9.2.1 Zero-Field Measurements

The majority of measurements employing the beam-foil light source use a presumably amorphous carbon foil oriented with its normal parallel to the incident beam direction. Such a source has cylindrical symmetry and reflection symmetry in any plane passing through the beam axis. In the absence of external fields, the intensity is then given by (9.9)

$$I = \tfrac{1}{3}CS\{1 - \tfrac{1}{2}h^{(2)}(j_i,j_f)A_0^{col}[\tfrac{1}{2}(3\cos^2\theta - 1) - \tfrac{3}{2}\sin^2\theta\cos 2\psi]\}\exp(-t/\tau) \quad ,(9.13)$$

Only He-like, Li-like, and hydrogen-like ions have been extensively studied. We consider He-like systems first.

a) Non-hydrogenic species. Since nuclear spin presumably plays no role in the excitation process, and since most measurements are made with the nuclear spins initially randomly oriented, we may factor the time-dependence from $\langle T_q^{[k]}\rangle$ quite easily. The average $\langle (I|\exp(iHt/\hbar)T_q^{[k]}\exp(-iHt/\hbar)|I)\rangle_{nuc}$ over nuclear spins is an irreducible tensor $\bar{T}_q^{[k]}(t)$ acting on electronic variables only.

The Wigner-Eckart theorem enables us to write

$$\langle (J|\bar{T}_q^{[k]}(t)|J)\rangle_{elec} = [(J\|\bar{T}^{[k]}(t)\|J)/(J\|T^{[k]}\|J)]\,\langle (J|T_q^{[k]}(0)|J)\rangle_{elec} \quad , \quad (9.14)$$

where the mean value of $T_q^{[k]}(0)$ is evaluated at t=0, the instant of collision, and pertains to the electronic state only. The factor $G^{(k)} = (J\|\bar{T}^{[k]}(t)\|J)/(J\|T^{[k]}\|J)$ expresses completely the time evolution of the alignment and orientation. It is expressed in terms of recoupling coefficients as (see Refs.[9.8], [9.6,Eq.(37)], and [9.3,Eq.(32)]).

$$G^{(k)}(t) = \sum_{FF'}[(2F'+1)(2F+1)/(2I+1)]\begin{Bmatrix} F' & F & k \\ J & J & I \end{Bmatrix}^2 \cos\omega_{FF'}t \quad , \tag{9.15}$$

where F and F' represent the total angular momentum quantum numbers of the initial hyperfine levels with electron angular momentum quantum number J. The frequencies $\omega_{FF'}$ relate to the initial hyperfine energy levels according to $\omega_{FF'} = (E_{F'} - E_F)/\hbar$. If I=0 but the levels are split by small fine-structure splitting, $G^{(k)}(t)$ is given by

$$G^{(k)}(t) = \sum_{JJ'}[(2J'+1)(2J+1)/(2S+1)]\begin{Bmatrix} J' & J & k \\ L & L & S \end{Bmatrix}^2 \cos\omega_{JJ'}t \quad , \tag{9.16}$$

with the assumption that electronic spin plays no role in the collision excitation process. In this case $T_q^{[k]}(0)$ is constructed from components of orbital angular momentum operator $\underset{\sim}{L}$ rather than $\underset{\sim}{J}$. Correspondingly, j_i and j_f in the expression for $h^{(k)}$ are taken to be L_i and L_f. When fine and hyperfine splittings are comparable, (9.13) still applies, having $T_q^{[k]}$ constructed from components of $\underset{\sim}{L}$, and $\bar{T}_q^{[k]}$ is defined as an average over both electronic and nuclear spins. Since J is not normally a good quantum number for such levels, calculation of $G^{(k)}(t)$ requires a diagonalization of the matrix H for each specific set of levels. Equations (9.9) to (9.16) completely disentangle the geometrical factors from the time-dependent factor and the single source parameter $A_0^{col}(0)$ so that we have $A_0^{col} = G^{(2)}(t)A_0^{col}(0)$. Note that the internal interactions modulate the initial alignment but do not change its tensor character. This follows because $\underset{\sim}{F} = \underset{\sim}{J} + \underset{\sim}{I}$, for example, is conserved, but $\underset{\sim}{J}$ and $\underset{\sim}{I}$ are not. The modulation reflects the reversible exchange of alignment and orientation from electronic to nuclear variables, a process that cannot change the tensor character of $T_q^{[k]}$.

For our purposes here, (9.13) and (9.16) provide three alternative means of measuring $A_0^{col}(0)$. We could measure the modulation amplitudes relative to the non-modulated part of I and thereby extract $A_0^{col}(0)$ from the modulation amplitude. We could also measure the angular distribution of the radiation, with no polarization analysis, say, and obtain $A_0^{col}(0)$ as the coefficient of $\frac{1}{2}h^{(2)}G^{(2)}(t)\frac{1}{2}(3\cos^2\theta - 1)$ in (9.13). Finally, one could vary the orientation of a polarization analyzer, thereby varying ψ at fixed θ to determine $A_0^{col}(0)$ as the coefficient of $\frac{3}{4}h^{(2)}G^{(2)}(t)\sin^2\theta$ in (9.13). In beam-foil studies the first method has been principally used.

WITTMANN et al.'s [9.9] observation of the $3\ ^3P$ $2\ ^3S$ transition in ^4He represents the most extensive measurement of a perturbation pattern by the beam-foil technique, showing 49 periods of oscillation. This measurement is a more refined version of

ANDRÄ's [9.10] earlier measurement, which first demonstrated anisotropy in the beam-foil source. WITTMANN's observation is shown in Fig.9.3.

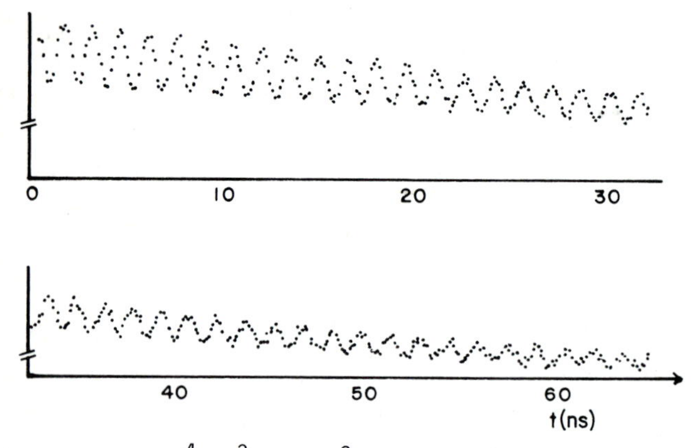

Fig.9.3 He^4 4 $^3P_1$ - 3 $^3P_2$ zero-field quantum beats at 659 MHz. The lower curve is the continuation of the upper curve. The data, from WITTMANN et al. [9.9], show the line intensity as a function of time after excitation

A similar measurement by WITTMANN et al. [9.11] with higher time resolution shows a high-frequency modulation superposed on the low frequency curve; see Fig. 9.4. When account is taken of the finite time window [9.11] the relative amplitudes of the oscillation terms are in good agreement with (9.16). The alignment was found to be negative, but its absolute value was not reported.

The sign of A_0^{col} was deduced by noting that with $h^{(2)}(1,0) = -2$, $\theta=90°$ and $\psi=0°$ in (9.13), the intensity was maximum at t=0. This can only happen if A_0^{col} is negative. Measurements for initial n=3,5, and 7 ^3P levels were also made and are shown in Fig. 9.4. The excellent agreement between theory and experiment confirmed the correctness of (9.13) and (9.16), with all of their implicit assumptions of spin independence, cylindrical symmetry, and impulsive excitation, for application to the beam-foil source. This experiment, and another by BURNS and HANCOCK [9.12], were quite important in placing the theory on a sound experimental foundation.

BURNS and HANCOCK [9.12] measured a relative modulation amplitude of 0.13 for the J = 2,1 frequency in the 3 $^3p \rightarrow 2\ ^3S$ transitions in He I excited by collisions of 200 keV He^+ with thin carbon foils. This amplitude corresponds to an alignment of -0.014. They also determined the initial phase of the modulation and found it to be (+ 0.06, -0.09). This confirms the hypothesis that the electron spin is indeed randomly oriented, as assumed in (9.14).

Measurements of the hyperfine structure [9.13-15] in ^3He and $^{6,7}Li$ II indicate that the n ^3P levels of these atoms are aligned in the beam-foil collisions, but the degree

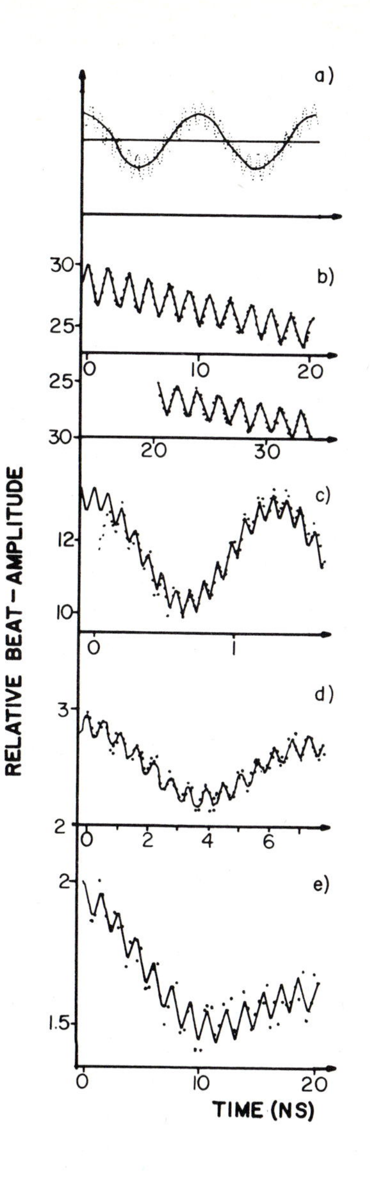

Fig. 9.4 a) Calculated quantum beat of $n\ ^3P\ ^4He$
b) Time calibration derived from the $3\ ^3P_2 - ^3P_1$ 658.55 MHz frequency
c) to e) Experimental data for n=3,5,7 respectively, with fitted functions. (Data of WITTMANN et al. [9.11])

of alignment has not been reported, although it is expected that the alignment does not depend upon nuclear spin. No obvious departures from this expectation are apparent.

POULSEN and SUBTIL [9.16] have measured the alignment of some ^2D and ^2F levels in Be II for 50-500 keV, ions using the zero-field technique. They observed the unresolved $^2D_{J=5/2,3/2}$ and $^2F_{7/2,5/2}$ multiplets. The alignments given in Table 9.1, deduced from the polarization they report, accordingly refer to an average over the unresolved fine structure. Specifically, we have from (9.13),

$$h^{(2)}(L_i,L_f)A_0^{col} = 4P/(3-P) \quad , \qquad (9.17)$$

where P is the polarization of the light emitted perpendicular to the beam at t=0, as defined by POULSEN and SUBTIL [9.16]. In this case, the observed oscillations are due to the hyperfine splitting. The fine-structure splitting is

Table 9.1 Alignment of some levels in Be II deduced from measurements of [9.16]

Transition	$h^{(2)}$	$\overline{A_0^{col}}(0)$
4d $^2D \rightarrow$ 3p ^2P	-1	-0.106 ± 0.027
5d $^2D \rightarrow$ 3p ^2P	-1	-0.045 ± 0.013
4f $^2F \rightarrow$ 3d ^2D	-4/5	-0.128 ± 0.034
5f $^2F \rightarrow$ 3d ^2D	-4/5	-0.078 ± 0.034

averaged over and contributes to a reduction in the alignment according to

$$\overline{A_0^{col}} = A_0^{col}(0)\ \overline{G^{(2)}(t)}$$

$$= A_0^{col}(0) \sum_J (2J+1)^2 (2S+1)^{-1} \begin{Bmatrix} J & J & 2 \\ L & L & S \end{Bmatrix}^2 \quad , \tag{9.18}$$

where we have supposed $\tau/\omega_{JJ'} \approx 0$.

b) Hydrogen and Hydrogen-Like Species. BASHKIN and BEAUCHEMIN [9.17] used a photographic technique to examine the emissions from beam-foil-excited He and He^+ in the spectral region from 4300 to 6800Å. In He II the emissions associated with the ^2G levels of the n=7,8 and 9 terms showed an interesting temporal behavior. In each case, the ions appeared to cease to radiate at some time after exit from the foil. Further, the time involved was different for each transition. It is difficult to draw conclusions concerning the origin of these particular beats. It is probable that in the strict zero field that there will be at least four frequency components (and probably more if we include the possibility of coherence between states with different total angular momentum) associated with the light intensity from each of these states. It is unlikely that these oscillations could have been resolved in this experiment. Further, an electric field of approximately 2 V/cm is sufficient to mix $^2P_{3/2}$ with $^2D_{3/2}$ of the n=7 term of He II, while 0.8 V/cm will mix the $^2D_{5/2}$ and $^2F_{3/2}$ levels of the same term. In the experiment for an energy of 0.7 MeV, the smallest used in the experiment, the motional electric field due to the earth's magnetic field alone is at least 3 V/cm, in excess of that needed to cause the mixing of the above levels. Consequently, it is likely that the effect is due to an interference associated with the Stark effect, as originally suggested by BASHKIN and BEAUCHEMIN [9.17].

Following MACEK's [9.18,19] suggestion of the existence of zero-field beats SELLIN et al. [9.20] performed an experiment on H atoms of 50-150 keV energy. Using photoelectric detection and observing a specific polarization of the radiation, they obtained the time-dependence of the Balmer emissions H_β and H_γ while sampling a beam length of 6.4 mm, giving a frequency window of 490 and 860 MHz for detection of oscillations in the emission from H atoms with energies of 50 and 150 keV respectively. In addition, the Ly-α emission was monitored by sampling a beam length of 1.6 mm, giving equivalent frequency windows of 2 and 3.4 GHz. Although it was thought that oscillations of magnitude ≤ 5% could be identified, none was observed.

The experiment could have recorded oscillations associated with beats between the $D_{3/2}$, $D_{5/2}$ levels of the n=4 term for H_γ. This beat has a frequency of 457 MHz, but it is now known that the alignment is such that the amplitude of the oscillations is less than 1% of the average background in this case and therefore smaller than the sensitivity of the experiment. Other levels do interfere strongly in the case of H_γ, but the frequencies are all in excess of 1200 MHz and were unfortunately outside the range of the experiment. A frequency resolution of approximately 11 GHz is required to resolve the fine-structure beats between $P_{3/2}$ and $P_{1/2}$ for the n=2 term of H. Thus it was not surprising that no beats were detected in the case of the Ly-α emission, since the frequency window for the experiment was 3.4 GHz.

ANDRÄ [9.10] made the first definitive experiment on zero-field quantum beats in both hydrogen and helium. Protons of energy 133 keV were sent through a foil, and the subsequent emissions from a 1 mm section of the beam were examined. This is equivalent to a frequency of 4 GHz. It was shown that relatively large amplitude beats were found in the n=3 and 4 terms of hydrogen. Radiation polarized parallel to the beam was compared with that perpendicular to the beam, and was shown to be approximately phase reversed by π as required by (9.9) with $\theta=90°$ and $\psi=0$ and $\pi/2$. Because of the ℓ-degeneracy in hydrogen and the relatively complex nature of the ensuing beats, it was impossible to draw definite conclusions regarding the magnitude of relative magnetic substate populations and hence the actual alignment. Nevertheless the existence of the beats was sufficient to confirm that states were indeed aligned as the result of collisions within the foil. It was clear that a new technique now existed for the measurement of fine-structure detail of levels in atoms and ions. In confirmation of this ANDRÄ [9.10] showed that, e.g., the H_α emission had a beat frequency of 1063 ± 30 MHz corresponding to the $D_{3/2}$ - $D_{5/2}$ fine-structure splitting. Similarly in H_β frequencies of 1390 ± 20 MHz and 440 ± 20 MHz were decomposed from the data and corresponded to the fine-structure separations of $P_{1/2}$-$P_{3/2}$ (1371.07 MHz) and $D_{3/2}$-$D_{5/2}$ (457.03 MHz).

LYNCH et al. [9.21] made the first measurement of zero-field quantum beats in the vacuum ultraviolet. They observed the 1026Å Ly-β emission from the n=3 term of hydrogen. The experiment was performed in an environment where the magnetic field was less than 0.03G perpendicular to the beam axis, to reduce the motional electric field experienced by the H atoms and eliminate the possibility of Stark mixing of the S and D with the P states of the n=3 term. A specific polarization was not detected. Instead a signal was recorded which was proportional to the sum of the radiation polarized perpendicular to the beam plus 3.4 times that of radiation polarized parallel to the beam. In this case it is relatively simple to write down the explicit dependence of the light intensity. Ignoring the hyperfine splitting, since it is small compared to the fine structure, we have from (9.9) with $h^{(2)} = -2$

$$I = \tfrac{1}{3}CS\{1 + A_0^{col}(0)G^{(2)}(t)\ [-\tfrac{1}{2} - \tfrac{3}{2}\cos 2\psi]\}\ \exp(-t/\tau) \quad . \tag{9.19}$$

Here

$$A_0^{col}(0) = (\sigma_1 - \sigma_0)/(2\sigma_1 + \sigma_0) \quad ,$$

$$G^{(2)}(t) = \frac{1}{3}(1 + 2\cos\omega t) \quad . \tag{9.20}$$

Substituting ψ=0 for radiation polarized parallel to the beam,

$$I_{\parallel} = C' \{4\sigma_1 + 5\sigma_0 + 4(\sigma_0 - \sigma_1)\cos\omega t\} \exp(-t/\tau) \quad . \tag{9.21}$$

With $\psi = \pi/2$,

$$I_{\perp} = C'\{7\sigma_1 + 2\sigma_0 - 2(\sigma_0 - \sigma_1)\cos\omega t\} \exp(-t/\tau) \quad , \tag{9.22}$$

where σ_0 is the cross section for exciting the magnetic substate m_ℓ=0, σ_1 is the cross section for exciting the magnetic substate $|m_\ell| = 1$, $\omega = [(E_{3/2} - E_{1/2})/\hbar]$ is the frequency separation of the $P_{1/2}$-$P_{3/2}$ levels, and τ is the lifetime of the P state.

The data exhibited weak but reproducible oscillations and were analyzed in terms of the intensity $I_\perp + 3.4\ I_\parallel$. From the relative magnitudes of the oscillatory and non-oscillatory components of this expression it is possible to evaluate the relative magnitude of the cross sections σ_0 and σ_1. Consequently from this experiment it is possible to obtain the alignment A_0^{col}. LYNCH et al. [9.21] thus obtained the negative of the polarization (i.e., $(\sigma_1-\sigma_0)/(\sigma_1+\sigma_0)$) between approximately 200 and 500 keV/atom. We have calculated the equivalent alignments; they are shown in Fig. 9.5 as dots.

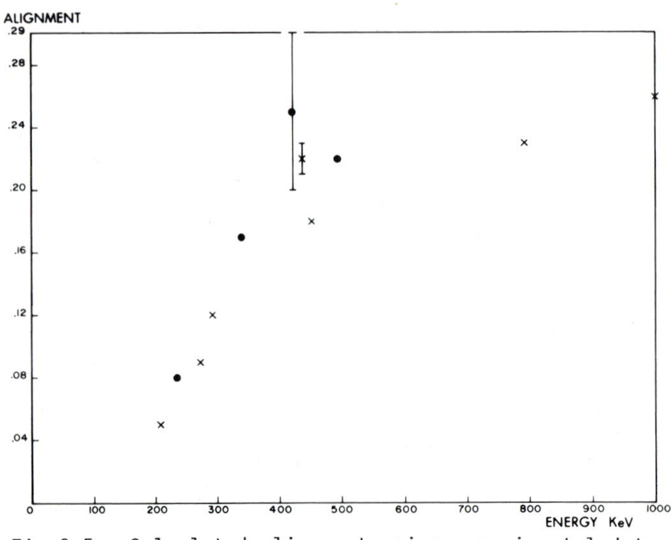

Fig.9.5 Calculated alignment using experimental data on zero-field beats for n=2 (x, based on DOBBERSTEIN et al. [9.23]) and n=3 (●, based on LYNCH et al. [9.21]). Note the apparent peak in the alignment near 420 keV

A maximum alignment of approximately 0.25 was found at $\sim$ 420 keV.

BURNS and HANCOCK [9.22] examined the H_γ emission of hydrogen with an ultimate resolution of 10 GHz and with a magnetic field such that the motional electric field experienced by an atom in the beam was 0.19 V/cm. The resultant decay for radiation polarized parallel to the beam was shown to consist of four separate frequencies corresponding to the $P_{1/2}$-$P_{3/2}$ separation (± 6% oscillations) and the $D_{3/2}$-$D_{5/2}$ separation (± 0.8% oscillations). Somewhat surprisingly, there were two other components corresponding to the S-$D_{3/2}$ and S-$D_{5/2}$ separations with respective beat amplitudes of ± 2.5% and ± 3.7%. This was the first confirmation of the possibility of the existence of these interesting beats alluded to by MACEK [9.18,19] which require a coherence between the excitation amplitudes for the $S_{1/2}$ state and the $D_{3/2}$ and $D_{5/2}$ states. In addition BURNS and HANCOCK [9.22] showed that it was the alignment rather than the total intensity which was modulated. In (9.12) above, if θ is set equal to 90° and ψ is set equal to 54°44', then the expression contains no oscillatory component. Figure 9.6 shows the result of setting the polarizer at this "magic" angle, where

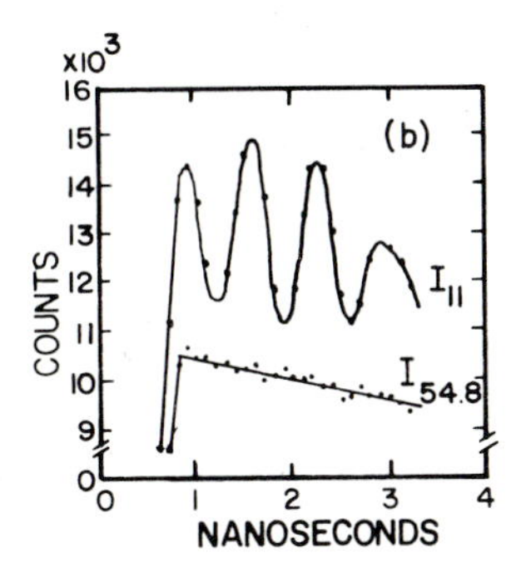

Fig.9.6 Oscillations associated with radiation polarized parallel to the beam axis for H_γ. No modulations are observed with the polarizer at an angle ψ = 54°44'. (Data are from BURNS and HANCOCK [9.22])

no oscillation is recorded in the emitted intensity. This is an indirect confirmation of the fact that the excitation and radiation processes are axially symmetric, and that the oscillations are not due to a field in the viewing region.

DOBBERSTEIN et al. [9.23] observed beats in Ly-α radiation associated with the 10.969 GHz splitting of the n=2 term of H. The experiment was performed in the earth's magnetic field, but in this case Stark-mixing effects associated with the motional field are negligible. The data were analyzed in an analogous manner to the similar experiment of LYNCH et al. [9.21], except that the detection system was assumed to have no intrinsic polarization. From these results, it is possible to obtain the alignment again, and this is shown as a function of beam energy in Fig.9.5. It is interesting to note that both this experiment and that of LYNCH et al. find a peak in the alignment at approximately the same energy for the n=2 and n=3 terms of H. These are the only experiments which have obtained specific values for the alignment by using zero-field quantum beats in H.

9.2.2 Electric Field

There has been considerable development in both theory and experiment since BASHKIN et al. [9.24] first reported Stark-effect-related interferences associated with four members of the Balmer series of hydrogen. Beam-foil-excited hydrogen was observed in emission in an electrostatic field created by two condenser plates parallel to the beam axis. The Arizona group used mostly photographic detection techniques in this and similar early work, with the result that it was impossible to make careful analysis of the beat patterns. BICKEL and BASHKIN [9.25] and BICKEL [9.26] observed fluctuations in four emissions from the n=7, 8, 9 and 10 terms of He II (the same three lines at 5411, 4859 and 4541Å were also observed to have intensity fluctuations in the "zero-field" experiment of BASHKIN and BEAUCHEMIN [9.17] described above.) This and further work illustrated that the two-level weak-field Stark effect was not adequate to explain the experimental observations.

SELLIN et al. [9.27,28] made a considerable improvement in interpreting data by calculating a multi-frequency beat pattern using a formulation of the Bethe-Lamb theory of the Stark effect. In hydrogen, for the n=5 level, it was found possible to explain the prominent frequencies in the beat pattern by assuming a preferential population of certain angular momentum substates, with particular emphasis on the D state. This was discovered by reproducing a series of different beat patterns by assigning different weights to the components associated with the S, P, D, F, and G fine-structure levels. Thus the so-called triangular weighting with the above ratios 1:2:3:2:1 produced a pattern which in appearance was similar to that recorded photographically. In addition to this work on Balmer radiation, intensity oscillations with close to the predicted frequency were found in the case of Ly-α. A single frequency of 1340 MHz was recorded for an electric field of 197 V/cm. As expected, this is slightly larger than the Lamb-shift frequency of 1058 MHz and is due to the electric-field-induced coherence between the $S_{1/2}$ and $P_{1/2}$ states.

ANDRÄ [9.29] investigated Stark-induced beats in Ly-α, using electric fields both perpendicular and parallel to the beam. He also used a magnetic field to obtain an equivalent motional field. ANDRÄ draws a careful distinction between coherence between $S_{1/2}$ and $P_{1/2}$ produced by the collision processes within the foil and that which is produced by impulsive entry of the atom into an electric field. In this experiment, with observation times limited to 5 ns, only the single frequency between the Stark components of the $S_{1/2}$ and $P_{1/2}$ levels was detected. Excellent agreement was obtained between the predicted beat frequencies and those observed, for a number of electric field strengths parallel to the beam. There was agreement between these results and those for beats induced by the motional electric field. However, no beats could be detected with a field less than 240 V/cm with an electric field perpendicular to the beam axis. Above 240 V/cm relatively weak beats were observed. ANDRÄ explains this reduction of the beat amplitude as being due to the existence of excitation coherence between the $S_{1/2}$ and $P_{1/2}$ levels, since in the case of the Stark effect the observed intensity fluctuations should be independent of the field direction. In retrospect, this conclusion is correct, since such coherence is found to exist. However, it is likely that ANDRÄ's results for the perpendicular electric field were severely affected by fringing fields associated with the design of the condenser plates used to produce the field. A later experiment [9.30] extends the analysis of Ly-α beats induced by the magnetic-field-induced Stark oscillations.

GAUPP [9.31] has observed Stark beats associated with the perturbation of an initially pure $^2S_{1/2}$ state in hydrogen. A beam of metastable atoms prepared in the $2\ ^2S_{1/2}$ state was made to enter a longitudinal electric field impulsively, which produced a coherent superposition of the perturbed S- and P-state wave functions. The resultant emission of Ly-α radiation showed large-amplitude oscillations (see Fig.9.7) at a frequency related to the field strength, permitting a measurement of the zero-field splitting or Lamb shift.

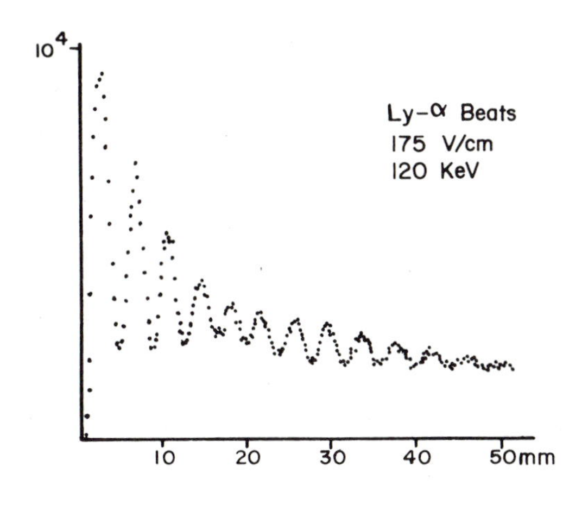

Fig.9.7 Stark-induced beats in Ly-α produced from metastable $^2S_{1/2}$ hydrogen atoms in a longitudinal electric field. (Data from GAUPP [9.31])

ALGUARD and DRAKE [9.7] have extended the study of Stark-induced Ly-α beats, while including work on Ly-β beats. This work was performed with a motional electric field to try to reduce some of the problems with fringing fields outlined above. In all, 36 possible frequencies were considered for Ly-β, being associated with the Stark-split S, P and D states. Six separate patterns were calculated for the six possible states denoted by ℓ, $|m_\ell|$. These were used to fit the data (see Fig.9.8) with coefficients proportional to the

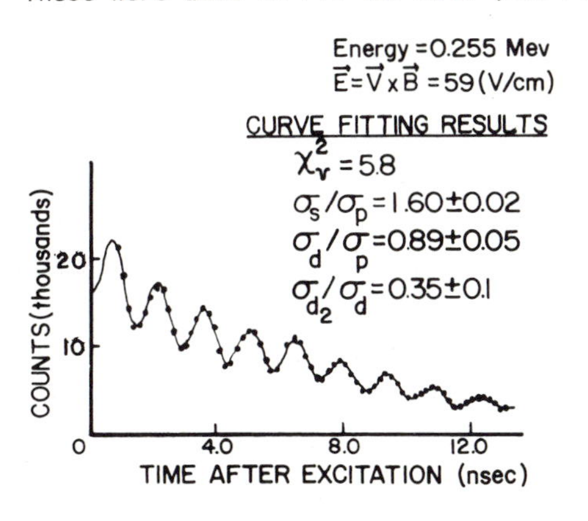

Fig.9.8 Stark-induced beats in Ly-α from beam-foil-excited hydrogen atoms at an energy of 255 keV and a motional electric field of 59 V/cm. Cross-section ratios are obtained from analysis of the data. (Data from ALGUARD and DRAKE [9.7])

cross-sections for exciting the various substates. Thus one could obtain the relative cross-sections. Figure 9.8 illustrates the curve fitting to the Ly-β transition with the computer-generated relative cross-sections. The data show consistently that the cross-section for exciting the S-state is greater than for the P- and D-states; a statistical distribution would be 1:3:5 for $\sigma_s:\sigma_p:\sigma_d$. In the case of Ly-α the presence of a beat minimum close to the foil implies that $\sigma_S=3.3\sigma_{P_{1/2}}$, in disagreement with SELLIN et al. and the first results of ANDRÄ.

ECK [9.32] has commented on these results and on the possibility that S- and P-states could be excited coherently as a result of the excitation process itself rather than because of impulsive entry into an electric field. Since the S-state in the case of Ly-α cannot decay radiatively it is not possible to observe this coherence in the manner of BURNS and HANCOCK. In this case, the Stark effect must be used. With an electric field along the beam axis, the $S_{1/2,1/2}$ level couples with $P_{1/2,1/2}$ while $S_{1/2,-1/2}$ couples with $P_{1/2,-1/2}$. MACEK [9.18,19] has pointed out that these are the possible combinations which may be excited coherently (different L, but same m_J). Ignoring the effect of the $P_{3/2}$ state, the total beat signal is then proportional to

$$(V/\omega)^2[\tfrac{1}{3}(\sigma_{P_0} + 2\sigma_{P_1}) - \sigma_S] \cos\omega t + \sqrt{1/3}\,(V\omega_0/\omega^2)\,\langle |f_S||f_{P_0}|\cos\alpha\rangle \cos\omega t$$
$$+ (V/\omega)\sqrt{1/3}\,\langle |f_S||f_{P_0}|\sin\alpha\rangle \sin\omega t \quad , \qquad (9.23)$$

and is exponentially damped by the lifetime of the P-state. V is the Stark matrix element coupling the $S_{1/2,1/2}$ and $P_{1/2,1/2}$ states, $\omega = (\omega_0^2 + 4V^2)^{1/2}$, where ω_0 is the Lamb shift in angular frequency units, $|f_S|$ and $|f_{P_0}|$ represent the amplitudes for exciting the magnetic substates $m_\ell=0$ of the S- and P-states, and α is the relative phase of the above amplitudes.

The first term in (9.23) represents the contribution to the incoherent signal, while the next two terms represent the contribution associated with coherent excitation of the S- and P-states. Since V is proportional to the component of the electric field along the beam axis, the last two terms change sign on reversal of the field direction. Thus the sum of the signals associated with E_z parallel and anti-parallel with the beam direction gives twice the incoherent contribution, while the difference yields twice the coherent contribution.

SELLIN et al. [9.33] performed an experiment at a beam energy of 186 keV, and an electric field strength of 525 V/cm to obtain the data shown in Fig.9.9. In this figure (a) and (c) represent the sum and difference curves for the electric field alternately parallel and anti-parallel with the beam, and (c) confirms the existence of coherence between the S- and P-states. Further, the average magnitude of the S-P phase coherence angle α was found to be π/2. Thus SELLIN et al. [9.33] conclude there is initial charge isotropy within the atom, but in time this changes, with the electron charge density initially moving toward the backward hemisphere.

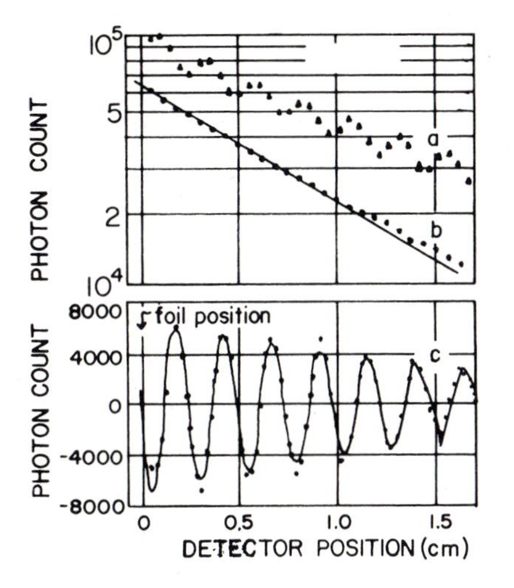

Fig.9.9 Ly-α emission in zero field is shown in (b). The sum of the signals for the electric field parallel and anti-parallel to the beam is shown in (a). The coherent part of the emission equal to the difference between parallel and anti-parallel components is shown in (c). (Results of SELLIN et al. [9.33])

GAUPP et al. [9.34] have performed a similar experiment at 110, 165 and 210 keV with a smaller electric field of 295 V/cm. A more complete analysis of the data was made, which, together with the results of DOBBERSTEIN et al. [9.23] and ALGUARD and DRAKE [9.7], yields the density matrix $\sigma_{ij}(i,j = S, P_1, P_0, P_{-1})$ for n=2 hydrogen at 210 keV.

$$\sigma = \begin{vmatrix} 0.56 & 0 & 0.22e^{2.06i} & 0 \\ 0 & 0.15 & 0 & 0 \\ 0.22e^{-2.06i} & 0 & 0.14 & 0 \\ 0 & 0 & 0 & 0.15 \end{vmatrix} \qquad (9.24)$$

In this case the relative SP phase is 2.06 radians, indicating that the electron charge density is originally concentrated in front of the proton. The measured value of the SP coherence $\sigma_{SP} \propto |f_S||f_{P_0}|$ is 0.22, which is almost the maximum allowable value of 0.28 (i.e., $\{\sigma_S\sigma_P\}^{1/2}$). This is consistent with the similarly large coherence observed by BURNS and HANCOCK [9.22].

9.2.3 Magnetic Field

We consider here only the normal Zeeman effect, i.e., only fields B sufficiently weak that J and J_z are good quantum numbers. Thus the field does not change the value of $\underset{\approx}{A}^{col}$ in a frame rotating about the field axis at a rate $-g_iB$ equal in magnitude to the Larmor frequency ω, where g_i is the gyromagnetic ratio of the initial level. Two geometries are of particular importance: $\underset{\sim}{B} \parallel \hat{z}$ and $\underset{\sim}{B} \perp \hat{z}$, where z is along the ion beam axis. In the first instance, the field does not alter A_0^{col} since the z axes of the precessing frame and the collision frame coincide. In the latter case, θ_p, ϕ_p and ψ_p defined in the precessing frame relate to θ, ϕ and ψ measured in the collision frame by

$$\cos\theta_p = \cos\theta \cos\omega t - \sin\theta \sin\omega t \cos\phi$$

$$\sin\theta_p \cos\phi_p = \cos\theta \sin\omega t + \sin\theta \cos\omega t \cos\phi$$

$$\cos(\psi_p - \psi) = \cos\phi \cos\phi_p + \sin\phi \sin\phi_p \cos\omega t \quad . \tag{9.25}$$

In (9.25) the y axis is along the field direction. Experiments reported to date place the detector in the xz plane, and at right angles to the beam axis z. Hence $\phi=0$ and $\theta=90°$. For this arrangement, we have $\theta_p = 90 + \omega t$, $\phi_p = 0$, and $\psi_p = \psi$. Using these values, (9.9) becomes

$$I = \frac{1}{3}CS\{1 - \frac{1}{2}h^{(2)}(j_i,j_f)A_0^{col}(0)[\frac{1}{4} - \frac{3}{4}\cos2\omega t - \frac{3}{4}(1+\cos2\omega t)\cos2\psi]\}\exp(-t/\tau) \quad . \tag{9.26}$$

Since $\omega = g_i B$, the intensity observed at a fixed time t will exhibit oscillations as a function of magnetic field strength B. This provides a convenient means to measure $A_0^{col}(0)$ as well as g_i. LIU et al. [9.35] measured the polarization of the $2p_9$ levels in Ne I and the $4f\ ^4D$ level in Ne II at 425 keV incident energy by this technique. THe alignment of the $2p_9$ level in Ne II is particularly interesting since a high value, equal to -0.20 ± 0.04, is deduced from the reported 12% polarization. This represents one of the largest alignments observed in a non-hydrogenic atom.

The g_i values of a series of levels in Ar II and III have also been measured with this technique by CHURCH and LIU [9.36] and by CHURCH et al. [9.37]. Although the alignments were not reported, the maximum and minimum amplitudes of the modulation term were. Using (9.26), we find that this "beat" amplitude C relates to A_0^{col} according to

$$h^{(2)}A_0^{col}(0) = -4C/(3+C) \quad . \tag{9.27}$$

In general, $A_0^{col}(0)$ cannot be calculated from (9.27) unless the sign of C is known. The sign of C was not measured in [9.36,37], but in all cases C was less than 0.07, which is small compared to 3. We can therefore calculate the magnitude of $A_0^{col}(0)$ from (9.27) to within 2% by neglecting C compared to 3 in the denominator. Thus we have $|A_0^{col}(0)| \simeq \frac{4}{3}|C|/|h^{(2)}|$. The values of $|A_0^{col}(0)|$ thus calculated from the measured beat amplitudes of the unblended transitions are given in Table 9.2.

Averaging $\cos2\omega t \exp(-t/\tau)$ gives the usual Hanle-effect denominator $[1+4\omega^2\tau^2]^{-1}$, which peaks at zero field. A zero-field level-crossing measurement [9.39] on $5\ ^1D$ state in He indicates that this level is aligned; the degree of alignment was not reported.

Table 9.2 Alignment of Ar II and Ar III levels excited by 425-725 keV Ar^+ collisions with thin carbon foils, deduced from the data of [9.36,37]

Charge State	Transition	$h^{(2)}$	$\lvert A_0^{col} \rvert$
II	$4p\ ^2D^o_{5/2} \rightarrow {}^2P_{3/2}$	-7/8	0.051 ± 0.001
II	$4p\ ^2P_{3/2} \rightarrow {}^2P_{1/2}$	-5/4	0.034 ± 0.001
II	$4p\ ^2F^o_{7/2} \rightarrow {}^2D_{5/2}$	-3/4	0.116 ± 0.004
II	$4p'\ ^2F^o_{5/2} \rightarrow {}^2D_{3/2}$	-7/8	0.073 ± 0.009
III	$4p'\ ^3F^o_4 \rightarrow {}^3D_3$	-5/7	0.041 ± 0.003

YELLIN et al. [9.40] extended the zero-field level-crossing measurements to include the measurement of coefficient of the Hanle-effect term for a series of singlet and triplet D states in He excited by 40 keV He^+ collisions with foils. In their arrangement, the signals at a sequence of distances from the foils were added to get a zero-field level crossing signal. Owing to the finite integration time, the observed signal oscillates in the wings of the line. They express the results of their measurement in terms of polarization of radiation that would be observed at t=0. From their data and (9.17), we obtain the alignment parameters given in Table 9.3.

Table 9.3 Alignment parameter $A_0^{col}(0)$ for some D states of He deduced from the measurement of [9.40]. The incident He^+ energy is 40 keV. In all cases $h^{(2)}(j_i j_f) = -1$

n	$A_0^{col}(0)$	n	$A_0^{col}(0)$
^1D 4	-0.17 ± 0.023	^3D 3	-0.048 ± 0.020
5	-0.13 ± 0.018	4	-0.039 ± 0.021
6	-0.058 ± 0.025	5	-0.054 ± 0.021
		6	-0.068 ± 0.023

A possible complication arises when the Hanle-effect is applied to rapidly-moving atoms. Owing to the motional electric field, $\underset{\sim}{E} = \underset{\sim}{v} \times \underset{\sim}{B}/c$, the atom is also subjected to an external electric field; indeed, this is often used to apply electric fields to hydrogenic species to study the linear Stark effect. The linear effect is absent in non-hydrogenic species, but the quadratic effect is present and could perturb the alignment.[9.41,42].

The perturbation depends upon the polarizability tensor of the excited state and the product vB/c. In the experiments discussed above, the effect is small, owing to the low value of the polarizability, except possibly in 5 $^1D_2$ in He.

9.3 Orientation and Circular Polarization

By orientation we mean that a single magnetic substate is overpopulated with respect to an equilibrium value. This type of non-equilibrium population distribution spread occurs commonly in many beam-foil experiments. The decays of the states so populated give rise to circularly-polarized light. As in the case of alignment, the effects are present both with and without external fields, as we shall now show.

9.3.1 Zero Field

A beam-foil source with the foil tilted at an angle α with respect to the beam axis could, in general, exhibit orientation O_{1-}^{col} as well as an alignment tensor with $A_{1+}^{col} \neq 0$ and $A_{2+}^{col} \neq 0$. It is by no means obvious *a priori* that such a tilted-foil arrangement has a lower degree of symmetry on a microscopic scale, since the foil surface is generally wrinkled, and is also subject to sputtering and impurity build-up under ion bombardment. Nonetheless BERRY et al. [9.43] showed that as the foil is tilted the excited atoms exhibit an increasing orientation, implying that the macroscopic foil-surface direction influences the anisotropy of the excited atom. The one orientation parameter and the three alignment parameters were extracted from the data obtained for 3p ^{1}P at 130 keV by measuring the light intensity with three different settings for a linear polarizer ψ=0°, 45° and 90° with and without a quarter-wave plate. The measured source parameters for the 3 ^{1}P level in He at 130 keV beam energy are given in Table 9.4.

Table 9.4 Alignment and orientation parameters for 5016Å, ^{4}He I, 2 ^{1}S-3 ^{1}P transition at 130 keV beam energy. From [9.43]

Foil Tilt Angle (°)	A_0^{col}	A_{2+}^{col}	A_{1+}^{col}	O_{1-}^{col}
0	-0.090(36)	0.016(9)	...	...
20	-0.081	0.012	-0.024(7)	-0.013(11)
30	-0.072	0.008	-0.021	-0.038
45	-0.054	-0.0002	-0.040	-0.040

The error bars are rather large, but, even so, there is definite proof of non-zero orientation at non-zero tilt angles. At α=0°, all parameters except A_0^{col} should be zero on the basis of cylindrical symmetry. That A_{2+}^{col} is non-zero possibly indicates experimental error larger than quoted, or some departure of carbon foils from strict cylindrical symmetry. A later measurement [9.44] shows that the orientation was also observed in 3p' ^{1}F Ne III and 3p' ^{2}D Ne II. This is illustrated below in

Fig.9.10. It was also found that the anisotropy of the excited atom increases with increasing tilt angle. This could be of practical importance, since levels with no

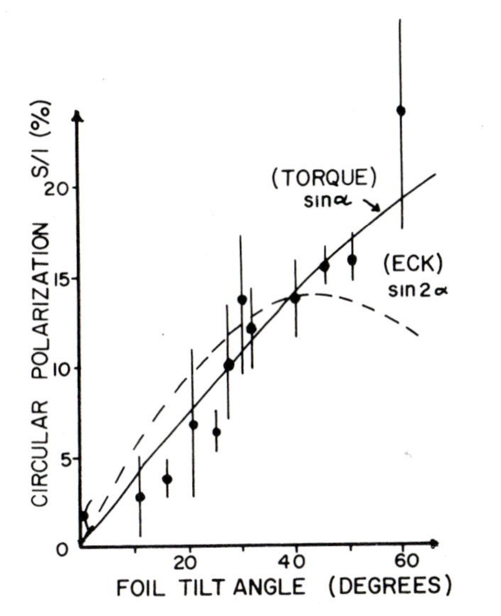

Fig.9.10 The circular polarization of the 3 $^1P \rightarrow 2\ ^1S$ emission of ^{4}He I as a function of foil tilt angle α, at an energy of 130 keV. Data compared with ECK's theory. (Data from BERRY et al. [9.44])

anisotropy at $\alpha=0$ might show anisotropy at $\alpha\neq0$, thereby making them accessible to study by level-crossing and other high-resolution techniques.

9.3.2 Magnetic Field Measurements

LIU et al. [9.45] detected orientation in beam-foil excited O II, Ar II, and He I states at energies between 540 and 1350 keV. They measured the linear and circular polarization of the emitted light for $\underset{\sim}{B}\parallel\hat{z}$, where z is along the beam axis, and $\underset{\sim}{B}\perp\hat{z}$. They do not extract the alignment and orientation components, but they clearly detect orientation via their measurement of circular polarization. There is a tendency for the circular polarization to decrease with increasing ion energy.

CHURCH et al. [9.46] extensively studied the alignment and orientation of the 4d $^1D_2$ level in ^{4}He excited by collision of 40 keV He^+ ions with tilted foils. They measured the light intensity, at a fixed distance downstream from the foil, emitted in a direction perpendicular to the beam as a function of the orientation of both a linear and a circular polarization analyzer and the strength of an applied magnetic field for two directions, $\underset{\sim}{B}\parallel\hat{z}$ and $\underset{\sim}{B}\perp\hat{z}$, of the applied field. Here $\hat{z}$ is along the beam axis. Substituting $\theta_p = \theta + \omega t$ for $\underset{\sim}{B}\perp\hat{z}$ and $\phi_p = \phi + \omega t$ for $\underset{\sim}{B}\parallel\hat{z}$ into (9.10) gives equations for the field-dependent intensity. For a specific polarization-analyzer setting they measured an oscillatory curve for intensity vs. B which they fitted to determine two frequencies and three amplitudes. The amplitudes for a series of polarization-analyzer settings provide the input to determine O^{col}_{1-}, A^{col}_0, A^{col}_{1+}, A^{col}_{2+}. The data, in fact, over-determine these parameters and thereby check the conjecture that the tilted-foil geometry does indeed have a plane of symmetry on the macroscopic scale. If no plane of symmetry exists, additional parameters, O_0^{col}, O_{1+}^{col}, A_{1-}^{col} and A_{2-}^{col} are in general non-zero. These results confirm the observation by BERRY et al. [9.43] that the collisions with tilted foils produce oriented and aligned atoms.

Some perturbation of the orientation by the motional electric field was reported and allowed for in the data analysis.

9.3.3 The Quadratic Stark Effect

An electric field acting crosswise on an aligned atom perturbs the alignment owing to the quadratic Stark effect, as demonstrated by LOMBARDI [9.41]. A non-zero orientation arises from the coupling of the alignment $\underset{\approx}{A}$ with the electric field. This effect is of potential importance for beam-foil measurements since the dynamic evolution of atomic states perturbed via the second-order Stark effect could be directly observed, thus providing a measure of the polarizabilities of excited states. Furthermore, fields near the foil surface can perturb the initial alignment in the tilted-foil geometry; indeed ECK [9.47] has advanced this as an explanation of the observed orientation in such geometries. He postulates that fields near the foil, due to the charging up of the foil, are, on the average, normal to the foil surface. Such fields then perturb the initial alignment and induce a net orientation. The alignment and orientation of initially aligned but unoriented atoms, subjected to an electric field $\underset{\sim}{F}$ at an angle α to the beam direction, changes with time according to [9.48]:

$$O_{1-}^{col} = -A_0^{col}(0)\,\frac{1}{2}\sin\chi(t)\,\sin 2\alpha \quad ,$$

$$A_0^{col} = A_0^{col}(0)\,[1 - \frac{3}{2}\sin^2 2\alpha\,\sin^2\frac{\chi(t)}{2}] \quad ,$$

$$A_{1+}^{col} = A_0^{col}(0)\,\frac{1}{2}\sin 4\alpha\,\sin^2\frac{\chi(t)}{2} \quad ,$$

$$A_{2+}^{col} = A_0^{col}(0)\,\frac{1}{2}\sin^2 2\alpha\,\sin^2\frac{\chi(t)}{2} \quad , \qquad (9.28)$$

for the special case of $^1P \rightarrow {}^1S$ transitions, which we use to illustrate the theory. The phase $\chi(t)$ is given by

$$\chi(t) = \hbar^{-1}\int_0^t \{E_0(t') - E_1(t')\}\,dt' \quad , \qquad (9.29)$$

where $E_0(t') - E_1(t')$ is the splitting of the initial m=1 and m=0 levels due to a quasi-static electric field $\underset{\sim}{F}(t)$ in a frame with z axis along $\underset{\sim}{F}$. By quasi-static we mean that the field causes no transitions between the m=0 and 1 levels or between these levels and other atomic levels. In ECK's picture, the foil excitation produces only aligned atoms with an alignment given by A_0^{col} for α=0. A microscopic electric field normal to the foil then converts some of the alignment to orientation as the atoms leave the foil region. Since $\underset{\sim}{F}(t)$, and, hence, $\chi(t)$ is not known for large t, the exact degree of orientation cannot be predicted. However, regarding $\chi(t)$ for large t as a parameter, the theory predicts the dependence of $\underset{\sim}{A}$ and $\underset{\sim}{O}$ on α. Recent measurements by BERRY [9.44] and by CHURCH et al. [9.49] show that the observed orientation and alignment are not in accord with (9.28). It thus appears that microfields near the foil do not account for the observed orientation. Nonetheless any attempt to calculate the orientation must take into account possible electric fields in the foil region since they do perturb the alignment and orientation if they are present and of sufficient magnitude.

Equation (9.28) also serves to estimate the Stark-effect perturbation to aligned atoms due to the motional $\underset{\sim}{v} \times \underset{\sim}{B}$ electric fields for ^1P states. Note that fields perpendicular to the beam axis ($\alpha=\pi/2$) do not alter A_0^{col}. However, the magnetic field causes the alignment to precess so that at some time after the foil excitation the motional electric field acts crosswise on the alignment and the alignment will be perturbed due to the second-order Stark effect. Thus if the product of the motional electric field, the atomic polarizability, and the time during which the field acts is large enough that $\chi(t)$ is of the order of a few degrees, the alignment is perturbed in a complicated way. It is best to avoid such high motional fields.

References

9.1 H. J. Andrä, M. Gaillard, L. Henke, M. Kraus, J. Macek, W. Wittmann: In Proceedings of 4th Intl. Conf. on Atomic Physics, Heidelberg 1974, Abstracts 168

9.2 J. T. Park, F. D. Schowengerdt, D. R. Schoonover: Phys. Rev. A 3, 679 (1971)

9.3 H. J. Andrä: Physica Scripta 9, 257 (1974)

9.4 G. Breit: Rev. Mod. Phys. 5, 91 (1933)

9.5 R. Kornblith, J. H. Eberly: to be published

9.6 U. Fano, J. H. Macek: Rev. Mod. Phys. 45, 553 (1973)

9.7 M. J. Alguard, C. W. Drake: Phys. Rev. A 8, 27 (1973)

9.8 K. Adler: Helv. Phys. Acta 25, 235 (1952)

9.9 W. Wittmann, K. Tillmann, H. J. Andrä: Nucl. Instr. and Meth. 110, 305 (1973)

9.10 H. J. Andrä: Phys. Rev. Letters 25, 325 (1970)

9.11 W. Wittmann, K. Tillmann, H. J. Andrä, P. Dobberstein: Z. Physik 257, 299 (1972)

9.12 D. J. Burns, W. H. Hancock: J. Opt. Soc. Am. 63, 241 (1973)

9.13 H. G. Berry, J. L. Subtil: Phys. Rev. Letters 27, 1103 (1971)

9.14 H. G. Berry, J. L. Subtil, E. H. Pinnington, H. J. Andrä, W. Wittmann, A. Gaupp: Phys. Rev. A 7, 1609 (1973)

9.15 K. Tillmann, H. J. Andrä, W. Wittmann: Phys. Rev. Letters 30, 155 (1973)

9.16 O. Poulsen, J. L. Subtil: Phys. Rev. A 8, 1181 (1973)

9.17 S. Bashkin, G. Beauchemin: Can. J. Phys. 44, 1603 (1966)

9.18 J. Macek: Phys. Rev. Letters 23, 1 (1969)

9.19 J. Macek: Phys. Rev. A 1, 618 (1970)

9.20 I. A. Sellin, J. A. Biggerstaff, P. M. Griffin: Phys. Rev. A 2, 423 (1970)

9.21 D. J. Lynch, C. W. Drake, M. J. Alguard, C. E. Fairchild: Phys. Rev. Letters 26, 1211 (1971)

9.22 D. J. Burns, W. H. Hancock: Phys. Rev. Letters 27, 370 (1971)

9.23 P. Dobberstein, H. J. Andrä, W. Wittmann: Z. Physik 257, 272 (1972)

9.24 S. Bashkin, W. S. Bickel, D. Fink, R. K. Wangsness: Phys. Rev. Letters 15, 284 (1965)

9.25 W. S. Bickel, S. Bashkin: Phys. Rev. 162, 12 (1967)

9.26 W. S. Bickel: J. Opt. Soc. Am. 58, 213 (1968)

9.27 I. A. Sellin, C. D. Moak, P. M. Griffin, J. A. Biggerstaff: Phys. Rev. 184, 56 (1969)

9.28 I. A. Sellin, P. M. Griffin, J. A. Biggerstaff: Phys. Rev. A 1, 1553 (1970)

9.29 H. J. Andrä: Phys. Rev. A 2, 2200 (1970)

9.30 H. J. Andrä, P. Dobberstein, A. Gaupp, W. Wittmann: Nucl. Instr. and Meth. 110, 301 (1973)

9.31 A. Gaupp: Diploma Thesis (I. Phys. Inst. Free Univ. of Berlin) unpublished See also [9.3]

9.32 T. G. Eck: Phys. Rev. Letters 31, 270 (1973)

9.33 I. A. Sellin, J. R. Mowat, R. S. Peterson, P. M. Griffin, R. Lambert, H. H. Haselton: Phys. Rev. Letters 31, 1335 (1973)

9.34 A. Gaupp, H. J. Andrä, J. Macek: Phys. Rev. Letters 32, 1335 (1974)

9.35 C. H. Liu, S. Bashkin, W. S. Bickel, T. Hadeishi: Phys. Rev. Letters 26, 222 (1971)

9.36 D. A. Church, C. H. Liu: Phys. Rev. A 5, 1031 (1972)

9.37 D. A. Church, M. Druetta, C. H. Liu: Phys. Rev. Letters 27, 1763 (1971)

9.38 C. H. Liu, D. A. Church: Phys. Rev. Letters 29, 1208 (1972)

9.39 J. Yellin, T. Hadeishi, M. C. Michel: Phys. Rev. Letters 30, 417 (1973)

9.40 J. Yellin, T. Hadeishi, M. C. Michel: Phys. Rev. Letters 30, 1286 (1973)

9.41 M. Lombardi: J. Phys. Radium 30, 631 (1969)

9.42 M. Lombardi, M. Giroud: Compt. Rend. B 266, 60 (1968)

9.43 H. G. Berry, L. J. Curtis, D. G. Ellis, R. M. Schectman: Phys. Rev. Letters 32, 751 (1974)

9.44 H. G. Berry, L. J. Curtis, R. M. Schectman: Phys. Rev. Letters 34, 509 (1975)

9.45 C. H. Liu, S. Bashkin, D. A. Church: Phys. Rev. Letters 33, 993 (1974)

9.46 D. A. Church, W. Kolbe, M. C. Michel, T. Hadeishi: Phys. Rev. Letters 33, 565 (1974)

9.47 T. G. Eck: Phys. Rev. Letters 33, 1055 (1974)

9.48 J. H. Macek: In Proceedings of Symposium on Electron and Photon Interactions with Atoms, ed. by M. R. C. McDowell (to be published by University of Sterling)

9.49 D. A. Church, M. C. Michel, W. Kolbe: private communication

10. The Measurement of Autoionizing Ion Levels and Lifetimes by Fast Projectile Electron Spectroscopy

Ivan A. Sellin

With 13 Figures

Inspection of such venerable reference sources as the U.S. National Bureau of Standards volumes on atomic energy levels [10.1] creates the misleading impression that the most probable fate of excited atomic systems is decay by photon emission. For at least the first third of the periodic table, quite the opposite is true: the preferred decay mode of most excited atomic systems in most states of excitation and ionization is the one first observed by AUGER [10.2]. For example, a singly-charged neon ion lacking one K electron is more than 50 times as likely to decay by electron emission as by photon emission! Hence atomic spectroscopy can be said to be in its infancy in that the vast preponderance of data concerning levels of excited atomic systems concerns only optically-allowed, single-particle, valence-shell excitations in low states of ionization. Since most of the mass in nature is found in stars, and most of the elements therein rarely occupy such ionization-excitation states, it can be argued that standard reference sources provide a very unrepresentative description of the commonly-occurring excited atomic systems in nature. When the mean lives of excited atomic systems are considered, the relative rarity of lifetime measurements on Auger electron-emitting states is even more striking. The experimentally inconvenient typical lifetime range (10^{-12} - 10^{-16}s) accounts for this lack.

The optical branch of what has become known as the method of beam-foil spectroscopy has done much to expand knowledge of the structure and lifetimes of highly ionized and excited states of the light elements insofar as they are manifested in photon-decay processes. The sheer bulk of such optical data compared to electron emission data once again obscures the fact that Auger emission is the dominant decay process of light-element excited states. The knowledge of autoionizing energy levels and transition rates that does exist is, furthermore, heavily concentrated on neutral atoms and their negative ions.

The present discussion will mainly be devoted to describing progress in acquiring heretofore unavailable structure and lifetime data on Auger-emitting levels of highly ionized and excited states of the light elements. A novel variation on electron spectroscopy (DONNALLY et al. [10.3], SELLIN et al. [10.4]) has been most useful, particularly in acquiring lifetime data, and will be referred to as the method of fast-projectile electron spectroscopy (FPES). Since the method is relatively new and has some kinematic attributes it does not share with some more common electron spectroscopic methods, much of our discussion will be concerned with these attributes.

As was the case with the optical branch of beam-foil spectroscopy, decay-in-flight methods for the study of the level structures and lifetimes of projectile particles were invented and exploited much earlier in nuclear physics than in atomic physics. Beta-ray spectroscopy has been pursued for many years (cf. the reviews of SIEGBAHN [10.5] and SEVIER [10.6]. Early work on internal conversion electron spectra from moving recoil ions has, for example, been reviewed fairly recently by SEVIER [10.6]. The first discussion known to the author of atomic electron emission from fast atomic systems was given substantially later by RUDD et al. [10.7] in characterizing electrons emitted in keV Ar^+ on Ar collisions.

We will use the terms Auger emission and autoionization (autoejection) interchangeably, since for highly-ionized atomic systems, inner-shell vacancies, core excitations, and valence-shell excitations are often indistinguishable. The terms atom, ion, and atomic system will also be used interchangeably. There are many excellent recent articles and reviews concerning the electron spectroscopy of neutral atoms and their singly-charged ions, some of which contain data on electron emission from fast atoms (e.g., SEVIER [10.6], EDWARDS et al. [10.8], CROOKS et al. [10.9], RUDD and MACEK [10.10], EDWARDS and CUNNINGHAM [10.11], GROENEVELD et al.[10.12], RUDD et al.[10.13], CUNNINGHAM and EDWARDS [10.14], STOLTERFOHT et al. [10.15], EDWARDS and CUNNINGHAM [10.16], RISLEY et al. [10.17], PATTERSON et al. [10.18], EDWARDS and CUNNINGHAM [10.19], STOLTERFOHT et al. [10.20], STOLTERFOHT et al. [10.21], and PEGG et al. [10.22]). Among these articles and reviews, only those which overlap the main themes of this chapter (SEVIER [10.6], RUDD and MACEK [10.10], GROENEVELD et al. [10.12], STOLTERFOHT et al. [10.15,20,21], PEGG et al. [10.22]) will receive any detailed attention. High-resolution electron spectroscopic studies of multiply-ionized target atoms are rare and will also receive limited attention. An exception is the interesting Auger electron spectrum of multiply-ionized neon atoms produced by impact of highly-charged oxygen ions (MATTHEWS et al. [10.23]).

In the context of target-atom spectroscopy it is known from the work of MACDONALD et al. [10.24], MOWAT et al. [10.25], and MOWAT et al. [10.26] that highly-ionized, energetic, heavy projectiles produce remarkable degrees of target ionization and excitation, even under single collision conditions. For example, for Ar^{17+} ions incident on neon gas at 80 MeV, copious production of Lyman-alpha photons from one-electron neon atoms was observed by MOWAT et al. [10.26]. Hence removal of nine electrons and simultaneous excitation of the tenth to the 2P state were observed in single collisions. The higher the incident ion charge state, the greater is the degree of target ionization. For a given incident projectile energy, it is well known that the highest states of projectile ion charge state are reached by passage through thin foils. Even in experiments on multiply-ionized target atoms, then, projectile excitation in foils will be a dominant technique. See Fig.10.1.

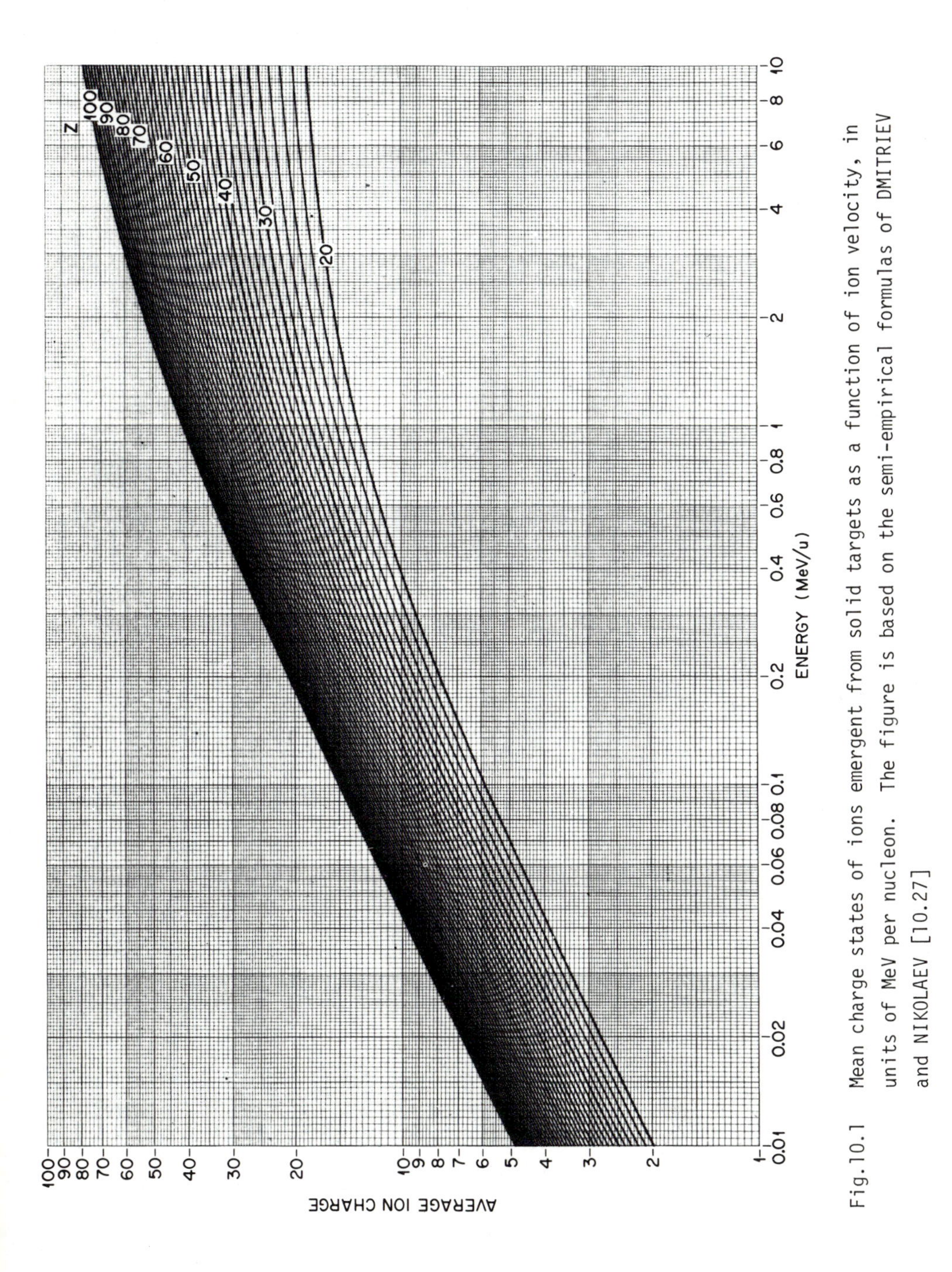

Fig.10.1 Mean charge states of ions emergent from solid targets as a function of ion velocity, in units of MeV per nucleon. The figure is based on the semi-empirical formulas of DMITRIEV and NIKOLAEV [10.27]

Figure 10.1 provides a compact description of the approximate mean charge states of ions emergent from solid targets as a function of ion velocity in units of MeV per nucleon, and is based on the semi-empirical formulas of DMITRIEV and NIKOLAEV [10.27]. This description is most useful in picking a projectile beam energy at which a given state of ionization and excitation is likely to be observed. For $Z < 20$, the differences between solid and gas target emergent charge-state distributions are minor.

For both radiating and Auger-emitting states, it is also well known that foil excitation creates states of not only high ionization and excitation, but also high angular momentum. Copious production occurs of excited states which cannot relax under the selection rules governing allowed radiative and Auger processes. A discussion of higher-order radiative processes in such states is given in Chapter 8 of this volume. Table 10.1 exhibits selection rules appropriate to Auger processes.

Table 10.1 Autoionization Selection Rules (LS Coupling)

	ΔL	ΔS	ΔJ	Parity
Coulomb	0	0	0	Conserved
Spin-Orbit	$0,\pm1$	$0,\pm1$	0	Conserved
Spin-Spin	$0,\pm1,\ \pm2$	$0,\pm1,\ \pm2$	0	Conserved

Example: $(1s\ 2s\ 2p)\ ^4P^o_{5/2} \rightarrow (1s^2)\ ^1S_0 + k\ ^2F^o_{5/2}$

To conserve parity, L must be odd
To conserve J, L = 2,3
Hence, $k\ ^2F^o_{5/2}$ is the final state;
Only the spin-spin interaction is effective

As can be seen, ordinary allowed Auger processes caused by the e^2/r_{ij} Coulomb interactions among excited electrons are frequently forbidden when the excited-state angular momentum is sufficiently large. Higher-order magnetic interactions then govern the decay rates. The mean lives of metastable states are especially sensitive to the matrix elements for these more subtle interactions because of selection rule orthogonalities in the matrix elements for normally-dominant allowed processes. Measured mean lives can be used as a probe of the relativistic magnetic interactions in excited states in a way similar to that in which the analogous radiative mean lives are used. Generally, for Auger processes, somewhat different electronic structural properties (e.g., spin-spin interactions and aspects of electron correlation) are probed, and interactions with the radiation field usually need not be considered.

Single-particle, valence-shell excitation states do not, of course, autoionize. For autoionization to be energetically possible, an autoionizing state must be degenerate with a continuum state, i.e., lie higher in energy than the ground state (or other excited states) of an ion in the next higher adjacent charge state. An autoionizing state may then be adjacent to more than one continuum if it lies above not only the ground but also one or more excited states of the adjacent ion. RUDD and MACEK [10.10] give a number of examples of states which can autoionize: (a) multiple-excitation states in which two or more electrons are simultaneously excited; (b) inner-shell excitation states, in which an inner electron in an atom having more than one shell is excited to a higher energy orbit; (c) single-particle excitation states together with rearrangement of the core (same core configuration but different core angular momentum coupling); (d) inner-shell vacancy states, in which an inner-shell electron in an atom with more than one shell is completely removed; (e) single-particle excitation accompanied by some other internal energy change (e.g., in molecules, vibration-rotation energy may be present).

Much of the following discussion will be concerned with KLL processes, in which two L electrons interact to fill one K vacancy and yield one Auger electron, and LMM processes, in which two M electrons interact analogously. The total electronic energy of the autoionizing state is usually quoted relative to the situation in which the atomic nucleus and all electrons of the system are infinitely removed from one another and are in a state of rest. The kinetic energy of the Auger electron in the rest frame of the emitting ion will then be given by this total energy minus the total energy of the final state of the residual ion. Neglecting recoil in a KLL process in a three-electron ion, the rest-frame Auger electron energy is given by the sum of the binding energies (ionization potentials) of the two electrons in the resultant ground-state helium-like ion minus the sum of the binding energies of the three electrons in the excited lithium-like ion. For example, KLL processes in oxygen generate Auger electrons of roughly 500 eV energy. The laboratory energy will be higher by an amount corresponding to the translational velocity of the emitter and the angle of emission of the Auger electron. A 500-eV electron could easily be kinematically shifted to, say, 900 eV. It will be found that relativistic kinematic corrections are not necessarily negligible.

10.1 The Fast-Projectile Electron Spectroscopy (FPES) Method

The kinematic shifts referred to above are, of course, accompanied by corresponding kinematic broadening of even the sharpest Auger lines in the laboratory coordinate system. The ratio v/v_p of laboratory electron emission velocity v to projectile velocity v_p and the angle θ between $\vec{v}$ and $\vec{v}_p$ determine the broadening. The velocity $\vec{v}_p$ is usually confined to a range of polar angles of $\sim$ 0.1 to 1° about the projectile beam direction (the z direction) and has a magnitude which is generally easily established by accelerator analyzing-magnet systems to an amount which often renders the spread in v_p in the emergent beam negligible. Projectile passage through a thin foil leads to a reduction in v_p: the reduction can be measured or established by foil-thickness measurements combined with the use of energy-loss tables (cf. NORTHCLIFFE and SCHILLING [10.28]. Energy straggling is very often found to be negligible. Thus $\vec{v}_p$ may be regarded as having a transverse spread but a negligible axial spread. It is found that even the thinnest foils increase the transverse spread in $\vec{v}_p$ by a significant amount through multiple scattering. For a sharp Auger line in the laboratory system, the energy of an electron emitted at a fixed polar angle $\overline{\theta}$ will be spread because of the transverse spread in $\vec{v}_p$, but will be independent of the azimuthal angle of emission ϕ. The output of an energy analyzer observing electrons emitted in some range $\delta\overline{\theta}$ of polar angles near $\overline{\theta}$ will exhibit a broadening due both to kinematic broadening described by the transformation between rest and laboratory frames and the transverse spread in $\vec{v}_p$.

10.1.1 Choice of an Analyzer

An analyzer recently constructed in our laboratory is shown in Fig.10.2. We chose

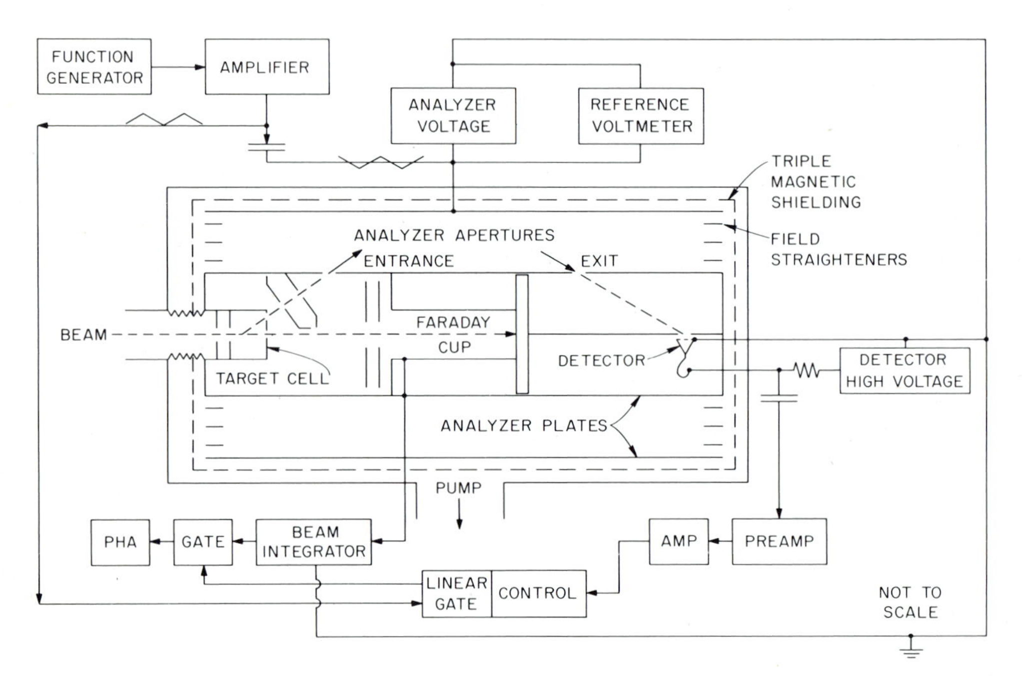

Fig.10.2 Schematic diagram of an electrostatic cylindrical-mirror analyzer suitable for use in FPES experiments

this design for the following reasons. The narrowest projectile Auger lines and hence best energy resolution will result from the smallest possible values for the analyzer polar-angle acceptance range and the smallest possible spread in $\vec{v}_p$ consistent with needed line intensities. Because of the consequent loss in solid angle subtended by an analyzer entrance slit at the site of the emitter, it is imperative to take advantage of the cylindrical symmetry in ϕ to regain intensity. It is cheaper and more convenient to use electrostatic as opposed to magnetic analyzers. Hence the analyzer of choice is the electrostatic cylindrical-mirror analyzer, whose collection geometry permits small values of $\delta\bar{\theta}$ and an azimuthal range $\geqslant \pi$. Such an analyzer was apparently first described by BLAUTH [10.29], although its original inventor is not established. Of the many articles which have subsequently described its use and optimal design, the definitive papers of ZASHKVARA et al. [10.30] and of SAR-EL [10.31,32] best characterize its electron optical properties as they relate to the present discussion.

Cylindrical-mirror analyzers possess a unique advantage for FPES over spherical-sector, flat-plate, or other instruments of comparable intrinsic resolution. Spherical-sector instruments, for example, typically have intermediate angular acceptances in two planes and are nearly useless for FPES due to the resulting severe kinematic broadening. If the polar angle bite were reduced to compensate for this broadening, then spherical-sector or flat-plate instruments would still compare unfavorably to the cylindrical-mirror analyzer because of the much larger azimuthal range of the latter.

10.1.2 Properties of a Cylindrical-Mirror Analyzer Suitable for FPES

The first FPES lifetime experiments with a cylindrical-mirror analyzer appear to be those done in our laboratory, using a small instrument (resolving power $\sim$ 150) designed and built by SMITH [10.33]. Its success stimulated the design and construction of the larger device shown in Fig.10.2 (inner-cylinder radius a=5.71 cm, outer-cylinder radius b=12.22 cm, demonstrated resolving power $\geqslant$ 2500), built by HASELTON and SELLIN [10.34]. In Fig.10.2, the dashed arrow indicates the central ray emerging from the target cell. Electrons which are emitted from either projectile or target ions at points near the intersection of the axis with the central ray are deflected according to their energy by a retarding electrostatic field and detected by a quiet, channel electron multiplier (noise counting rate $\leqslant$ 0.04 Hz) placed at the axial location of the conjugate focus. The multiplier is baffled by a 0.25 mm effective-diameter aperture in a molybdenum sheet coplanar with an axial plane. The configuration of Fig.10.2 is appropriate to the study of projectile collisions in gases. When foil excitation is used, the projectile beam-defining apertures are mounted behind a foil on a longitudinal drive rod which can adjust the separation of the foil and the spectrometer viewing region over the region 0 to $\sim$ 50 cm. The mean angle of launch θ

into the analyzer is chosen to be 42.3°, for which angle there is for non-relativistic particles a second order focus at the axial image point. For a radius a of the inner cylinder, the axial distance L_0 between object and image points is chosen to be $L_0 \simeq 6.1a$; for our analyzer $L_0 = 35.03$ cm. Second-order focusing corresponds to $\partial L/\partial\theta = \partial^2 L/\partial\theta^2 \simeq 0$ for $\theta = 42.3°$, and occurs for $L = L_0$. HAFNER et al. [10.35] give a convenient reduction of a formula of ZASHKVARA et al. [10.30] from which the linear dispersion D can be derived

$$\Delta L \cong 5.6\, a(\Delta E/E_0)\ (1 + 1.84\ \delta\bar{\theta}) - 15.4\ a(\delta\bar{\theta})^3 \quad , \tag{10.1}$$

$$D \equiv \Delta L/(\Delta E/E_0) \simeq 5.6\ a \simeq 0.92\ L_0 \quad . \tag{10.2}$$

Here ΔL is the range of axial distances near L_0 corresponding to a band of energies of width ΔE in electron kinetic energy E_0. For the spectrometer sketched in Fig. 10.2, $\delta\bar{\theta}$ (for the moment ignoring the transverse spread in $\vec{v}_p$) is limited for kinematic reasons to 1.2 mrad $\simeq$ 0.07°. Hence the terms in (10.1) in $\delta\bar{\theta}$ are negligible (e.g., the angular aberration term $(\delta\bar{\theta})^3$ is $< 10^{-3}$ times the leading term) and _in the FPES application, the cylindrical-mirror analyzer can often be said to be aberration-free, insofar as axial rays are concerned_. For our analyzer, $D \simeq 0.92\ L_0 \simeq 32.2$ cm.

The source volume in Fig. 10.2 is determined by the intersection of the cylindrical beam (typical radius c = 0.8 mm) with the field of view of the defining slits. The latter consist of a semi-circular molybdenum slit of mean radius d = 5.12 mm, width 0.25 mm, and thickness 0.15 mm forming the rear wall of the target cell and a $2\pi/3$ molybdenum slit in the inner cylinder wall, located on a radius a = 5.71 cm, of width 0.18 mm and thickness 0.15 mm. Because $d \simeq 0.09$ a, the ratio of the axial extent of the penumbra to that of the umbra region is $\leqslant 0.1$. Taking into account the finite widths of the limiting slits, the collision volume can be approximately described as the region common to a cylinder of radius $c \simeq 0.8$ mm and the space between two cones of half-angle $\cong$ 42.3°, whose axis coincides with the beam axis and whose vertices are separated by a distance $w_1 \cong 0.15$ mm. The combined axial widths of both source points and detection aperture of diameter w_2 is then $\Delta L \simeq 0.15$ mm + 0.25 mm = 0.4 mm, and using (10.2) with D = 32.2 cm, an estimate can be derived of the expected peak base width $\Delta E/E_0$ (when the analyzer voltage is scanned) for a sharp line of energy $E = E_0$: $\Delta E/E_0 \cong \Delta L/D \simeq 1.2 \times 10^{-3}$. As discussed by RUDD [10.36], the FWHM/E_0 of such a peak depends only on the wider of the two widths w_1, w_2, and is given in this case by FWHM/$E_0 \simeq w_2/D = 0.25$ mm/$D \simeq 8 \times 10^{-4}$.

Thus far, our discussion has been limited to axial rays, as would emerge from a line source on axis. ZASHKVARA et al. [10.30] indicate that for a disc source of radius u and negligible axial extent located at the focus, true imaging is obtained at unit magnification in the conjugate focal plane. Non-axial rays can, however, have a screw-shaped trajectory through the analyzer, and, according to these authors, image in the focusing plane in a circle of radius $\sim 1.7u$, with decreasing intensity in the region between u and 1.7u. They state without proof that the smallest energy interval that can be resolved is $\Delta E/E_0 \cong 0.5\ u/a$. For our collision volume, the annular-shaped source elements would have a characteristic transverse dimension $\geqslant w_i = 0.15$ mm, corresponding to a value for $\Delta E/E$ of $\geqslant 1.3 \times 10^{-3}$ FWHM. The apparent conclusion that non-axial rays can be critically limiting even for transverse source dimensions ~ 0.1 mm can, in practice, be mitigated by the presence of other slit edges which limit non-axial rays but do not further limit axial rays. Nonetheless, it is of interest to compare the results of the convolution in quadrature of the expected axial and non-axial FWHM widths with our measurements on the electron impact excited Ne KLL line at 804 eV. The estimated and observed values of FWHM/E_0 are 1.5×10^{-3} and 1.3×10^{-3}, respectively. We conclude that further reductions in source dimensions could materially improve resolution, at corresponding expense in line intensity. An often employed and better solution is to use a retarding lens system to reduce the electron energy in some ratio R before analysis so that the effective resolution is $\Delta E/E_0|_{eff} = (\Delta E/E_0)\ R^{-1}$. The conical-electrode lens system indicated in Fig.10.2 has been used in this way, and has achieved values of $R \geqslant 5$ at about the same factor loss intensity. Hence our present FWHM/E_0 can be made as low as 4×10^{-4}. We have not as yet found it advisable to use this particular retarding system regularly, because of the further limitations (to be discussed) associated with beam collimation and foil-induced transverse spreads in $\vec{v}_p$.

Our measured analyzer constant (defined as the slope of the analyzer voltage vs. electron energy characteristic in the non-relativistic limit) has the value 0.582 ± 0.002. The theoretical estimate of ZASHKVARA et al. [10.30] is $0.77\ \ln\ (b/a) = 0.586$. The sharp-edged fringe-field correction rings which are spaced at logarithmic radial intervals are found to be essential in establishing agreement to two significant figures. A triple layer of annealed conetic magnetic shielding (thickness 1.27 mm) reduces the ambient magnetic field inside the spectrometer vacuum chamber to less than 7 mG, an important feature for low-energy electron spectroscopy. Proximity of the spectrometer to a 4000Å, 18 kG, 1 megawatt cyclotron magnet explains our preference for shielding to Helmholtz coils. The metal analyzer parts are made mainly of non-magnetic stainless steel; the insulators are machined out of a partially-fired ceramic called remcolox. The analyzer is housed in a non-magnetic stainless steel vacuum enclosure maintained at pressures on the 10^{-8} torr scale. As indicated in Fig.10.2, an electron spectrum is generated by cycling the analyzer voltage through some chosen range of interest with a repetitive linear ramp voltage, a constant fraction of which is routed to the linear input of a multichannel analyzer through a linear gate. The gate is opened by detector pulses, thereby generating a spectrum of counts vs. a linear voltage proportional to electron energy (in the non-relativistic limit). We have very recently found it preferable to replace the linear ramp generator indicated in Fig.10.2 with a 12-bit digital-to-analog converter system devised by THOE [10.37]. This system is used for simultaneous addressing of the multiscaling input of a multichannel analyzer and the input of a high-loop-gain, well-regulated operational power supply which provides the analyzer voltage. Channel advance is controlled by a pulse triggered by a preset number of ion-beam-current digitizer pulses. The latter system permits analyzer voltage setting accuracy of ~ 1 part in 2^{13} and has the additional advantage that the voltage can be repetitively reset to the same value within this accuracy.

10.1.3 Kinematic Modification of Analyzer Optimization Criteria

RUDD [10.36] has given a number of criteria for selecting entrance and exit slit widths to maximize analyzer current output for a desired peak FWHM and dispersion D in the presence of angular aberrations. In FPES applications, angular aberrations are often unimportant. RUDD [10.36] suggests that for a fixed FWHM, the greatest analyzer output results when entrance and exit slit widths w_1 and w_2 are equal. This conclusion will sometimes require modification to $w_1 < w_2$ in FPES experiments. The analyzer input current is proportional to $\ell_{EFF} d\Omega_{EFF}$, where ℓ_{EFF} is the effective length of beam viewed by the entrance slit system, and $d\Omega_{EFF}$ is the solid angle subtended by the limiting slit nearest the analyzer entrance ($\ell_{EFF} d\Omega_{EFF} \cong 2.32 \times 10^{-4}$ mm sr in the present configuration). The kinematic width of a sharp Auger line is proportional to the polar angle acceptance bite and hence to w_1. It is advantageous to specify $w_1 < w_2$ to reduce the laboratory system width of a line $\delta E/E_0$ to a level comparable to the intrinsic analyzer FWHM/E_0 peak width for a sharp line. This intrinsic width is given by w_g/D, where w_g is the wider of the slit widths w_1, w_2. Considering first a flat-topped line of width $\delta E/E_0$ appreciably less than the intrinsic FWHM/E_0, the case $w_1 < w_2$ gives an approximately trapezoidal peak shape (when the spectrometer voltage is scanned) of FWHM/$E_0 = w_2/D$, of height $\propto w_1$ (the input current is proportional to w_1), of (base width)/$E_0 = w_2/D + (w_1/D + \delta E/E_0)$, and of (top width)/$E_0 = w_2/D - (w_1/D + \delta E/E_0)$. The area under the peak and thus current output is proportional to $E_0 w_1 w_2/D$. To maximize current output for a given FWHM/E_0, w_1 should be as large as possible, but may not exceed $(w_2 - D\delta E/E_0)$ because then the FWHM/E_0 becomes limited by w_1. Since $\delta E/E_0$ is proportional to w_1, we may define g such that $g \equiv D\delta E/(E_0 w_1)$ in which case for a given FWHM/E_0 - $w_2 D$, maximum current output arises if $w_1 = w_2/(1+g)$.

Considering next a wide, flat-topped line of width $\delta E/E_0$ such that $(w_1 + D\delta E/E_0) > w_2$, the peak FWHM/$E_0$ becomes $(w_1/D + \delta E/E_0)$, the peak height becomes $w_1[w_2/(w_1 + D\delta E/E_0)]$, the (base width)/$E_0$ becomes $w_2/D + (w_1/D + \delta E/E_0)$, and the (top width)/$E_0 = [(w_1/D + \delta E/E_0) - w_2/D]$. The area under the peak and thus current output will be once again proportional to $E_0 w_1 w_2/D$. To maximize current output for a given FWHM, w_2 should be as large as possible and may be as large as $w_1(1 + g)$ without degrading the FWHM/E_0.

Again, maximum current output results in $w_1 = w_2/(1+g)$, for a given $FWHM/E_0$. Summarizing the situation for a flat-topped line of any kinematic width $\delta E/E_0$, the maximum output will result when $w_2 = w_1(1+g)$, the $FWHM/E_0$ of a peak will then be w_2/D, the peak height will be proportional to w_1, and the peak area will be proportional to $E_0 w_1 w_2/D$.

10.1.4 Relativistic Corrections to Analyzer Performance

Relativistically, the plate voltage on an electrostatic analyzer is not proportional to the kinetic energy of a transmitted electron, even if a relativistically correct expression for the electron kinetic energy is used. SAR-EL [10.32] derives relativistic corrections which can be rewritten in a simple form appropriate to electrostatic analyzers. We define the analyzer constant K as the limit as $T \to 0$ of the ratio eV/T_0, where e is the electron charge, V the analyzing plate voltage, and T_0 is the relativistic kinetic energy of the entrant electron ($T_0 \to E_0 \equiv \frac{1}{2}mv^2$ as $v/c \to 0$). Then the relation between V and T_0 can be written

$$K^{-1}V = (T_0/2) + (T_0/2)/(1 + T_0/mc^2) \quad . \tag{10.3}$$

For a 1000-eV electron, the denominator in (10.3) $\cong$ 1.002, so that its neglect would occasion a fractional error of 1×10^{-3} in the energy one would assign a line in the basis of the more approximate equation $K^{-1}V \cong E_0$. Since this error has the same order as the intrinsic analyzer resolution, and the difference between T_0 and E_0 is also of this order, this denominator should not be neglected for electrons in the range of a few hundred eV and above. Relativistic corrections to atomic electron binding energies in the range of a few hundred eV and above typically have a relative magnitude of this order or larger, so that consideration of both experimental and theoretical line shifts is sometimes important.

The correctness of (10.3) is easy to demonstrate for the simple case of a parallel-plate analyzer. If an electron of velocity v enters such an analyzer through a grounded plate at a launch angle $\theta = \pi/4$ and the analyzing plate voltage V is adjusted so as to permit reverse passage of the trajectory through the grounded plate a distance L_0 away from the entrance point, then the transit time τ is given by $L_0/(v\cos\theta)$. Non-relativistically, half the transit time $\tau/2$ corresponding to the maximum potential energy is related to the plate separation H by

$$\tau/2 = (eV/Hm)^{-1}\, v\sin\theta \quad . \tag{10.4}$$

Simultaneous solution of these equations for τ gives the standard result (for $\theta = \pi/4$) that $eV = KE_0 = (2H/L_0)(mv^2/2)$. Relativistically, the only change is the insertion of the extra factor γ on the right hand side of (10.4), where $\gamma = (1-v^2/c^2)^{-1/2}$, and we find

$$eV = (2H/L_o)E_o\gamma \quad . \tag{10.5}$$

Using the relations $T_o = mc_2(\gamma - 1)$ and $T_o^2 + 2mc^2T_o = p^2c^2 = m^2v^2\gamma^2c^2$, it is easy to transform (10.4) into the form of (10.3).

10.1.5 Broadening from Transverse Velocity Spread

Significant broadening results from any source of transverse spread in $\vec{v}_p$. Since the bite in polar angle of the instrument under discussion has a value 1.2 mrad $\cong$ 0.07°, ion-beam collimation to comparable extent is desirable. We have often accomplished this by placing a 1-mm aperture a distance of approximately 6 m upstream of the spectrometer, yielding a transverse spread in $\vec{v}_p$ due to ion-beam collimation of about 1.5 mrad $\cong$ 0.09°.

A more seriously limiting consideration when foils are used is the angular spread in $\vec{v}_p$ due to multiple scattering. This subject has been very recently reviewed and treated theoretically by SIGMUND and WINTERBON [10.38] and has been experimentally investigated by GROENEVELD et al. [10.12]. Figure 10.3 exhibits the comparison between theory and experiment concerning the half-angle to be expected for multiple

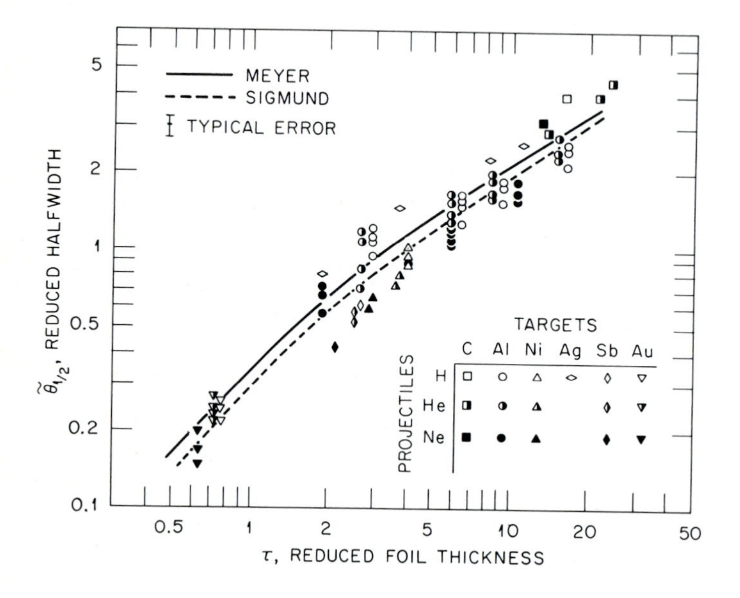

Fig.10.3 Comparison between theory and experiment concerning the half-angle to be expected for multiple scattering in foils. The figure shows a universal plot of reduced half-angle vs. reduced foil thickness. Experimental data from GROENEVELD et al. [10.12]

scattering on a suitable universal plot of reduced half-angle vs. reduced foil-thickness. Also shown in Fig.10.3 is a theoretical plot by MEYER [10.39]. Reduced foil-thickness τ_F is defined as $\pi a_{TF}^2 Nx$, where a_{TF} is the Thomas-Fermi screening radius, which is assumed to describe screened Coulomb scattering in the energy range of interest. The radius $a_{TF} = 0.885 a_o (Z_p^{2/3} + Z_t^{2/3})^{-1/2}$, N is the number of target atoms (nuclear charge Z_t) per unit volume, and x is the foil thickness. Here Z_p is the projectile nuclear charge and a_o is the Bohr radius. The reduced multiple scattering half-angle $\tilde{\theta}_{1/2}$ is related to the actual half-angle $\theta_{1/2}$ by $\tilde{\theta}_{1/2} = \theta_{1/2} (E_o a_{TF}/2Z_p Z_t e^2)$. As can be seen, the theoretical curve for $\theta_{1/2}$ provides a useful estimate for the transverse spread in $\vec{v}_p$ due to multiple scattering, although there is considerable question whether the theory ought to be applied to heavy-ion penetration in matter in the MeV range. Taking a nominal $5\mu g/cm^2$ C foil, the reduced thickness is $\cong 2.5 \times 10^{17}\ \pi a^2\ cm^{-2}$. For the typical case of 5 MeV oxygen ions in such a foil, $a \cong 1.73 \times 10^{-9}$ cm, so that $\tau_F \cong 2.4$, and $\theta_{1/2} = \tilde{\theta}_{1/2} \times 1.6 \times 10^{-3} \cong 0.9$ mrad $\cong 0.05°$. Actual measurements of CLINE et al. [10.40] for 10 MeV oxygen ions in $50\mu g/cm^2$ C foils give $\theta_{1/2} \sim 0.33°$. A scaling proportional to $1/E_o$ and based on the N-dependence of Fig.10.3 would predict a value of $\sim 0.12°$ for 5 MeV ions in $5\mu g/cm^2$ foils. Hence a value $\sim 0.2°$ for $2\theta_{1/2}$ is a reasonable estimate of the transverse spread in $\vec{v}_p$ from multiple scattering of 5 MeV ions in $5\mu g/cm^2$ C foils.

It might be assumed that an appreciable gain would result from using, say $2\mu g/cm^2$ C foils, the thinnest foil commonly available. However, we have observed a beam dose-dependent foil thickening of the irradiated beam spot that tends to saturate at large doses. For 5 MeV oxygen ions in C foils, the apparent thickness of a $2\mu g/cm^2$ foil is approximately $5\mu g/cm^2$ after a dose of a few hundred μC of beam. The initial thickening is so rapid that it has not proved possible to get a complete useful spectrum at thicknesses near $2\mu g/cm^2$. It nevertheless pays to begin with the thinnest foils possible, since the saturation thickness seems always to be appreciably larger than the initial thickness (e.g., $15\mu g/cm^2$ C foils quickly approach $\gtrsim 30\mu g/cm^2$ thickness). These changes in thickness are easily measured by monitoring the peak shifts in electron spectra arising from additional ion energy loss in the foils.

The total spread $\delta\bar{\theta}$ in launch angle into the analyzer may be estimated from adding in quadrature the estimated values for finite acceptance angle bite, transverse spread in $\vec{v}_p$ due to beam collimations and transverse spread due to multiple scattering. For the typical case of 5 MeV oxygen beams traversing $5\mu g/cm^2$ C foils in the geometry of Fig.10.2, the resulting value $\delta\bar{\theta} = (0.07^2 + 0.09^2 + 0.2^2)^{1/2} = 0.23°$.

10.1.6 Further Kinematic Considerations: Sample Estimates of Net Line Widths Observed in FPES

Non-relativistically, the laboratory electron velocity is obtained by adding the beam velocity $\vec{v}_p$ to the electron rest frame velocity $\vec{v}_r$; the resultant vector $\vec{v}$ has a magnitude given by the vector triangle relation

$$v_r^2 = v^2 + v_p^2 - 2v_p v \cos\theta \quad . \tag{10.6}$$

When $v_p > v_r$, v will in general be double-valued. The rest-frame emission angle will be related (non-relativistically) to θ by

$$v \cos\theta = v_p + v_r \cos\theta \quad . \tag{10.7}$$

For given values of v_p and v, and $v_p > v_r$, there will be a maximum angle at which such electrons can be observed in the laboratory, corresponding to kinematic peaking of the angular distribution in the forward direction, given by $\sin \overline{\theta}_{MAX} = v_r/v_p$. This situation results when $\vec{v}$ and $\vec{v}_r$ form the legs of a right triangle of which $\vec{v}_p$ is the hypotenuse. Implicit differentiation of (10.6,7) easily allows computation of the relationship between solid angles in the rest frame and the laboratory in the form

$$d\Omega_R/d\Omega_L = (v/v_r)^2[v/v_r - (v_p/v_r) \cos \overline{\theta}]^{-1} \quad , \tag{10.8}$$

which grows with acute angles $\overline{\theta}$ and approaches ∞ for $\overline{\theta} = \overline{\theta}_{MAX}$. Hence the range of solid angles in the laboratory system is generally larger than the corresponding range in the rest frame, and becomes very large near the limiting angle of emission (if any) $\overline{\theta}_{MAX}$. Best resolution will result for as low values of v_p as possible consistent with excitation of states of interest, and with necessary restrictions on $\delta\overline{\theta}$ due to multiple scattering.

Relativistically, the relevant vector triangle involves the addition of the four-vector components $\vec{v}_p\gamma_p$ and $\vec{v}_r\gamma_r$. Standard relativistic formulas can be used to express the resultant $\vec{v}$ in terms of T_0. The analyzer voltage V corresponding to a given peak centroid will be given by (10.3).

As an example of the combined effects of all of the kinematic effects discussed so far, we consider KLL transitions in highly-ionized oxygen ions (e.g., lithium-like ions) which are excited by passage of $\sim$ 5 MeV oxygen ions through 5μg/cm^2 C foils and have rest-frame energies $\sim$ 500 eV. Energy loss in such foils reduces the ion energy by about 58 keV (NORTHCLIFFE and SCHILLING [10.28]), and we will take 5 MeV to be the post-foil ion energy.

Solution of (10.6) with $\overline{\theta} = 42.3 \pm 0.12°$ gives a value for E_0 of 914.2 ± 1.7 eV or $\sim$ 0.4% FWHM. Convolution in quadrature with the intrinsic analyzer FWHM/E_0 value of 1.9×10^{-3} gives a net FWHM/E_0 of $\sim$ 0.44%. This resolution is very close to that realized in practice. Relativistically, E_0 is replaced by T_0 = 913.8 eV, and V = 531.3 eV (note that $V/T < K = 0.582$).

As a second example we consider a monoenergetic 70-keV Li beam undergoing collisions in a light gas target, and neglect both relativistic and single-collision recoil effects. Doubly-excited states are easily created under these conditions and give rise to electrons with energy near 55 eV in the rest frame. It is convenient, because gas target thicknesses are so much smaller than foil thicknesses, to open up the beam collimation (polar angle spread) to $\sim \pm 0.4°$ to gain intensity. For a 55-eV electron, E_0 becomes 74.37 eV $\pm$ 0.24 eV or $\sim$ 0.6% FWHM. This resolution is also very close to what is realized in practice.

In both examples, the resolution is very much limited by the spread in $\vec{v}_p$ associated with $\delta\overline{\theta}$. In the Li example, much better resolution would be achieved at a corresponding loss in intensity by better ion-beam collimation to below 0.1° spread. In the first example, gains are harder to achieve. Beam energy E must first be chosen to allow population of ionization-excitation states of interest and to reduce multiple scattering as much as possible ($\theta_{1/2} \propto E^{-1}$). These two requirements are often opposed to keeping kinematic spread as low as possible. A compromise in which v_p/v is $\leqslant 1/\sqrt{2}$ allows some reduction in multiple scattering loss while avoiding the resolution disaster near $\overline{\theta}_{MAX}$. Where states of ionization $\leqslant$ 10 are of interest, it may in the future be possible to gain greatly in resolution by using keV-energy sources of multiply-charged ions as opposed to foil-transmitted high-energy beams.

10.1.7 Summary of the Advantages of FPES

Having noted the difficulty in limiting kinematic resolution loss to a level comparable to intrinsic spectrometer resolution, it is reasonable to consider compensating advantages of FPES. The ability to measure lifetimes by the time-of-flight technique is certainly one clear-cut advantage. The ability to create nearly any ionization-excitation state of interest is much more easily achieved in projectiles than in target particles. Even when such ionization-excitation states are created in target atoms, the spectrum of ionization states is typically much broader than in a typical projectile beam, so that FPES experiments have a charge-state resolution advantage. The principle of differential metastability is a unique advantage of the FPES method. Peaks which are often unresolved in energy can be resolved temporally through choice of the separation distance between the point of excitation and the spectrometer viewing region. Knowledge of even very approximate relative state lifetimes is often a valuable aid in identifying emitting states, whose matrix elements leading to available final states can vary by orders of magnitude. We have noted the efficiency of highly-charged projectile ions in creating high states of ionization and excitation in target atoms. Foil excitation of fast heavy projectile ions is thus a valuable technique in both projectile and target-atom electron spectroscopy. Finally, we note that FPES will have more nearly comparable resolution to target-atom electron spectroscopy when keV-energy multiply-charged ion beams are sufficient to reach ionization ($\leqslant$ +10) and excitation states of interest.

10.2 Examples of FPES

Various examples of spectra obtained by the FPES method are discussed in detail below.

10.2.1 Spectra of Long-Lived States of the Li-Like, Be-Like, and B-Like Ions

Figures 10.4 and 10.5 display spectra of KLL, KLM, ... electrons from fluorine ions

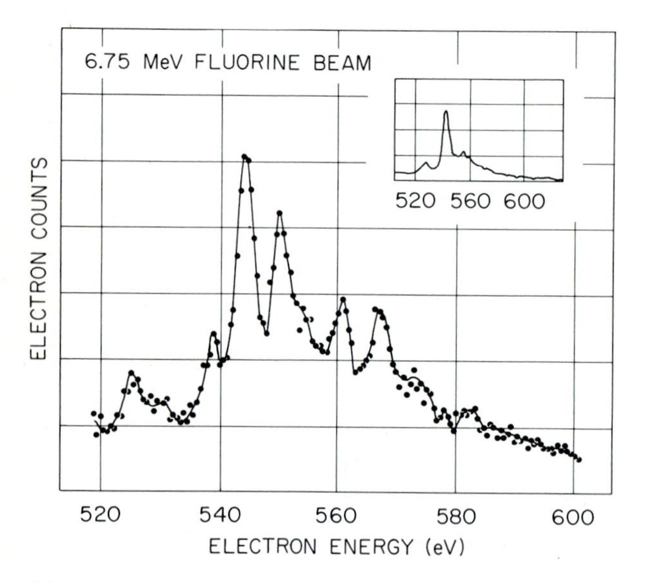

Fig.10.4 Spectrum of autoionization electrons emitted from (∼ 2μg/cm^2) foil-excited 6.75 MeV fluorine ions undergoing decay in flight. Electron energies are plotted in the rest frame of the emitting ion. The inset shows a lower-resolution spectrum by PEGG et al. [10.41]

of 6.75 MeV, ionized and excited in nominally 2μg/cm^2 C foils, and observed after a time delay of approximately 0.1 ns after excitation. The inset to Fig.10.4 displays a similar but lower resolution spectrum by PEGG et al. [10.41]. That article contains a much more comprehensive discussion of such spectra than is possible here. The additional resolution permits separation of a number of previously-unresolved states, whose decay in time was also discussed in the article of PEGG et al. The feature near 525 eV may result from the optically-forbidden $1s2s^2\ ^2S$ state of the lithium-like ion which has not been observed previously. The smaller feature near 530 eV is actually the dominant feature of the spectrum at time delays > 0.4 ns and corresponds to the metastable $1s2s2p\ ^4P_J$ states. Hence this spectrum provides a good illustration of the principle of differential metastability, and indicates how the levels of states having low Auger transition probabilities can be established in the presence of strongly emitting states by means of such cross-over spectra. Because

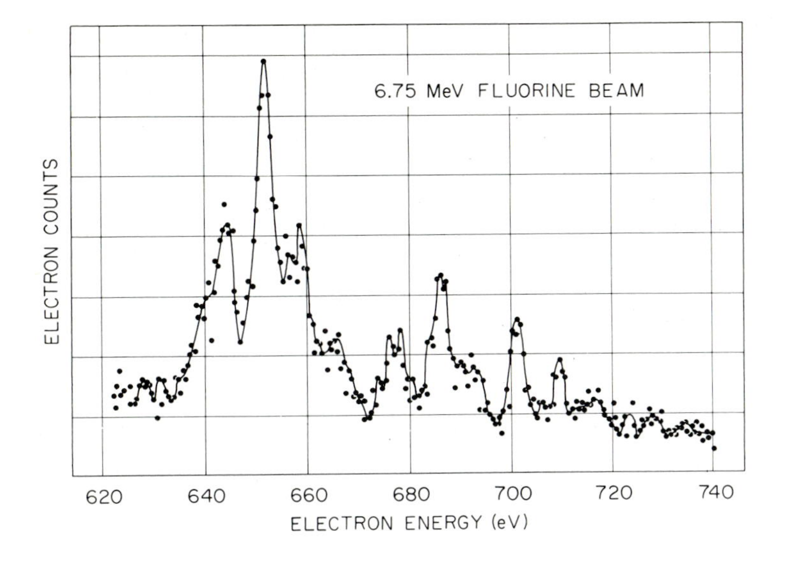

Fig.10.5 As in Fig.10.4, but for higher-energy electrons

the 1s2s2p ^4P_J states have been the subject of two extensive variational calculations by HOLØIEN and GELTMAN [10.42] and by JUNKER and BARDSLEY [10.43], and because absolute peak locations are uncertain due mainly to the dose-dependent foil-energy-loss phenomenon we have found it useful to relate our absolute energy scale to the relativistically-corrected Auger-electron energy value of 530.0 eV for this state derived from the work of HOLØIEN and GELTMAN [10.42]. The alternate variational calculations for this state were performed with a smaller number of terms in the variational wave function and hence led to a state energy about 1.3 eV higher. A central point of the work of JUNKER and BARDSLEY [10.43] was not so much high variational energy accuracy but rather approximate location of large numbers of metastable lithium-like, beryllium-like and boron-like states, many of which are in fact found in our spectra. In fact, for some lithium-like states, the latter authors' eigenvalues are slightly lower than those of HOLØIEN and GELTMAN [10.42]. Thus 530.0 eV should be regarded as an upper limit to the Auger transition energy, but one which is probably correct to a few tenths of an eV.

The largest feature of the spectrum in Fig.10.4, near 544 eV, contains Auger electrons from the $1s2s2p^{2,\,4}P^e$ state, which is also observed to survive at delays of a few ns, but must at the delay shown contain at least one short-lived state yet to be assigned. Both sets of variational calculations appear to overestimate the Auger energy of the $1s2p^2\ ^4P^e$ state.

The KLM and higher-series members are in the range of Fig.10.5. The largest feature of this spectrum overlaps a pair of metastable lithium-like states designated by

$^4P^e(2)$ and $^4S^e(2)$, which are also observed at centimeter distances downstream. (The parenthetic number specifies the numerical ordering of states of a given symmetry as a function of excitation energy). The relatively long $^4P^e(2)$ state lifetime has been measured in lithium-like oxygen by DONNALLY et al. [10.3] and is discussed in Section 10.3. For the $^4P^e(2)$ state, the respective variational calculations of HOLØIEN and GELTMAN [10.42] and JUNKER and BARDSLEY [10.43] give the rather closely corresponding values for Auger electron energies, of 654.1 and 654.2 eV. However, for the $^4S^e(2)$ state the alternative calculations are 0.5 eV apart and there are other instances of states involving s electrons where the gap is nearly 3 eV. Thus, even at experimental resolution levels of $\sim$ 0.4%, the available theoretical level calculations sometimes have uncertainties about equal to a line width. More precise theoretical level calculation would clearly aid in the spectral interpretation. As will be noted, a similar situation prevails concerning core-excited states of the nearly neutral alkali systems.

Many of the short-lived features in the spectra of Figs.10.4 and 10.5 arise from allowed doublet or semi-allowed quartet states of the lithium-like, beryllium-like, and boron-like ions. There are instances for other ions in these isoelectronic sequences for which optical radiation (representing a competitive decay mode to autoionization) has been observed. Examples are radiation from the $1s2p^2$ doublet and quartet states for ions like C, N, O, and other astrophysically-important elements. Optical lines originating in the upper levels appear as satellites to the helium resonance lines and have, for example, been studied by GABRIEL [10.44]. Corresponding level calculations have been provided by SUMMERS [10.45] and by SAFRONOVA and KHARITONOVA [10.46] to accuracy $\sim$ 1 eV. For the specific case of fluorine, such lithium-like states as $1s2s2p\ ^{2,3}P$ and $2p^2\ ^2P,^2D$ have also been identified through their soft X-ray decay channel by RICHARD et al. [10.47]. Optical intersystem lines at longer, often visible, wavelengths can be expected to arise, coupling any pair of opposite-parity excited states in the same spin system in such spectra as those in Figs.10.4 and 10.5. In some cases, better resolution may thus be obtained by optical means. FPES still enjoys a significant advantage in that the dynamic range of electron energies analyzable by an electron spectrometer vastly exceeds the corresponding wavelength range of any optical spectrometer, and avoids such uncertainties as wavelength-dependent grating or crystal reflectivity values.

10.2.2 Spectra of Long-Lived Core-Excited States of Sodium-Like Chlorine

The interpretation of the spectra of very long-lived ($\sim$ 50 ns) autoionizing levels of core-excited states of sodium-like chlorine ions (with perhaps some contributions from adjacent charge states) studied in our laboratory by PEGG et al. [10.48] is still far from clear. The needed theoretical calculations on these levels seem not to be available.

Figure 10.6 displays a more recently-acquired spectrum for 6.75 MeV Cℓ ions excited in nominally 2µg/cm^2 C foils, at time delays of a few tenths of a nanosecond. It can

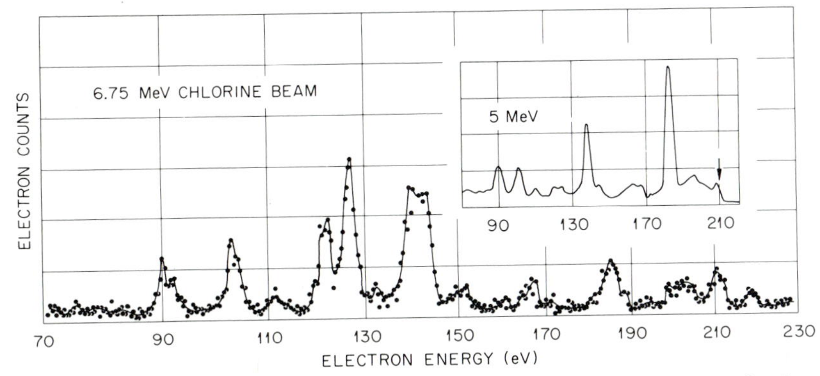

Fig.10.6 Spectrum of autoionization electrons emitted by (∿ 2µg/cm^2) foil-excited 6.75 MeV chlorine ions undergoing decay in flight. The electron energies are expressed in the rest frame of the emitting atom. The inset spectrum contains lower-resolution data by PEGG et al. [10.48], obtained at a relative delay of about 6 ns

be compared to the lower-resolution inset spectrum obtained by PEGG et al. [10.48], at a time delay of about 6 ns. Some of the Auger lines are thought to arise from such configurations as $2p^5n\ell n'\ell'$; the metastability arises if the spins of the two-valence electrons are parallel to the spin of the L-shell hole. Taking the difference in resolution into account, the spectra are quite similar, despite the difference in time delay, but one clear difference is that the series limit marked by the arrow in the inset does not occur in the higher resolution spectrum. This series limit corresponds to the limiting-energy Auger electrons one would find from configurations like $2p^53sn'\ell'$, with $n'\ell'$ very large. It thus appears that decays from states with parent terms like $2p^53p$ and $2p^53d$ appear in the complete spectrum, but have mean lives too short to appear in the inset spectrum. What is surprising is the large number of very long-lived states (on an Auger time scale of $\sim 10^{-14}$ s). Once again, the principle of differential metastability is illustrated in the comparison of the features between 120 and 130 eV in the two spectra.

10.2.3 Core-Excited States of the Neutral and Nearly-Neutral Alkali Metals

FPES also has applications to the electron spectroscopy of atoms in low states of ionization, especially in those areas where the corrosive nature of the species of interest deters their use in target cells, and where the levels of interest are optically forbidden. An example is the optically-forbidden, core-excited states of the alkali metals, which have recently been studied in our laboratory by PEGG et al. [10.22]. Specifically, core-excited states of Li, Na, K and Mg^+ have been studied.

The energies of a number of core-excited levels of several alkali systems have been the subject of calculations of WEISS [10.49], HOLØIEN and GELTMAN [10.42], and GARCIA and MACK [10.50], and typically have uncertainties on the order of 0.1 eV. Our present results for resolved peaks have estimated absolute accuracies of ± 0.03 eV. The fortunate occurrence in the same spectra of optically-accessible doublets, for which the accurate photoabsorption energies measured by EDERER et al. [10.51] are available, was indispensable in establishing absolute energy scales. Hence our data are interpretable at theoretically-useful levels of accuracy. Theoretical difficulties arise from the strong state mixing due to the spin-orbit interaction associated with the p vacancy in the noble gas core of the heavier alkalis. The strong spin-orbit interaction is manifested in the large spin-orbit splittings in core-excited states of Na.

In these experiments, an alkali beam (∿ 70 keV) was collisionally excited by passage through a thin gas target in the target cell of Fig.10.2. Examples of the alkali FPES spectra are shown in Figs.10.7 and 10.8.

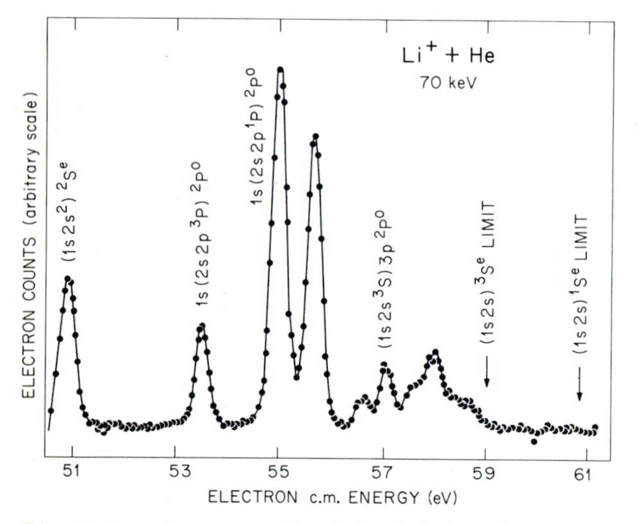

Fig.10.7 Spectrum of autoionization electrons emitted by 70-keV lithium atoms undergoing decay in flight after collisional excitation in a He gas target. The electron energies are expressed in the rest frame of the emitting atom. Data from PEGG et al. [10.22]. Excitation energies are obtained by adding the lithium ionization potential of 5.392 eV

The lowest-energy feature of the Li spectrum shown can be associated with the allowed (Coulomb) autoionizing decay of the previously-unobserved, optically-inaccessible $(1s2s^2)\ ^2S$ state. We measure the excitation energy of this state to be 56.31 ± 0.03 eV. The observed energies of the optically-accessible $1s(2s2p^3P)^2P^o$, $1s(2s2p^1P)\ ^2P^o$,

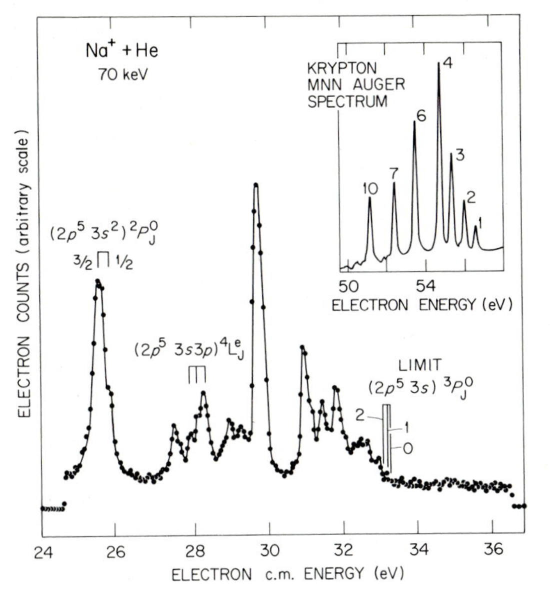

Fig.10.8 Spectrum of autoionization electrons emitted by 70-keV sodium atoms undergoing decay in flight after collisional excitation in a He gas target. The electron energies are expressed in the rest frame of the emitting atom. The inset shows a calibration spectrum MNN Auger electrons from Kr obtained by electron-impact excitation. Data from PEGG et al. [10.22]. Excitation energies are obtained by adding the sodium ionization potential, 5.139 eV

and $(1s2s\,^3S)3p\ ^2P^o$ states are in good agreement with the very accurate photoabsorption data of EDERER et al. [10.51] and such states serve to establish an accurate absolute energy scale. The intense line in the spectrum corresponding to an excitation energy of 61.04 ± 0.03 eV may be associated with the decay of the previously-unobserved, optically-inaccessible doublet, $(1s2p^2)\ ^2D$. The energy of this state has been estimated by NICOLAIDES [10.52] to be at 62.0 eV. The other features of the spectrum are attributable to the autoionizing decay of doublets and quartets associated with core-excited configurations of the type $1s2sn\ell(n \geq 3)$ and $1s2pn\ell(n \geq 2)$. A rather well-defined drop-off in intensity occurs near the $(1s2s)\ ^3S$ series limit, as is to be expected.

Calculations involving the superposition of configurations have been made for some core-excited states of Na by WEISS [10.49]. The lowest-lying features of the Na spectrum shown in Fig.10.8 can be associated with the decay of the optically-accessible $(2p^5 3s^2)\,^2P_{1/2,3/2}$ states, and their excitation energies are found to be in good agreement with recent photoabsorption data of WOLFF et al. [10.53]. The group of lines in the range 32.78 to 33.38 eV (excitation energy) can be attributed to optically-inaccessible quartet states associated with the $2p^5 3s3p$ configuration, based upon WEISS' predictions. It is also likely that the group of lines in the range 33.88 -34.79 eV (excitation energy) is associated with doublets of the same configuration. A few other lines in the spectrum can be attributed to the decay of optically-accessible states observed in the previously-mentioned photo-absorption studies, but many remain unidentified and presumably are associated with optically-inaccessible core-excited configurations. It is conceivable that core-excited states of Na^+, such as $2s2p^6n$ ($n \geq 3$), may also be present in the spectra. Further identification of many states in both Li and Na require calculations as yet unavailable.

10.2.4 Electron Background in FPES with Foil Targets

Returning once again to the subject of the FPES of KLL, KLM ... transitions in highly-ionized ions, as in Figs.10.4 and 10.5, the lines of interest typically are superposed on a continuous background of free electrons ejected from the target. The peak-to-background ratio is sensitively dependent on spectrometer resolution. From another perspective, it is of interest to inquire about the cross section for production of free electrons per eV-sr in solid targets. Such a spectrum has been obtained for Ne ions in C foils by GROENEVELD et al. [10.12] and is shown in Fig.10.9.

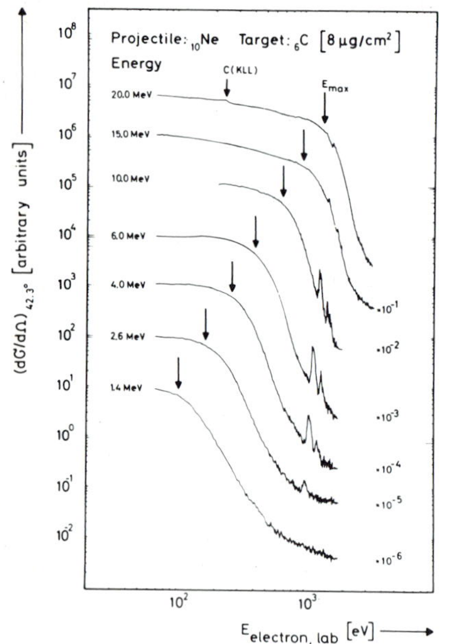

Fig.10.9 Cross-section for production of free electrons per eV-sr in solid targets following ion impact. The figure shows electron production by Ne ions in C foil targets. The KLL Auger transitions at ∿ 255 eV are only observable at the highest energies. With increasing beam energy the Doppler-shifted Auger lines of the projectile temporarily rise above the continuously rising continuum electron spectrum. (Data from GROENEVELD et al. [10.12])

Since the spectrometer used was also of the cylindrical-mirror type, their results are especially important in FPES. The free electrons arise from a number of processes, an important one of which is simply the knock-on of target electrons from the foil target. They are not sharply cut off at the expected limit of $E_{MAX} \cong 4(m/M_{Ne})\, E \cos^2 42.3°$ because of the non-zero momentum distribution of the electrons in the target atom. In addition to Ne KLL electrons, a small amount of C KLL Auger activity is visible. For still another reason, then, it is desirable to choose the beam energy so as to optimize both signal strength and resolution, the latter being especially important in providing appropriate peak-to-background ratios.

Surprisingly little is known of the systematics of the elementary processes governing ejection of electrons from the front and back surfaces of thin foils by heavy projectiles. A number of surprises occur. For example, as a function of incident proton energy, it has been found by MECKBACH et al. [10.54] that the ratio of forward-emitted to backward-emitted electrons for 5 $\mu g/cm^2$ C foils rises linearly from a value $\sim$ 1.1 at proton energies $\sim$ 20 keV to a value $\sim$ 1.5 at 125 keV, and then stays approximately constant as a function of increasing proton energy up to at least 250 keV. Even more surprising, the energy spectrum of backward-emitted and forward-emitted electrons has essentially the same shape, which shape is nearly independent of beam energy. Some typical results of MECKBACH et al. [10.54] are shown in Fig.10.10.

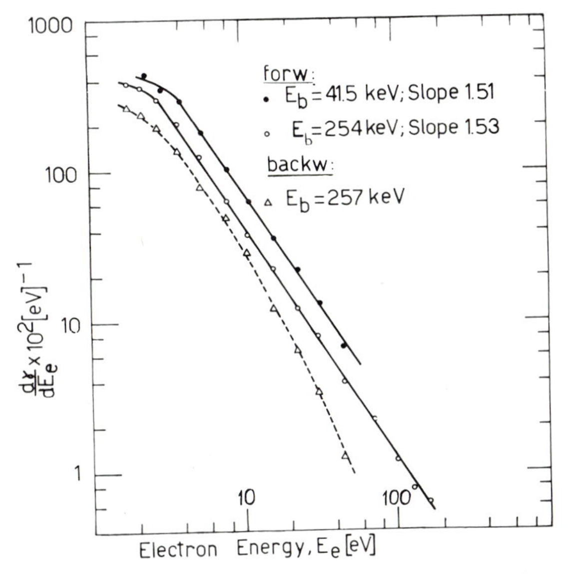

Fig.10.10 Differential emission coefficient $d\gamma/dE_e$ for secondary electrons emitted in the backward and forward directions from thin carbon foils transversed by 25-250 keV protons. (Data from MECKBACH [10.54])

The emergent-electron energy-spectrum appears to follow approximately an $(E_e)^{-3/2}$ power law. From both the proton and neon experiments, one concludes there is much to be done in FPES to understand the energy spectrum of electrons created by the passage of heavy particles through foils.

10.2.5 Electron Background in FPES with Gas Targets

Qualitatively, there exist both similarities and differences in the electron spectra obtained with heavy-ion bombardment of gas as opposed to solid targets, as can be seen in Fig.10.11, which is taken from the work of STOLTERFOHT et al. [10.20].

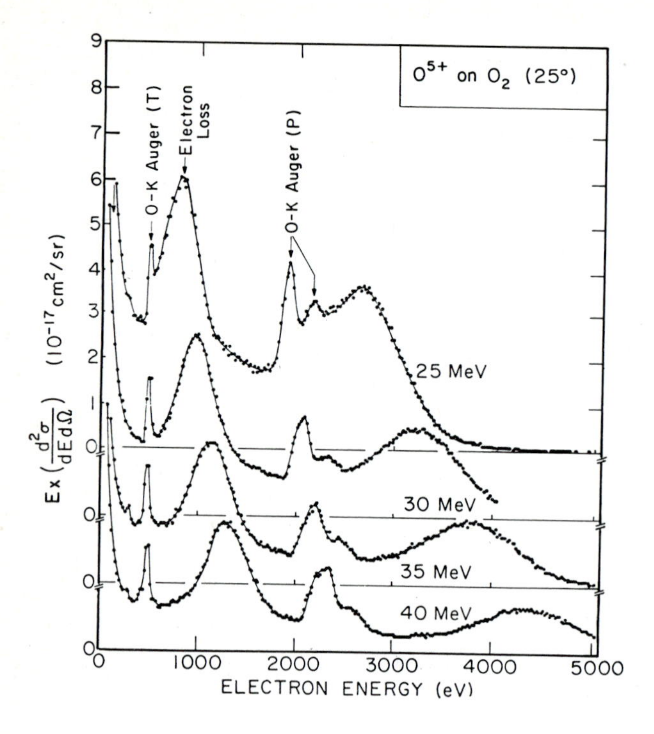

Fig.10.11 Doubly-differential cross-section for electron emission at 25° for O^{5+} on O_2 at various energies. (Data from STOLTERFOHT et al. [10.20])

What is plotted is the low resolution, doubly differential cross-section for electron emission at 25° for O^{5+} on O_2 at various energies. The large low-energy tail exhibits ionization of target electrons which are nearly at rest (near minimum momentum transfer) after the collision. Oxygen K Auger electrons from the target appear next, followed by the kinematically-shifted electrons from the projectile L shell in the peak near 1000 eV. The oxygen K Auger peaks (analogous to those shown in higher resolution in Figs.10.4 and 10.5) are seen at energies $\geqslant$ 2000 eV. Finally, the tail of hard knock-on electrons from the target is seen at energies which sensitively reflect incident ion energy, with the cut-off extending over a region analogous to the one observed for foils in Fig.10.9. The O^{q+} on O_2 cross-sections were studied also as a function of angle of observation and as a function of incident-ion charge-state q by STOLTERFOHT et al. [10.20].

In a lower beam energy study by STOLTERFOHT et al. [10.15,21], projectile and target K Auger-electron spectra for Ne^{q+} on Ne collisions (q = 0,1,2) in the range 0.05-2 MeV were accumulated. It was found that both projectile and target K Auger electron emission were essentially isotropic, and that the K vacancy was on the average equally shared by the projectile and target.

A much higher-resolution Ne K Auger spectrum (0.04%) (but for target neon ions only) was obtained by MATTHEWS et al. [10.23] using O^{5+} ion bombardment, and is shown in Fig.10.12, together with a comparison spectrum obtained by electron impact. Five

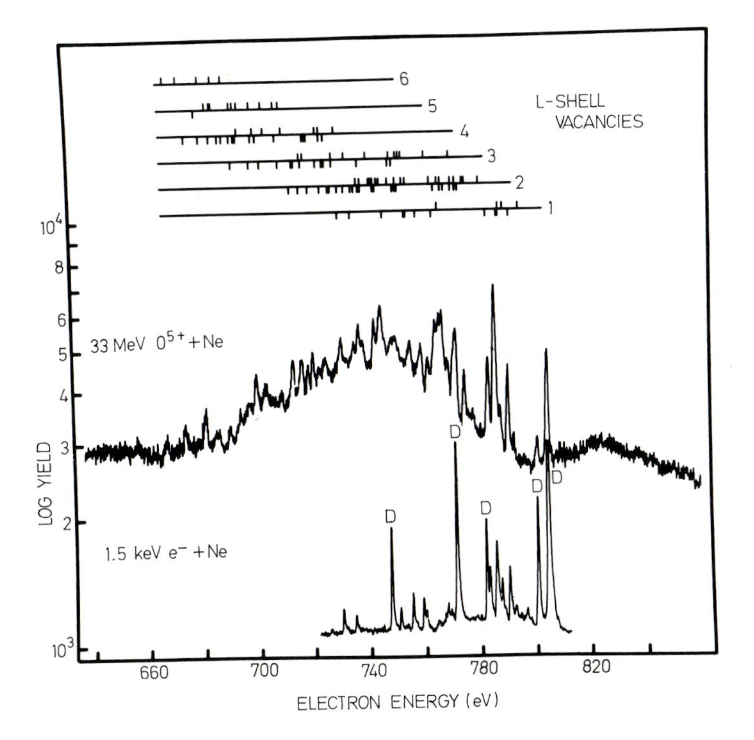

Fig.10.12 Comparison of Ne K-Auger spectra produced by heavy-ion and electron bombardment. In the e^- + Ne spectrum the diagram Auger lines are marked D. The calculated Auger satellite transition energies are shown on the axes which label the number of L-shell vacancies. (Data from MATTHEWS et al. [10.23])

"dominant" (D) lines and some 58 satellites were observed in the heavy-ion-induced spectrum. This spectrum at once provides an illustration of the power of energetic heavy ions in high charge states in producing highly-ionized target ions as well as the dual limitation of target atom spectroscopy: (i) the lack of excited-target charge-state specificity, and (ii) the lack of the differential metastability advantage. The KLX Auger lines of oxygen fall below the energy range of Fig.10.12, and, due to kinematic broadening and the large polar bite of the spectrometer viewing system, would in any case have widths many times the width of the target lines.

10.3 The Measurement of Auger Lifetimes by FPES

The technique of making optical lifetime measurements by determining the decay length of spectrally-selected radiation is a well-known and powerful tool of the optical branch of beam-foil spectroscopy. The analogous technique in FPES experiments is so nearly identical in every respect that there is no need to describe details of the method here. Problems relating to cascades, to normalization, and to backgrounds are indistinguishable from their optical counterparts. Instead of stepping a grating rotation to scan wavelengths, one steps an analyzer plate voltage. Normalization can be made to digitized beam charge, to optically filtered light from some transition in a beam-excited ion, or to another Auger transition. Flat-topping of spectral peaks to avoid intensity dependence on precise spectrometer tuning can be accomplished by enlarging the spectrometer exit-slit width. As with optical beam-foil experiments, decay-in-flight lifetime measurements are one of the most powerful features of the FPES method and one which is hard to achieve by other methods.

What is remarkable is the extremely small number of lifetime measurements ever made by the FPES technique, most of them in fact occurring in our own laboratory. Because of the special physical interest which attaches to the measurement of lifetimes of metastable states which decay only by violation of the normal selection rules; because of the convenient range of the metastable state lifetimes for light, highly-ionized ions; and because sufficiently precise theoretical calculations have not progressed beyond about the two- and three-electron systems, much of this initial lifetime work has been concerned with the metastable lithium-like ions of elements up to argon.

10.3.1 Auger Lifetimes of Metastable Lithium-Like Ions

Interest in the quartet states of three-electron systems stems historically from the early observations (HIBY [10.55]; DOPEL [10.56]) of the existence of a bound state associated with the three-electron helium ion, He^-. WU [10.57] suggested that the (1s2s2p) $^4P^o_{5/2}$ state of He^- would be bound, and since it was also expected to be metastable against both autoionization and radiation, it accounted for the observations. Subsequent studies such as those by HOLØIEN and GELTMAN [10.42], JUNKER and BARDSLEY [10.43], GARCIA and MACK [10.50], WU and SHEN [10.58], and HOLØIEN and MIDTAL [10.59] were directed at the study of those highly-excited states which are metastable against autoionization.

The lifetimes and energies of the lowest-lying quartet states in three-electron systems have been investigated extensively both experimentally and theoretically because they are metastable against Coulomb autoionization (as well as against radiative decay, in the case of the lowest quarter state (1s2s2p) $^4P^o_{5/2}$). These states autoionize only through the spin-orbit and spin-spin interactions in three-electron systems (for J = 5/2, spin-orbit matrix elements also vanish). Hence their study has provided a sensitive test of the spin-orbit and spin-spin interactions in three-electron systems, including the effects of electron correlation. Comparison of experiments with theoretical predictions shows that accuracy in wave functions for the three-electron system is critical for predicting level lifetimes.

FELDMAN and NOVICK [10.60,61] experimentally studied the forbidden autoionizing decay of metastable autoionizing states in several of the alkali atoms. The binding energy of the (1s2s2p) $^4P^o_{5/2}$ level in lithium was determined and its lifetime measured by observing the change in the charge states of an electron-impact-excited lithium beam following autoionization in flight. BLAU, NOVICK, and WEINFLASH [10.62] used a similar method to measure the lifetime of the corresponding state in He^-. The (1s2s2p) $^4P^o_{5/2}$ state was studied in other three-electron ions (Z=4-8) by DMITRIEV et al. [10.63], who used a time-of-flight technique to track the charge change in a fast, foil-excited, accelerator beam. MANSON [10.64], BALASHOV et al. [10.65], and, most recently, CHENG et al. [10.66] have calculated the lifetime of this lowest-lying quartet state in a number of ions of the lithium sequence.

The FPES experimental method is similar to that used by DMITRIEV et al., with the important exception that the FPES method includes energy analysis of the emitted autoionization electrons, which represents a considerable advance over the previous time-of-flight methods. In fact, the lifetimes measured by DMITRIEV et al. showed a residual dependence on beam velocity, which of course corresponded to a systematic error of unknown origin in their experiments. We are nonetheless indebted to these earlier experimenters for their being the first to demonstrate the possibility of measuring Auger lifetimes by applying time-of-flight methods to fast beams of excited heavy ions.

The decays of the J=1/2 and J=3/2 states are not currently resolvable spectroscopically from the J=5/2 decays but are resolvable through the principle of differential metastability; the J=5/2 state outlives the others and thus can be resolved temporally. Because of the requirements that both J and parity be conserved in autoionizing transitions to the final state $(1s^2)\ ^1S_0 + k$, only $^2F_{5/2}$ states are possible for the emitted electron, requiring $\Delta S=1$, $\Delta L=2$. These changes rule out Coulomb, spin-orbit and spin-other-orbit autoionization processes, but the decay can be induced by the tensor part of the spin-spin interaction. Hence systematic measurements of the mean life of this state against autoionization in ions of the lithium sequence measures directly the size of the tensor part of the electronic spin-spin interaction in three-electron systems.

Results from our laboratory for J=5/2 state lifetimes for a number of ions in the lithium sequence in the range Z=8-18 have been reviewed recently (HASELTON et al. [10.67]) and compared with theory, especially with the very recent theoretical results of CHENG et al. [10.66]. Figure 10.13 provides a summary of the various experimental and theoretical results, including the charge-changing results.

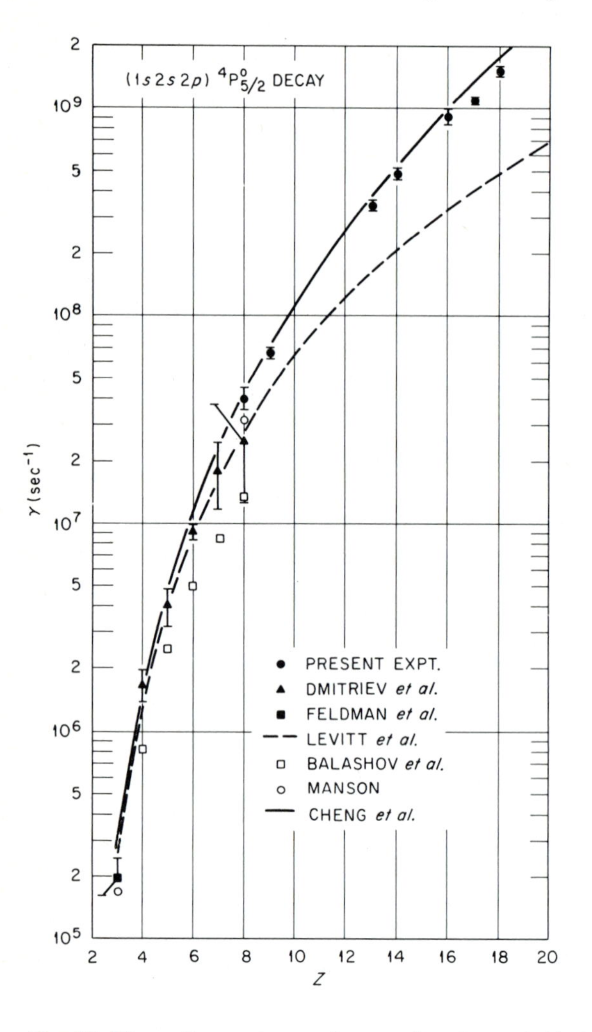

Fig.10.13 Comparison of experiments and theory for the total decay rate of the (1s2s2p) $^4P_{5/2}$ state vs. Z in three-electron ions. (Figure from HASELTON et al. [10.67])

In comparing the theoretical results of CHENG et al [10.66] with experiment, it was necessary to include in the theoretical curve a correction arising from the rapid onset of an M2 radiative-decay channel whose rate scales as $\sim Z^8$. The theoretical curve shown includes this correction, which will be discussed further in the subsequent subsection. Concerning the Auger rates above, there are two different aspects of such calculations to be tested: (i) the importance of relativistic effects and (ii) the importance of electron-correlation effects. The dashed curve gives an extrapolated semi-empirical fit $\propto (Z-\sigma)^3$ to the low-Z data. Such a Z-dependence was proposed by LEVITT et al. [10.68], who considered only the autoionizing channel; σ is a suitable screening constant. It is clear that the recent calculations of CHENG et al. represent a marked improvement over the semi-empirical extrapolation, and yield a Z-dependence nearer $(Z-\sigma)^4$ than $(Z-\sigma)^3$. It is also obvious that the most recent theoretical results systematically overestimate the experimentally-measured decay rates for the case $Z > 5$, while giving significantly better results for Z=8 than the calculations of MANSON or of BALASHOV et al.

The calculation of BALASHOV et al. employed simple screened-Coulomb functions for the bound electrons. This approach is not expected to yield particularly accurate results. MANSON used a single-configuration, separable, single-orbital type, variational wave function and a purely nonrelativistic Hamiltonian in his calculations. CHENG et al. [10.66] used Dirac-Hartree-Fock wave functions and a relativistic Hamiltonian. It is the opinion of the latter authors that the remaining discrepancy between their theory and our experiments may be due to the neglect of initial-state electron-correlation effects as well as distortion by the out-going final-state F-wave continuum electron. Though such correlation effects ought to be most significant at low Z, it appears that the lifetime measurements are sufficiently precise to permit sensitive measurements of such effects for ions as heavy as Ar^{15+}.

Table 10.2 summarizes some results on lifetime measurements and calculations.

Table 10.2 Lifetimes of the $(1s2s2p)\,^4P_{5/2}$ State in Some Three-Electron Ions

Ion	Lifetime (ns) Experiment	Theory	Autoionization Branching Ratio	Lifetime Ratio Expt/theory
O^{5+}	25 ± 3 [10.3]	23.1 [10.6]	0.993 [10.6]	1.08 ± 0.13
	40 ± 20 @10.3]	31 [10.4]		
		75 [10.5]		
F^{6+}	15 ± 1 [10.3]	13.5 [10.6]	0.988 [10.6]	1.11 ± 0.07
$A\ell^{10+}$	2.9 ± 0.2 [10.7]	2.58 [10.6]	0.947 [10.6]	1.12 ± 0.08
Si^{11+}	2.1 ± 0.1 [10.7]	1.84 [10.6]	0.929 [10.6]	1.14 ± 0.05
S^{13+}	1.1 ± 0.1 [10.9]			
	1.1 ± 0.2 [10.9]	0.993 [10.6]	0.883 [10.6]	1.11 ± 0.10
$C\ell^{14+}$	0.91 ± 0.04 [10.4]			
	0.95 ± 0.20 [10.9]	0.743 [10.6]	0.855 [10.6]	1.12 ± 0.05
Ar^{15+}	0.66 ± 0.04 [10.4]	0.563 [10.6]	0.823 [10.6]	1.17 ± 0.07

The last column of Table 10.2 shows the remarkably constant ratio of the present experimental to theoretical lifetimes. In a plot of this ratio vs. Z, it would be possible to represent the ratio as a constant of ~ 1.17 within the quoted error bars. The mean ratio for the cases cited is 1.14, and the standard deviation from this mean is $\pm$ 0.05. Hence it is possible to speak of a systematic difference between theoretical and experimental decay rates of 14 $\pm$ 5%.

The possibility of undiscovered systematic errors of this magnitude of course arises. However, this error would have to have remarkably constant-ratio properties in order to prevail for different experimental configurations ranging from those involving 2.5 MeV oxygen beams from the Oak Ridge Tandem Accelerator to those involving 80 MeV argon beams from the Oak Ridge Isochronous Cyclotron, and involving different target thicknesses, different laboratory decay-lengths, and different energy-loss corrections. Cascades, which often apparently lengthen beam-foil lifetime measurements, would also have to have remarkably coincidental behavior in order to lengthen the experimental lifetimes in the same ratio for such a wide variety of atomic systems and beam velocities.

A second example of such Auger lifetime measurements concerns the excited $^4P^e(2)$ state of lithium-like oxygen, whose basic configuration is 1s2p3p. The largest feature of Fig.10.5 contains the analogous state in lithium-like fluorine. The lifetime in O^{5+} was measured in our laboratory to be 1.00 $\pm$ 0.04 ns. Because the $^4P^e(2)$ state can decay to any lower state of opposite parity in the quartet system, including the lowest state (1s2s2p) $^4P^o(1)$, the lifetime is determined by a sum over both Auger and radiative decay channels. A consequence is the ability to measure the lifetime of an Auger emitting state by means of studying its optical decay channel. As will be seen in the subsequent subsection, such experiments have occasionally been done.

10.3.2 Examples of Lifetimes from Optical Decay Channels of Auger-Emitting Levels

A first example of the use of the optical decay channel to measure the lifetime of Auger-emitting levels has to do with the onset of an M2 radiative channel in the decay of the 1s2s2p $^4P_{5/2}$ state of the lithium-like ions mentioned in the previous subsection. It has been demonstrated that the autoionization branching ratio is reduced by the competing forbidden (M2) radiative transition $(1s^22s)\ ^2S_{1/2}$ - (1s2s2p) $^4P_{5/2}$. The decay in the flight of such radiation (X-ray) has been observed for both S^{13+} and $C\ell^{14+}$ ions by COCKE et al. [10.69], and this decay channel yields lifetimes in excellent agreement with the results depicted in Fig.10.13.

As a second example, the decay-in-flight of the shorter-lived blend of J=1/2 and 3/2 components of the 1s2s2p ^4P_J levels by soft X-ray emission has been studied for lithium-like oxygen by RICHARD et al. [10.70] and MOORE et al. [10.71]. The lack of agreement between the respective values of 1.87 $\pm$ 0.1 ns and 3.48 $\pm$ 0.8 ns may well have to do with different J-component mixtures (spectroscopically unresolved) in the two experiments. The decay-in-flight of a similar mixture of states in lithium-like fluorine was studied and assigned a lifetime of 2.00 ± 0.02 ns by RICHARD et al. [10.47].

Acknowledgments

The author is particularly grateful to D. J. Pegg for his help in completing the text of this manuscript and overseeing its final preparation. Thanks are due Mrs. Barbara Pack for her excellent typing assistance, and the staff of the Oak Ridge National Laboratory tandem and cyclotron for the several years of invaluable technical support that underlie the work in our laboratory. The direct support of the U. S. Office of Naval Research, the U. S. National Science Foundation, and the U. S. National Aeronautics and Space Administration in the work of our laboratory and in the preparation of this manuscript is gratefully acknowledged.

References

10.1 C. E. Moore: *Atomic Energy Levels*, NSRDS-NBS 35 (U.S.Govt. Printing Office Washington,D.C. 1971)

10.2 P. Auger: J. Phys. Radium 6, 205 (1925)

10.3 B. Donnally, W. W. Smith, D. J. Pegg, M. D. Brown, I. A. Sellin: Phys. Rev. A 4, 122 (1971)

10.4 I. A. Sellin, D. J. Pegg, M. D. Brown, W. W. Smith, B. Donnally: Phys. Rev. Letters 27, 1108 (1971)

10.5 K. Siegbahn, ed.: *Alpha-, Beta-, and Gamma-Ray Spectroscopy* (North-Holland Publ. Co., Amsterdam 1965)

10.6 K. D. Sevier, ed: *Low Energy Electron Spectroscopy* (Wiley Publ. Co., New York 1972)

10.7 M. E. Rudd, T. Jorgensen, Jr., D. J. Volz: Phys. Rev. Letters 16, 929 (1966)

10.8 A. K. Edwards, J. S. Risley, R. Geballe: Phys. Rev. A 3, 583 (1971)

10.9 G. B. Crooks, R. D. DuBois, D. E. Golden, M. E. Rudd: Phys. Rev. Letters 29, 327 (1972)

10.10 M. E. Rudd, J. Macek: Vol. III, No. 2, *Case Studies in Atomic Physics*. ed. by M. R. C. McDowell, E. W. McDaniel (North-Holland Publ. Co., Amsterdam 1972) pp 47-136

10.11 A. K. Edwards, D. L. Cunningham: Phys. Rev. A 8, 168 (1973)

10.12 K. O. Groeneveld, R. Mann, G. Spahn, W. Meckbach: 1973 Ann. Report of the Institüt für Kernphysik (Universität, Frankfurt am Main) pp 66-93

10.13 M. E. Rudd, B. Fastrup, P. Dahl, F. D. Schowengerdt: Phys. Rev. A 8, 220 (1973)

10.14 D. L. Cunningham, A. K. Edwards: Phys. Rev. A 8, 2960 (1973)

10.15 N. Stolterfoht, D. Burch, D. Schneider: Proceedings of the VIII Intl. Conf. on the Physics of Electronic and Atomic Collisions, Belgrade (Institute of Physics, Belgrade 1973) pp 731-732

10.16 A. K. Edwards, D. L. Cunningham: Phys. Rev. A 9, 1011 (1974)

10.17 J. S. Risley, A. K. Edwards, R. Geballe: Phys. Rev. A 9, 1115 (1974)

10.18 T. A. Patterson, H. Hotop, A. Kasdan, D. W. Norcross, W. C. Lineberger: Phys. Rev. Letters 32, 189 (1974)

10.19 A. K. Edwards, D. L. Cunningham: Phys. Rev. A 10, 448 (1974)

10.20 N. Stolterfoht, D. Burch, J. S. Risley, D. Schneider, H. Wieman: Ann. Report of the Nuclear Physics Laboratory (University of Washington, Seattle 1974)pp 157-172

10.21 N. Stolterfoht, D. Schneider, D. Burch, B. Aagaard, E. Børing, B. Fastrup: (to be published 1975)

10.22 D. J. Pegg, H. H. Haselton, R. S. Thoe, P. M. Griffin, M. D. Brown, I. A. Sellin: Phys. Letters 50A, 447 (1975)

10.23 D. L. Matthews, B. M. Johnson, J. J. Mackey, C. F. Moore: Phys. Rev. Letters 31, 1331 (1973)

10.24 J. R. Macdonald, L. Winters, M. D. Brown, T. Chiao, L. D. Ellsworth: Phys. Rev. Letters 29, 1291 (1972)

10.25 J. R. Mowat, D. J. Pegg, R. S. Peterson, P. M. Griffin, I. A. Sellin: Phys. Rev. Letters 29, 1577 (1972)

10.26 J. R. Mowat, I. A. Sellin, D. J. Pegg, R. S. Peterson, M. D. Brown, J. R. Macdonald: Phys. Rev. Letters 30, 1289 (1973)

10.27 I. S. Dmitriev, V. S. Nikolaev: Zh. Eksper, I. Teor. Fiz. 47, 615 (1964) [Sov. Phys. - JEPT 20, 409 (1965)]

10.28 L. C. Northcliffe, R. F. Schilling: In *Nuclear Data Tables A7*, No. 3-4 (Academic Press, New York 1970)

10.29 E. Blauth: Z. Physik 147, 228 (1957)

10.30 V. V. Zashkvara, M. I. Korsunskii, O. S. Kosmachev: Zh. Tekhn. Fiz. 36, 132 (1966) [Sov. Phys. - Tech. Phys. 11, 96 (1966)]

10.31 H. Z. Sar-El: Rev. Sci. Instr. 38, 1210 (1967)

10.32 H. Z. Sar-El: Rev. Sci. Instr. 39, 533 (1968)

10.33 W. W. Smith: University of Connecticut (unpublished results)

10.34 H. H. Haselton, I. A. Sellin: University of Tennessee (unpublished results)

10.35 H. Hafner, J. A. Simpson, C. E. Kuyatt: Rev. Sci. Instr. 39, 33 (1968)

10.36 M. E. Rudd: In *Low Energy Electron Spectrometry*, ed. by K. D. Sevier (Wiley Publ. Co., New York 1972)pp 17-32

10.37 R. S. Thoe: University of Tennessee (unpublished results)

10.38 P. Sigmund, K. B. Winterbon: Nucl. Instr. and Meth. 119, 541 (1974)

10.39 L. Meyer: Phys. Status Solidi B 44, 253 (1971)

10.40 C. K. Cline, T. E. Pierce, K. H. Purser, M. Blann: Phys. Rev. 180, 450 (1969)

10.41 D. J. Pegg, I. A. Sellin, R. S. Peterson, J. R. Mowat, W. W. Smith, M. D. Brown, J. R. Macdonald: Phys. Rev. A 8, 1350 (1973)

10.42 E. Holøien, S. Geltman: Phys. Rev. 153, 81 (1967)

10.43 B. R. Junker, J. N. Bardsley: Phys. Rev. A 8, 1345 (1973)

10.44 A. H. Gabriel: Mon. Not. R. Astr. Soc. 160, 99 (1972)

10.45 H. P. Summers: Astrophys. J. 179, L45 (1973)

10.46 U. I. Safronova, V. N. Kharitonova: Optics and Spectroscopy 27, 300 (1969)

10.47 P. Richard, R. L. Kauffman, F. F. Hopkins, C. W. Woods, K. A. Jamison: Phys. Rev. Letters 30, 888 (1973)

10.48 D. J. Pegg, I. A. Sellin, P. M. Griffin, W. W. Smith: Phys. Rev. Letters 28, 1615 (1972)

10.49 A. W. Weiss: private communication, 1973

10.50 J. D. Garcia, J. E. Mack: Phys. Rev. 138, A987 (1965)

10.51 D. L. Ederer, T. Lucatorto, R. P. Madden: Phys. Rev. Letters 25, 1537 (1970)

10.52 C. Nicolaides: Nucl. Instr. and Meth. 110, 231 (1973)

10.53 H. W. Wolff, K. Radler, B. Sonntag, R. Haensel: Z. Physik 257, 353 (1972)

10.54 W. Meckbach: private communication, 1975

10.55 J. W. Hiby: Ann. Phys. (Leipz) 34, 473 (1939)

10.56 R. Dopel: Ann. Phys. (N.Y.) 76, 1 (1925)

10.57 T-Y. Wu: Phys. Rev. 58, 1114 (1940)

10.58 T-Y. Wu, S. T. Shen: Chin. J. Phys. (Peking) 5, 150 (1944)

10.59 E. Holøien, J. Midtal: Proc. Phys. Soc. (London) A68, 815 (1955)

10.60 P. Feldman, R. Novick: Phys. Rev. Letters 11, 278 (1963)

10.61 P. Feldman, R. Novick: Phys. Rev. 160, 143 (1967)

10.62 L. M. Blau, R. Novick, D. Weinflash: Phys. Rev. Letters 24, 1268 (1970)

10.63 I. S. Dmitriev, V. S. Nikolaev, Y. A. Teplova: Phys. Letters 26A, 122 (1968)

10.64 S. T. Manson: Phys. Letters 23, 315 (1966)

10.65 V. V. Balashov, V. S. Senashenko, B. Tekou: Phys. Letters 48A, 487 (1967)

10.66 K. T. Cheng, C. P. Lin, W. R. Johnson: Phys. Letters 48A, 437 (1974)

10.67 H. H. Haselton, R. S. Thoe, J. R. Mowat, P. M. Griffin, D. J. Pegg, I. A. Sellin: Phys. Rev. A 11, 468 (1975)

10.68 M. Levitt, R. Novick, P. Feldman: Phys. Rev. A 3, 130 (1971)

10.69 C. L. Cocke, B. Curnutte, R. Randall: Phys. Rev. A 9, 1823 (1974)

10.70 P. Richard, R. L. Kauffman, F. F. Hopkins, C. W. Woods, K. A. Jamison: Phys. Rev. A 8, 2187 (1973)

10.71 C. F. Moore, W. J. Braithwaite, D. L. Matthews: Phys. Letters 44A, 199 (1973)

Appendix

The following papers have been selected by the editor from publications which appeared subsequent to completion of the various chapters. The editor apologizes in the event that some papers have been overlooked or that others are not entirely appropriate to their assigned chapters. Within each chapter the references are alphabetized by first author.

Chapter 1

P. Apard, S. Bliman, R. Geller: A multipurpose highly charged ion source. Nucl. Instr. and Method 129, 357 (1975).

M. G. Betigiri, M. S. Bhatia, T. P. David: Production of 50 keV intense heavy ion beams. Nucl. Instr. and Meth. 128, 29 (1975).

M. W. Chang, W. S. Bickel: Spectral-line profiles and wavelength measurements using the beam-foil light source. J. Opt. Soc. Am. 65, 1376 (1975).

K. R. Chapman: The inverted sputter source. Nucl. Instr. and Meth. 124, 299 (1975).

A. Chevallier, J. Chevallier, J. L. Gross, B. Haas, N. Schulz, J. Styczen, M. Toulemonde: Lifetime and g-factor measurements in ^{44}Sc. Z. f. Physik A 275, 51 (1975).

R. B. Clark, I. S. Grant, R. King, D. A. Eastham, T. Joy: Equilibrium charge state distributions of high energy heavy ions. Nucl. Instr. and Meth. 133, 17 (1976).

G. Clausnitzer, H. Klinger, A. Müller, E. Salzborn: An electron beam ion source for the production of multiply charged heavy ions. Nucl. Instr. and Meth. 128, 1 (1975).

G. Doucas, H. R. McK. Hyder, A. B. Knox: The brightness and emittance of negative ion beams from a Middleton source. Part II, Nucl. Instr. and Meth. 124, 11 (1975).

B. Dynefors, I. Martinson, E. Veje: A study of the beam-foil excitation mechanism using 60-360 keV Be^+ projectiles. Physica Scripta 12, 58 (1975).

J. M. Freeman, P. P. Kane, B. W. Hooton: Small angle multiple scattering of 12-40 MeV heavy ions from thin foils. Nucl. Instr. and Meth. 124, 29 (1975).

K. O. Groeneveld, G. Spahn: Angular straggling of heavy and light ions in thin solid foils. Nucl. Instr. and Meth. 123, 425 (1975).

J. Heinemeier, P. Hvelplung, J. O. Østgard, F. R. Simpson: Equilibrium charge-state distributions of heavy ion beams after passage through carbon-foils. Physica Scripta 10, 304 (1974).

R. Kirchner, E. Roeckl: A cathode with long lifetime for operation of ion sources with chemically aggressive vapors. Nucl. Instr. and Meth. 127, 307 (1975).

A. Latuszynski, V. I. Raiko: Studies of the ion source with surface-volume ionization. Nucl. Instr. and Meth. 125, 61 (1975).

F. W. Martin: Acceleration of doubly charged C, N, and O ions at a 3 megavolt Van de Graaff. Nucl. Instr. and Meth. 124, 329 (1975).

G. Newton, P. J. Unsworth, D. A. Andrews: Observation of fine-structure resonances in atomic hydrogen by the method of spatially periodic fields and variable beam velocity. J. Phys. B 8, 2928 (1975).

H. Oona, W. S. Bickel: Charge-state determination of beam-foil spectral lines. J. Opt. Soc. Am. 66, 278 (1976).

R. Saloman, S. Stenholm: Two-photon spectroscopy in a fast atomic beam. Optics Comm. 16, 292 (1976).

J. E. Sherwood, R. D. Zwicker: A hollow-beam prototype of the universal negative ion source for tandem accelerators. Nucl. Instr. and Meth. 129, 43 (1975).

H. V. Smith, Jr., H. T. Richards: A sputter pig source (SPIGS) for negative ions. Nucl. Instr. and Meth. 125, 497 (1975).

P. Thieberger, H. E. Wegner: Test of heavy-ion gas-foil stripping for improved foil lifetime in tandem Van de Graaff accelerators. Nucl. Instr. and Meth. 126, 231 (1975).

Chapter 2

L. Barrette, D. J. G. Irwin, R. Drouin: New identifications and lifetime measurements in Ne VIII. Physica Scripta 12, 113 (1975).

H. G. Berry: Multiply-excited states in beam-foil spectroscopy. Physica Scripta 12, 5 (1975).

H. G. Berry, R. Hallin, R. Sjödin, M. Gaillard: Beam-foil observations of Na I doubly-excited states. Phys. Letters 50A, 181 (1974).

W. S. Bickel: Molecular effects in beam-foil spectroscopy. Phys. Rev. A 12, 1801 (1975).

R. Bruch, G. Paul, J. Andrä: Metastable autoionizing three and four electron states in beryllium. J. Phys. B 8, L253 (1975).

M. S. Z. Chaghtai, S. P. Singh, S. Khatoon: Observation and classification of 4p-5d,6d and 4p-6s,7s transitions in Mo VIII. J. Phys. B 8, 1831 (1975).

C. L. Cocke, S. L. Varghese, J. A. Bednar, C. P. Bhalla, B. Curnutte, R. Kauffman, R. Randall, P. Richard, C. Woods, J. H. Scofield: X rays from foil-excited iodine beams. Phys. Rev. A 12, 2413 (1975).

J. Davidson: Relative and absolute level populations in beam-foil-excited neutral helium. Phys. Rev. A 12, 1350 (1975).

G. A. Doschek, U. Feldman, L. Cohen: Transitions $2s^2 2p - 2s2p^2$ in the B I isoelectronic sequence. J. Opt. Soc. Am. 65, 463 (1975).

B. Edlén: The oxygen spectrum below 200 Å and the high-limit terms of O IV. Physica Scripta 11, 366 (1975).

B. Edlén, J. W. Swensson: The spectrum of doubly ionized titanium, Ti III. Physica Scripta 12, 21 (1975).

J. O. Ekberg: Term analysis of Fe VI. Physica Scripta 11, 23 (1975).

J. O. Ekberg: Term analysis of Fe V. Physica Scripta 12, 42 (1975).

J. O. Ekberg: The spectrum of five-times-ionized vanadium. Physica Scripta 13, 111 (1976).

J. O. Ekberg, L. A. Swensson: Spectrum and term system of Ti XII. Physica Scripta 12, 116 (1975).

L. Gabła, M. Szymonski, M. Szulkin: The intensity ratios and polarization of spectral lines emitted by secondary particles during bombardment of zinc and copper targets by helium ions. Physica C 81, 193 (1976).

T. F. Gallagher, J. A. Edelstein, R. M. Hill: Collisional angular momentum mixing in Rydberg states of sodium. Phys. Rev. Letters 35, 644 (1975).

J. S. Hansen, W. Persson, A. Borgström: $3s3p^6\,3d \rightarrow 3s^2 3p^5 nf$ interaction in Ca III and identification of the $3s3p^6 3d$ configuration. Physica Scripta 11, 31 (1975).

L. Johansson: Additions to the displaced term system of Be I. Physica Scripta 10, 236 (1974).

S. Johansson, U. Litzén: Analysis of 4d-4f transitions in Fe II. Physica Scripta 10, 121 (1974).

E. Ya. Kononov, K. N. Koshelev, L. I. Podobedova, S. S. Charilov: Spectra of calcium in the vacuum ultraviolet. Part 1: Ca XVII, Ca XVI. Opt. and Spectr. 39, 458 (1975).

A. E. Livingston, J. A. Kernahan, D. J. G. Irwin, E. H. Pinnington: Beam-foil studies of phosphorus in the vacuum ultraviolet. Physica Scripta 12, 223 (1975).

C. E. Magnusson, P. O. Zetterberg: The spectrum and term system of P V. Physica Scripta 10, 177 (1974).

L. Minnhagen: The $2p^4\,4f$ configuration of doubly ionized sodium, Na III. Physica Scripta 11, 38 (1975).

O. Poulsen, T. Andersen, N. J. Skonbone: Fast-beam, zero-field level-crossing measurements of radiative lifetimes, fine and hyperfine structures in excited states of ionic and neutral beryllium. J. Phys. B 8, 1393 (1975).

A. V. Ryabtsev: Spectra of ions in the Ni I isoelectronic series. 1: Ga IV. Opt. and Spectr. 39, 239 (1975).

A. V. Ryabtsev: Spectra of ions of the isoelectronic Ni I series. Part 2: Ge V. Opt. and Spectr. 39, 455 (1975).

D. Schürmann: Atomic states of lithium by the beam-foil technique in the spectral range 400-1900 Å. Z. f. Physik 273, 331 (1975).

E. Veje: Phys. Rev. A. (in press).

J. R. Woodyard, Sr., P. L. Altick: Ab initio calculation of energy levels and transition probabilities in the spectra of rare gases II. Ne I. J. Phys. B 8, 718 (1975).

J. F. Wyart: Mélange de configurations dans les spectres du titane et du vanadium plusieurs fois ionisés. Physica Scripta 12, 33 (1975).

Chapter 3

T. Andersen, A. P. Petrkiev, G. Sørensen: Lifetimes of excited levels in Mg III, Ca III, K IV-V, Ca V, and Ge I. Physica Scripta 12, 283 (1975).

T. Andersen, O. Poulsen, P. S. Ramanujam, A. Petrakien Petkov: Lifetimes of some excited states in the rare earths: La II, Ce II, Pr II, Nd II, Sm II, Yb I, Yb II, and Lu II. Solar Phys. 44, 257 (1975).

A. Arnesen, A. Bengtsson, R. Hallin, S. Kandela, T. Noreland, R. Lidholt: Lifetime measurements of the Ba II 6p $^2P_{3/2}$ and 6p $^2P_{1/2}$ levels with the beam-laser method. Phys. Letters 53A, 459 (1975).

L. Barrette, D. J. G. Irwin, R. Drouin: New identifications and lifetime measurements in Ne VIII. Physica Scripta 12, 113 (1975).

Y. Baudinet-Robinet, P. D. Dumont, E. Biémont, N. Grevesse: Lifetimes and transition probabilities in N V. Physica Scripta 11, 371 (1975).

J. A. Bednar, C. L. Cocke, B. Curnutte, R. Randall: Lifetime of the 2 $^3S_1$ state in heliumlike sulphur and chlorine. Phys. Rev. A 11, 460 (1975).

F. Bely-Dubau, C. Camhy-Val, A. M. Dumont: Mean lifetimes of excited levels of Ar(II)-II. Theoretical considerations. J. Quant. Spectrosc. Radiat. Transfer 15, 375 (1975).

H. G. Berry: Experimental lifetimes in vanadium. Physica Scripta 13, 36 (1976).

W. J. Braithwaite, D. L. Matthews, C. F. Moore: Delayed x-ray emission in the Lyman and Lyman-like series of one- and two-electron oxygen. Phys. Rev. A 11, 465 (1975).

R. Bruch, G. Paul, J. Andrä, L. Lipsky: Autoionization of foil-excited states in Li I and Li II. Phys. Rev. A 12, 1808 (1975).

J. P. Buchet, M. Druetta: Beam-foil spectroscopy of neon between 80 and 350 Å. J. Opt. Soc. Am. 65, 991 (1975).

C. Camhy-Val, A. M. Dumont, M. Dreux, L. Perret, C. Vanderriest: Mean lifetimes of excited levels of Ar II. I. Time correlation measurements. J. Quant. Spectrosc. Radiat. Transfer 15, 527 (1975).

L. J. Curtis, B. Engman, I. Martinson: Lifetime measurements in Cu I and Cu II. Physica Scripta 13, 109 (1976).

K. E. Donnelly, P. J. Kindlmann, W. R. Bennet, Jr.: Radiative lifetimes and collisional deactivation rates of levels in the $4p^4$ 5p configuation of singly ionized krypton. J. Opt. Soc. Am. 65, 1359 (1975).

B. I. Dynefors: Lifetime measurements in Se I and Te I. Physica Scripta 11, 375 (1975).

B. Emmoth, M. Braun, J. Bromander, I. Martinson: Lifetimes of excited levels in Ca I-Ca III. Physica Scripta 12, 75 (1975).

B. Engman, A. Gaupp, L. J. Curtis, I. Martinson: Lifetime measurements for excited levels in Cr II. Physica Scripta 12, 220 (1975).

M. Gaillard, H. J. Andrä, A. Gaupp, W. Wittmann, H. J. Plonn, and J. O. Stoner, Jr.: Mean lives for the $5p_{3/2}[1/2]_1$ and $5p_{1/2}[1/2]_1$ levels in singly ionized rubidium $(Rb\ II)^+$. Phys. Rev. A 12, 987 (1975).

K. O. Groeneveld, R. Mann, G. Nolte, S. Schumann, R. Spohr: Lifetime measurements of the metastable 1s2s2p $^4P_{5/2}$ state in the lithium-like N, O and Ne. Phys. Letters 54A, 335 (1975).

H. H. Haselton, R. S. Thoe, J. R. Mowat, P. M. Griffin, D. J. Pegg, I. A. Sellin: Lifetimes of the metastable autoionizing (1s2s2p) $^4P_{5/2}$ states of lithiumlike $A\ell^{10+}$ and Si^{11+} ions: Comparisons with theory over the isoelectronic sequence Z = 8 - 18. Phys. Rev. A 11, 468 (1975).

F. Hopkins, P. v. Brentano: High n-state population and delayed photon emission from beam-foil interaction. J. Phys. B 9, 775 (1976).

D. Kaiser: Lifetime measurements of higher s- and d- levels in the Na I spectrum. Phys. Letters 51A, 375 (1975).

J. A. Kernahan, E. H. Pinnington, A. E. Livingston, D. J. G. Irwin: Mean-life measurements for levels in B I - B IV. Physica Scripta 12, 319 (1975).

J. Z. Klose: Mean life of the 27887-cm^{-1} level in U I. Phys. Rev. A 11, 1840 (1975).

E. J. Knystautas, R. Drouin: New identifications and lifetime measurements of excited states in highly ionized oxygen. J. Phys. B 8, 2001 (1975).

A. E. Livingston, Y. Baudinet-Robinet, P. D. Dumont: Radiative-lifetime measurements: effects of configuration mixing and singlet-triplet interaction in singly-ionized nitrogen. Phys. Letters 55A, 207 (1975).

A. E. Livingston, J. A. Kernahan, D. J. G. Irwin, E. H. Pinnington: Radiative-lifetime measurements in Si II - Si V. J. Phys. B 9, 389 (1976).

L. Maleki, C. E. Head: Radiative lifetimes for some 4p and 4d levels of singly ionized sulfur. Phys. Rev. A 12, 2420 (1975).

J. R. Mowat, P. M. Griffin, H. H. Haselton, R. Laubert, D. J. Pegg, R. S. Peterson, I. A. Sellin, R. S. Thoe: Heliumlike ^{19}F: 2 $^3P_2$ and 2 $^3P_0$ lifetimes. Phys. Rev. A 11, 2198 (1975).

A. L. Osherovich, E. N. Borisov, M. L. Burstein, Ya. F. Verolainen: Radiative lifetimes of levels of the mercury atom. Opt. and Spectr. 39, 466 (1975).

H. Panke, F. Bell, H. D. Betz, W. Stehling, E. Spindler, R. Laubert: Lifetimes of multiplet states in one-, two-, and three-electron sulphur ions. Phys. Letters 53A, 457 (1975).

E. H. Pinnington: Radiative lifetimes of some levels of Xe I and Xe II. J. Opt. Soc. Am. 65, 218 (1975).

O. Poulsen, T. Andersen, N. J. Skonbone: Fast-beam, zero-field level-crossing measurements of radiative lifetimes, fine and hyperfine structures in excited states of ionic and neutral beryllium. J. Phys. B 8, 1393 (1975).

V. P. Samoilov, Yu. M. Smirnov, G. S. Starihova: Transition probabilities and cross sections for excitation of Xe II. Opt. and Spectr. 38, 707 (1975).

W. Schlagheck: Lifetimes of ($2p^5\ 3s^{1,3}P$) levels of Na II. Phys. Letters 54A, 181 (1975).

D. Schürmann: Atomic states of lithium by the beam-foil technique in the spectral range 400-1900 Å. Z. f. Physik 273, 331 (1975).

Chapter 4

N. V. Afanaseva, P. F. Gruzdev: Lifetimes of the ns and np levels of the argon atom. Opt. and Spectr. 38, 450 (1975).

N. V. Afanaseva, P. F. Gruzdev: Lifetimes of nd and nf levels of the neon atom. Opt. and Spectr. 38, 583 (1975).

M. Aymar: Détermination théorique des durées de vie des niveaux $3s3p^3$ dans la séquence isoeléctronique de Si I. Physica 74, 205 (1974).

D. R. Beck, C. A. Nicolaides: Effect of electron correlation on atomic properties. Int. J. Qu. Chem. Symp. 8, 17 (1974).

D. R. Beck, C. A. Nicolaides: Theoretical lifetimes of the N II 2s2p, and $2s^2 2p3s\ ^1P^o$ states obtained by applying "fotos". Phys. Letters 56A, 265 (1976).

D. R. Beck, C. A. Nicolaides: Theoretical oscillator strengths for the nitrogen I and oxygen I resonance transitions. J. Quant. Spectrosc. Radiat. Transfer 16, 297 (1976).

D. R. Beck, C. A. Nicolaides: Absorption oscillator strengths to autoionizing states in lithium, nitrogen, and fluorine, and their isoelectronic sequences. Can. J. Phys. 54, 689 (1976).

F. Bely-Dubau, C. Camhy-Val, A. M. Dumont: Mean lifetimes of excited levels of Ar (II)-II. Theoretical considerations. J. Quant. Spectrosc. Radiat. Transfer 15, 375 (1975).

E. Biémont: Cancellation effects and trends of oscillator strengths in the potassium isoelectronic sequence. Physica B & C 81, 158 (1976).

J. L. Dehmer, M. Inokuti, R. P. Saxon: Systematics of moments of dipole oscillator-strength distributions for atoms of the first and second row. Phys. Rev. A 12, 102 (1975).

Th. M. Eℓ Sherbini: Calculation of Xe II line strengths and radiative lifetimes in intermediate coupling. Z. f. Physik A 275, 1 (1975).

Th. M. Eℓ Sherbini: Calculations of transition probabilities and radiative lifetimes for singly ionized krypton. J. Phys. B 8, L183 (1975).

C. Froese-Fischer: Theoretical oscillator strengths for nP → nD transitions in Mg. Can. J. Phys. 53, 184 (1975).

C. Froese-Fischer: Theoretical oscillator strengths for nS → mP transitions in Mg. Can. J. Phys. 53, 338 (1975).

S. Garpman, L. Holmgren, A. Rosen: Theoretical transition probabilities between the $np^3(n+1)s \rightarrow np^4$ configurations of Se I and Te I. Physica Scripta 10, 221 (1974).

O. Goscincki, G. Howat, T. Åberg: On transition energies and probabilities by a transition operator method. J. Phys. B 8, 11 (1975).

P. F. Gruzdev, A. V. Loginov: Radiation lifetimes of levels of the argon atom. Opt. and Spectr. 38, 234 (1975).

P. F. Gruzdev, A. V. Loginov: Radiation lifetimes of levels of the Kr I atom. Opt. and Spectr. 38, 611 (1975).

P. F. Gruzdev, A. V. Loginov: Neon transition probabilities. Part 2: $2p^5 4p$-$2p^5 ns$ (n = 3-6) transitions. Opt. and Spectr. 39, 464 (1975).

H. H. Haselton, R. S. Thoe, J. R. Mowat, P. M. Griffin, D. J. Pegg, I. A. Sellin: Lifetimes of the metastable autoionizing (1s2s2p) $^4P_{5/2}$ states of lithiumlike $A\ell^{10+}$ and Si^{11+} ions: Comparisons with theory over the isoelectronic sequence Z = 8-18. Phys. Rev. A 11, 468 (1975).

L. Holmgren: Theoretically calculated transition probabilities and lifetimes for the first excited configuration $nP_2(n+1)s$ in the neutral As, Sb, and Bi atoms. Physica Scripta 11, 15 (1975).

L. Holmgren: A relativisitic calculation of transition probabilities between the np(n+1)s and np^2 configurations for the elements of group IV. Physica Scripta 10, 215 (1974).

Y. K. Kim, J. P. Desclaux: Relativisitic f values for the resonance transitions of Li- and Be-like ions. Phys. Rev. Letters 36, 139 (1976).

G. L. Klimchitskeya, U. I. Safronova, L. N. Labzovskii: Relativistic calculations of transition probabilities in two-electron, multiply charged ions. Opt. and Spectr. 38, 480 (1975).

J. G. Leopold, M. Cohen: Bounds to transition integrals for the helium isoelectronic sequence. J. Phys. B 8, L369 (1975).

R. A. Lilly: Transition probabilities in the spectra of Ne I. J. Opt. Soc. Am. 65, 389 (1975).

R. A. Lilly: Transition probabilities in the spectra of Ne I, Ar I, and Kr. J. Opt. Soc. Am. 66, 245 (1976).

A. Lindgård, S. E. Nielsen: Numerical approach to transition probabilities in the Coulomb approximation: Be II and Mg II Rydberg series. J. Phys. B 8, 1183 (1975).

W. L. Luken, O. Sinanoğlu: Oscillator strengths for transitions involving excited states not lowest of their symmetry: nitrogen I and nitrogen II transitions. J. Chem. Phys. 64, 3141 (1976).

S. Lunell: Oscillator strengths for the lithium isoelectronic sequence from spin-optimized SCF wave functions. Physica Scripta 12, 63 (1975).

J. V. Mallow, P. S. Bagus: Ultraviolet oscillator strengths for carbon, nitrogen, and oxygen ions. J. Quant. Spectrosc. Radiat. Transfer 16, 409 (1976).

J. Migdałek: Theoretical oscillator strengths for some transitions in P III, As III, Sb III, and Bi III. J. Quant. Spectrosc. Radiat. Transfer 16, 385 (1976).

J. Migdałek: Theoretical relativistic oscillator strenths I. Transitions in principal, sharp, and diffuse series of aluminum I, gallium I, indium I, and thallium I spectra. Can. J. Phys. 54, 118 (1976).

J. Migdałek: Theoretical oscillator strengths II. Transitions in principal, sharp, and diffuse spectral series of aluminum III, gallium III, indium III, and thallium III spectra. Can. J. Phys. 54, 130 (1976).

C. A. Nicolaides, D. R. Beck: Length, velocity, and acceleration expressions for the calculation of accurate oscillator strengths in many-electron systems. Chem. Phys. Letters 35, 202 (1975).

C. A. Nicolaides, D. R. Beck: Approach to the calculation of the important many-body effects on photoabsorption oscillator strengths. Chem. Phys. Letters 36, 79 (1975).

C. A. Nicolaides, D. R. Beck: A comment on the effect of nonorthonormality on atomic transition probabilities. Can. J. Phys. 53, 1224 (1975).

J. S. Onello: $(1s)^2 3s\ ^2S$, $(1s)^2 3p\ ^2P$, and $(1s)^2 3d\ ^2D$ states of the lithium isoelectronic sequence. Phys. Rev. A 11, 743 (1975).

J. S. Onello, L. Ford: Magnetic quadrupole decay of the $(1s2s2p)\ ^4P_{5/2} - (1s)^2 2s\ ^2S^e_{1/2}$ transition of the lithium isoelectronic sequence. Phys. Rev. A 11, 749 (1975).

U. I. Safronova: Calculation of the term energies, dipole matrix elements, and oscillator strengths for isoelectric series of light atoms. Opt. and Spectr. 38, 477 (1975).

N. Spector, S. Garpman: Relative transition probabilities in Kr II. J. Opt. Soc. Am. 65, 1187A (1975).

D. K. Watson, S. V. O'Neill: 1/Z-expansion study of the $1s^2 2s^2\ ^1S$, $1s^2 2s2p\ ^1P$, and $1s^2 2p^2\ ^1S$ states of the beryllium isoelectronic sequence. Phys. Rev. A 12, 729 (1975).

A. Wells, K. J. Miller: Effect of a superposition of configurations on the generalized oscillator strengths and elastic and inelastic total cross sections for the 3p to 4s, 4p and nd ($n \leq 7$) transitions in aluminum. Phys. Rev. A 12, 17 (1975).

J. R. Woodyard, Sr., P. L. Altick: Ab initio calculation of energy levels and transition probabilities in the spectra of rare gases II. Ne I. J. Phys. B 8, 718 (1975).

Chapter 5

M. T. Anderson, F. Weinhold: Bounds to the lifetimes of the Ar XVII 2 ^3S state. Phys. Rev. A 11, 442 (1975).

M. Aymar: Détermination théorique des durées de vie des niveaux $3s3p^3$ dans la séquence isoélectronique de Si I. Physica 74, 205 (1974).

E. Biémont: Systematic trends of Hartree-Fock oscillator strengths along the sodium isoelectronic sequence. J. Quant. Spectrosc. Radiat. Transfer 15, 531 (1975).

E. Biémont: Theoretical oscillator strengths for ultraviolet lines of doubly-ionized elements of astrophysical interest. J. Quant. Spectrosc. Radiat. Transfer 16, 137 (1976).

R. T. Brown: Ultraviolet wavelengths and oscillator strengths for 3d-nf transitions in the helium isoelectronic sequence. Astrophys. J. 158, 829 (1969).

R. T. Brown, J-L. M. Cortez: Oscillator strengths for allowed nd-n′f transitions in the helium isoelectronic sequence. Astrophys. J. 176, 267 (1972).

D. K. Datta, S. K. Ghoshel, S. Sengupta: Hartree-Fock wave functions and oscillator strengths for the helium isoelectronic sequence. J. Quant. Spectrosc. Radiat. Transfer 16, 49 (1976).

J. L. Dehmer, M. Inokuti, R. P. Saxon: Systematics of moments of dipole oscillator-strength distributions for atoms of the first and second row. Phys. Rev. A 12, 102 (1975).

H. H. Haselton, R. S. Thoe, J. R. Mowat, P. M. Griffin, D. J. Pegg, I. A. Sellin: Lifetimes of the metastable autoionizing (1s2s2p) $^4P_{5/2}$ states of lithiumlike $A\ell^{10+}$ and si^{11+} ions: Comparisons with theory over the isoelectronic sequence Z = 8-18. Phys. Rev. A 11, 468 (1975).

J. G. Leopold, M. Cohen: Bounds to transition integrals for the helium isoelectronic sequence. J. Phys. B 8, L369 (1975).

A. Lindgård, S. E. Nielsen: Numerical approach to transition probabilities in the Coulomb approximation: Be II and Mg II Rydberg series. J. Phys. B 8, 1183 (1975).

G. A. Martin, W. L. Wiese: Atomic oscillator-strength distributions in spectral series of the lithium isoelectronic sequence. Phys. Rev. A 13, 699 (1976).

J. R. Woodyard, Sr., P. L. Altick: Ab initio calculation of energy levels and transition probabilities in the spectra of rare gases II. Ne I. J. Phys. B 8, 718 (1975).

Chapter 6

E. Biémont: Computation of oscillator strengths by a semi-empirical method for some elements of the iron-group and their solar photospheric abundance. III. Results for Mn I. Solar Phys. 44, 269 (1975).

E. Biémont, N. Grevesse: The solar photospheric abundance of iron. Solar Phys. 45, 59 (1975).

Chapter 7

W. E. Behring, L. Cohen, U. Feldman, G. A. Doschek: The solar spectrum: wavelengths and identifications from 160 to 770 Angstroms. Astrophys. J. 203, 521 (1976).

S. J. Czyzak, L. H. Aller, R. N. Euwema: Forbidden-line excitation data for certain coronal lines. Ap. J. Suppl. No. 272 (1974).

G. A. Doschek, U. Feldman, L. Cohen: Transitions $2s^22p$-$2s2p^2$ in the B I isoelectronic sequence. J. Opt. Soc. Am. 65, 463 (1975).

G. A. Doschek, U. Feldman, K. P. Dere, G. D. Sandlin, M. E. Van Hoosier, G. E. Brueckner, J. D. Purcell, R. Tousey: Forbidden lines of highly ionized iron in solar flare spectra. Astrophy. J. 196, L83 (1975).

S. O. Kastner, W. M. Neupert, M. Swartz: Observation of possible Fe XVII $2p^53p$ $(^1S_0)$ - $2p^53s$ $(^1P_1, {}^3P_1)$ transitions in spectra of a solar active region and flare. Solar Phys. 43, 111 (1975).

E. Ya. Kononov, K. N. Koshelev, L. I. Podobedova, S. V. Chekalin, S. S. Churilov: Identification of the solar spectra of multicharged iron ions on the basis of laboratory measurements. J. Phys. B 9, 565 (1976).

F. Magnant-Crifo: On the identification of Fe IX and Ni XI lines from coronal spectra. Solar Phys. 41, 109 (1975).

P. D. Noerdlinger, S. E. Dynan: Ultraviolet absorption lines arising on metastable states. Ap. J. Suppl. No. 283, 29, 185 (1975).

H. R. Rugge, A. B. C. Walker, Jr.: The relative abundance of neon and magnesium in the solar corona. Astrophys. J. 203, L139 (1976).

K. G. Widing: Fe XXIII 263 Å and Fe XXIV 255 Å emission in solar flares. Astrophys. J. 197, L33 (1975).

Chapter 8

M. T. Anderson, F. Weinhold: Bounds to the lifetime of the Ar XVII 2 ^3S state. Phys. Rev. A 11, 442 (1975).

C. K. Au: Non-relativisitic ground state Lamb shift of hydrogenic ions. Phys. Letters 56A, 186 (1976).

E. J. Kelsey, J. Sucher: 2 $^3S_1 \rightarrow 1\ ^1S_0$ + one photon transition in heliumlike ions: Exact result for the lowest-order effect of the electron-electron interaction. Phys. Rev. A 11, 1829 (1975).

H. Krüger, A. Oed: Measurement of the decay-probability of metastable hydrogen by two-photon emission. Phys. Letters 54A, 251 (1975).

Chapter 9

Y. B. Band: Alignment and orientation effects in beam-foil experiments. Phys. Rev. Letters 35, 1272 (1975).

H. G. Berry, S. N. Bhardwaj, L. J. Curtis, R. M. Schectman: Orientation and alignment of atoms by beam-foil excitation. Phys. Letters 50A, 59 (1974).

H. G. Berry, L. J. Curtis, D. G. Ellis, R. M. Schectman: Hyperfine quantum beats in oriented ^{14}N IV. Phys. Rev. Letters 35, 274 (1975).

W. W. Chow, M. O. Scully, J. O. Stoner, Jr.: Quantum-beat phenomena described by quantum electrodynamics and neoclassical theory. Phys. Rev. A 11, 1380 (1975).

J. S. Deech, R. Luypaert, G. W. Series: Determination of lifetimes and hyperfine structures of the 8, 9 and 10 $^2D_{3/2}$ states of ^{133}Cs by quantum-beat spectroscopy. J. Phys. B 8, 1406 (1975).

L. Gabła, M. Szymoński, M. Szulkin: The intensity ratios and polarization of spectral lines emitted by secondary particles during bombardment of zinc and copper targets by helium ions. Physica B&C 81, 193 (1976).

R. M. Herman: Theory of final-state coherence in beam-tilted foil interactions. Phys. Rev. Letters 35, 1626 (1975).

R. M. Herman, H. Grotch, R. Kornblith, J. H. Eberly: Quantum electrodynamic and semiclassical interference effects in spontaneous radiation. Phys. Rev. A 11, 1389 (1975).

R. Krotkov: Coherent excitation of the n = 3 states in hydrogen. Phys. Rev. A 12, 1793 (1975).

E. L. Lewis, J. D. Silver: Orientation and alignment effects in beam foil experiments with tilted foils. J. Phys. B 8, 2697 (1975).

M. Lombardi, M. Giroud: Orientation of hydrogen levels by stark effect and sp coherence resulting from direct excitation or molecular dissociation. Phys. Rev. Letters 36, 409 (1976).

M. Lombardi, M. Giroud, J. Macek: Coherently excited atoms in external electric fields. Phys. Rev. A 11, 1114 (1975).

I. R. Senitzky: Quantum beats and quantum mechanics. Phys. Rev. Letters 35, 1755 (1975).

J. D. Silver, J. Desequelles, M. L. Gaillard: Hyperfine structure measurements in the 2s3p ^3P levels of ^{14}N IV by zero-field quantum beats after beam foil excitation. J. Phys. B 8, L219 (1975).

H. Winter, H. H. Bukow: Alignment of H(2P) after beam-foil excitation. Z. Physik A 277, 27 (1976).

Chapter 10

R. Bruch, G. Paul, J. Andrä: Auger transitions in Li-like and Be-like ions. Phys. Letters 53A, 293 (1975).

M. M. Duncan, M. G. Menendez: Measurements of beam-foil electrons emitted near 0° from H^+ and H_2^+ on carbon in the energy range 0.175-1.0 MeV/amu. Phys. Rev. A 13, 566 (1976).

K. O. Groeneveld, G. Nolte, S. Schumann: Low-energy metastable autoionizing states in nitrogen, oxygen, and neon. J. de Physique Lettres 37, 7 (1976).

B. M. Johnson, D. Schneider, K. S. Roberts, J. E. Bolger, C. F. Moore: High resolution beam gas Auger electron spectra for oxygen ions excited by collisions with He, Me, and Ar. Phys. Letters 53A, 254 (1975).

M. G. Menendez, M. M. Duncan: Beam foil electron spectroscopy near the forward direction. Phys. Letters 53A, 409 (1975).

D. J. Pegg, H. H. Haselton, R. S. Thoe, P. M. Griffin, M. D. Brown, I. A. Sellin: Core-excited autoionizing states in the alkali metals. Phys. Rev. A 12, 1330 (1975).

D. Schneider, W. Hodge, B. M. Johnson, L. E. Smith, C. F. Moore: Auger electron emission spectra from foil and gas excited carbon beams. Phys. Letters 54A, 174 (1975).

D. Schneider, L. E. Smith, K. Roberts, W. Hodge, J. Whitenton, C. F. Moore: Relative satellite line intensities in Ar K auger electron spectra produced by 0.4 to 4 MeV proton impact. Phys. Letters 56A, 189 (1976).

Subject Index

absorption 50, 70
- lines or spectra 56, 179-190, 194
- 204, 217, 285, 286
- oscillator strength 65, 68, 69, 149
- 198, 205, 206
- , self 63, 181, 182
- , X-ray 15
accelerators 5-8, 10, 11, 34, 39, 41
- 45, 55, 72, 209, 212, 213, 216, 221
- 225, 231, 294
- , calibration of 75
alignment 68, 77, 83-86, 100-102, 228
- 237-263
- tensor 242-244
analyzer, electrostatic 11, 75, 239
- 270-279, 286
- , magnetic 10, 11, 270, 271
angular aberrations 272, 274
- correlations 218, 220, 239, 242
- momentum 33, 44, 68, 101, 217, 224
- 229, 230, 245, 250, 254, 268, 269
- momentum operator 242, 243, 247
anisotropy transfer 238
Auger or autoionization electrons 225
- 265-295
- levels 27, 48-56, 159
- lifetimes 280, 290-295
- lines or spectra 233, 269-271, 274
- 280-286, 288-290
- process 55, 63, 64, 232
- selection rules 49, 50, 53, 139, 232
- 268

A-values (also see lifetimes, oscillator strengths, transition probabilities) 8, 9, 69, 212, 217-220, 225-231
- as f(Z) 228-231, 293
- for intercombination decay 204, 231
- - M2 vs. E1 229

Balmer lines 20, 217, 250-254
beam, contamination of 9, 10, 34, 63
- , complex 9, 10
- foil properties 33-38, 47, 63, 64
- 70-72, 104, 211
- gas source 6, 26, 33, 34, 37-40, 42
- 43, 50, 67, 71, 77, 83, 92, 103, 279
- 283-289
- intensity 6-9, 34
- , pulsed 6, 64, 67, 68, 70, 71, 75, 77
- 83, 92
- scattering 5, 37, 71, 75
branches, choice of 184, 185
- identified 184
branching 65, 75, 86, 181, 185, 187, 190
- ratio 65, 70, 180-185, 198, 293, 294

cancellation effects 136, 137, 154, 157
- 158, 162, 164-166, 168
cascades 3, 63, 64, 66, 76-83, 86, 87
- 93-104, 112, 161, 162, 164, 174, 197
- 228, 238, 290, 294
- growing in 95, 96, 99
charge density 112, 116, 117, 139, 257
- expansion method 51, 53, 79, 111, 114
- 148, 151-153, 155, 165

charge distribution 26, 27, 34, 37, 39
- 114, 209, 266-268, 291
- supermultiplets 112, 119, 120, 133
- 135-141
- wave function 112, 114, 117-120, 123
- 126, 132, 135, 137-139, 149
coherence 69, 237-263
collisional de-excitation 38, 47, 63, 77
- 79, 195, 203, 205, 221
- excitation 38, 195, 197, 199, 202, 203
- 205, 239, 243, 247, 248, 261, 262, 266
- 284
- geometry 237, 243, 244, 257, 258
- orientation 237-263
configuration interaction or mixing 43
- 44, 46, 49, 51, 52, 54-56, 111, 113
- 116, 118, 119, 123, 130, 135, 137
- 139-141, 153-160, 162-166, 174, 231
- 284, 285
correlation effects 51, 116, 120, 123
- 136-138
- , all-external 114, 116, 120, 129
- 136-140
- , internal 113, 115, 135, 136, 140
- 141
- , non-closed-shell 112, 114, 140
Coulomb approximation 111, 155, 164, 165
- energy 216
- interaction 42, 45, 48-50, 53, 128
- 150, 151, 215, 232, 233, 268, 277, 284
- 290, 291
crossover 128, 130, 132, 166, 219
curve of growth 185-189

decay curves 74, 223
- , fits to 86-100
degeneracy 42, 47, 48, 68-70, 128, 149
- 153, 209, 215, 220, 251, 269
density effect 34
- matrix 69, 242, 245, 257
detection efficiency 8, 84, 86, 96, 97
- 183, 221-223
detection geometry 22-28, 34-37, 45, 48
- 74, 213, 237, 243, 244
detectors 19-22, 34, 74, 270
differential metastability 279, 280, 283
- 289, 291
dipole length 124, 125, 130, 135, 164
- moment 149, 220
- polarizability 44-47
- velocity 124, 125, 130, 135, 164
Doppler broadening 1, 22-24, 34, 36, 37
- - 63, 74, 75, 216
- shift 23, 26, 36, 37, 75, 239, 286
- tuning 103

electric dipole decay 84, 204, 212,
- - - 217-220, 225, 227, 229, 231, 232
- quadrupole decay 117
electron analyzer 270-279
- background 286-289
- capture 47
- correlation 111, 120, 268, 291-293
- density 195-198, 200-203, 205, 207
- excitation rate 197
- secondary 26, 28
- self energy 209, 210
- spectra 266, 280-291
- spectroscopy 265-295
- temperature 195-203, 206
- yield 287
element abundances 2, 33, 65, 133, 147
- - 179, 180, 185-189, 195, 196, 201-206
elements or ions studied
- Aℓ 6, 38, 42, 162, 187, 293
- Ar 38, 42, 43, 46, 47, 204, 216
- - 221-223, 225, 226, 228, 229, 258
- - 259, 261, 266, 293
- B 6, 9, 38, 40, 41, 44, 49, 53, 54
- - 112, 133, 134, 156
- Ba 103, 187
- Be 6, 38, 40, 41, 47-49, 53-56, 74, 96
- - 112, 117, 133, 134, 156, 159, 162, 249
- Bi 38

elements or ions studied (continued)
- C 5, 6, 9, 10, 34, 38, 39, 41, 45, 54
- - 112, 113, 117, 120, 125, 126, 128
- - 130, 133, 134, 156, 166, 187, 203
- - 286
- Ca 38, 56, 137, 187
- Ce 190
- Cℓ 9, 34, 38, 42, 43, 46, 56, 137, 228
- - 229, 282, 283, 293, 294
- Co 187
- Cr 35, 187
- Cu 187
- F 9, 10, 34, 38, 41, 120, 121, 131
- - 133, 134, 228, 231, 280-282
- - 293-295
- Fe 6-8, 10, 17, 38, 47, 180-182
- - 187, 189, 203, 206, 228, 229
- Ga 187
- H 5, 9, 10, 20, 47, 96, 210-212, 214
- - 215, 245, 246, 250-257, 287
- He 38, 40, 42, 50-53, 101, 137, 138
- - 220, 223, 225, 227-229, 231-233
- - 246-251, 254, 258-261, 282, 290
- Hg 187
- In 187
- K 38, 43, 187, 283
- Kr 9, 38, 209, 216
- Li 34, 38, 40, 42, 53, 54, 56, 75, 101
- - 102, 156, 165, 187, 211-213, 215
- - 233, 246, 248, 279, 283, 284, 286
- - 291
- Lu 190
- Mg 38, 42, 56, 164, 165, 187, 199, 201
- - 283
- Mn 187
- N 6, 9, 21, 24, 34, 36-39, 42, 44, 45
- - 54, 112, 117, 120, 125, 126, 128-131
- - 133-135, 157, 187, 215, 231
- Na 38, 40, 45, 55, 56, 134, 187, 283
- - 285, 286
- Nb 187

elements or ions studied (continued)
- Ne 6, 35, 38, 39, 44, 49, 77, 134, 156
- - 201, 260, 265, 266, 286-289
- Ni 38, 47, 184, 187, 188
- O 6, 10, 34, 38, 39, 41, 44, 45, 49, 53
- - 55, 112, 117, 120, 125, 126, 130-135
- - 187, 199-202, 204, 212, 231, 261
- - 266, 269, 277, 278, 282, 288, 289
- - 293, 294
- P 38, 42, 43, 117, 137
- Pb 38
- Pr 190
- Rb 187
- S 6, 34, 38, 42, 43, 45, 46, 50, 137
- - 187, 228, 229, 293, 294
- Sc 38, 122, 189
- Si 19, 34, 38, 42, 131, 134, 162-164
- - 187, 228, 293
- Sr 187
- Ti 34, 42, 187, 228, 229
- Tm 190
- U 6
- V 187, 228-230
- Xe 38, 216
- Y 187
- Zn 187
end-on geometry 35, 37, 45
equivalent width 181, 182, 185-189
excitation, coherent 69, 238, 239, 248-262
- functions 26, 37, 56
- mechanisms 33, 161, 195, 256
- rates 197-199, 201-205
external fields 14, 20, 28, 37, 42, 48,
- - 67, 69, 211-214, 237, 238, 246, 250
- - 253-259, 261-263

Faraday cup 13, 18, 25-27, 73, 75, 221
filters 11, 17, 27, 182-184
fine structure 42, 46, 64, 69, 72, 216
- - 239, 247, 249, 251, 254
foil characteristics 9, 14-17, 34, 71

foil, charge-equilibrating 25, 26, 221
- - 225
- mounts 12-16, 73, 74
- , scattering from 37, 276, 277
- thickness 9, 15, 16, 22, 25, 35, 75
- - 221, 225, 270, 276, 277, 279
- , tilted 244, 260-262
- , time-dependent changes in 27, 64
- - 73-75, 239, 260, 277
f-values (see oscillator strengths)

g-factor 258
Gamma-parameter 186, 187

Hanle effect 70, 77, 102, 155, 158, 164
- - 258, 259
hyperfine interaction or structure 42
- - 44, 64, 69, 72, 117, 123, 216, 230
- - 231, 239, 247-249, 251

image intensifier 22, 34
intensity change with energy 37, 39, 56
- - 186
- - field 48, 261
- enhancement 24, 29
- ratios 37, 39, 183, 195, 198-204
intercombination line 203, 204, 206, 224
- - 231, 282
intravalency 115, 116, 118, 119, 121
- 127, 130, 134
ion density 47, 196, 198-201, 204
ionization energy, limit, potential 41
- - 48-50, 116, 127, 128, 159, 169, 186
- - 187, 232, 269, 284, 285
- equilibrium 195, 196, 205
ion sources 5-8
isoelectronic sequences 37, 39, 41-44, 46
- - 46, 47, 49, 55, 65, 71, 78, 79, 113
- - 114, 120-122, 127-129, 133, 134, 136
- - 137, 141, 147-175, 198-204, 209-233
- - 269, 280-283, 291

kinematic broadening 269-271
- 274-279, 289
- corrections 269
- shifts 269, 288

Lamb shift 41, 42, 64, 209-217, 239
- - 254-256
- , discrepancy in 210
- , limit to measurement of 214
- in one-electron ions 209-215
- - two-electron systems 41, 42, 215, 216
- , scaling with Z 212, 213, 216, 217
- , uncertainty in 214-216
laser excitation 102-104
level calculations 41, 50, 51, 53, 54
- - 281, 282, 284-286, 290
level(s) (also see states, term) defined
- 68
- , dependence on n 41
- , displaced 38, 48-50
- , doubly (or multiply) excited 1, 38
- - 40-42, 48, 50-56, 63, 72, 103, 138
- - 139, 232, 233, 269, 279
- , hydrogenic 1, 28, 38, 43-48, 101, 211
- - 250-257
- , populations of 33, 38, 47-49, 56, 63
- - 68, 69, 80, 83-86, 96, 97, 100, 103
- - 181, 197, 199, 202, 203, 232, 238, 260
lifetime(s) (also see transition probabilities)
- and oscillator strength 65, 149, 150
- - 165, 179
- as $f(\ell)$ 47, 48
- background corrections 226
- calculations 22, 51, 67, 214, 225, 291
- - 292
- for autoionization 52, 265, 268, 290-295
- - state identification 38, 48, 49, 52
- - 279
- - two-electron states 52, 216, 232
- , disagreement in 225, 228, 293, 294
- in He-like systems as f(Z) 216, 224, 225

lifetime(s) (continued
- in hydrogenic systems 17, 47, 48,
- - 220, 224, 225, 238, 256
- - Li-like systems 282, 291-295
lifetime measurements 6, 17, 38-40, 43
- 49, 63-104, 179, 180, 221, 223, 225
- 226, 228-233, 282, 290-295
- , ambiguities in 93-96
- , distortions of 85, 86, 97
- , effect of field on 48
- , errors in 64, 73, 75, 78, 79, 111
- - 294
- , review of 65
- , need for 65, 174, 179, 189, 190, 206
- - 207
- , range of 71, 72, 76, 290
light sources 35, 38-43, 47, 49, 50, 52
- - 53, 55, 63, 65, 77, 181-184
line blending 64, 71, 75-77, 97, 101
- - 185, 294, 295
- broadening (also see Doppler and kinematic broadening) 47, 53
- , defined 68
- identifications 37, 42, 46-49, 55, 56
- - 195, 232, 237
- intensity (see intensity)
- strength 70, 124, 149, 152, 169
- - 170-172, 243
- width 22-24, 35-37, 52, 203, 212, 214
- - 232, 272-277, 289

M1 radiation 217-219, 227, 228, 231
M2 radiation 229, 230, 292-294
magic angle 86, 253
magnetic interaction 268
- quantum number 28, 242, 243, 252, 256
mass analysis 9-11, 34
- energy product 10, 11
- resolving power 10
- spectrum 10
mean lives (see lifetimes)
metastable levels or states 50, 53, 77
- 103, 166, 197, 198, 202, 203, 206, 207
- 209, 211, 213, 214, 216, 217, 221, 223
- 225, 232, 255, 280-283, 289-291
molecular ions 5-10, 34
monitor or normalizer 12, 13, 17, 21
- 25-28, 73, 182, 183, 212, 214, 290
multiplet 37, 68, 75, 149, 194, 195, 249

neutral currents 220

operator, electric dipole 212, 214, 218
- - 242
- , projection 124
- , raising or lowering 123
- , line strength 124
- , magnetic moment 217
- , parity 220
- , radial 217
orbital, penetrating 128
- relativistic effects 173
- , semi-internal 115, 116, 122, 123, 125
- - 129, 132, 138
- , Slater-type 125, 129
orientation 237-263
- vector 242, 244
oscillator strength or f-value (also see transition probability, lifetime) 49
- - 65, 66, 68-70, 186, 198, 204, 206
- , anomalous 154, 159, 174
- , error bounds for 79, 160, 161
- calculations 111-142
- , differential 168
- errors in 112, 129, 156, 160-162, 187
- , generalized 117, 136, 142
- , maxima in 154-164, 174
- , minima in 154, 164-166, 174
- , multiplet 114, 117, 127, 135, 153, 155
- - 169
- needed 189, 206
- , recommended 133, 134

oscillator strength or f-value(cont'd)
- regularities 147-175
- , relativistic effects in 113, 149
- - 156, 169-175
- sum rule 169
- , Z-expansion for 111, 113, 114
- - 150-153, 155

parity violation 220
perturbation theory 111, 148, 150-153
- - 158, 172, 218
plasma gradients 200, 201
- impurities 33, 147
- instabilities 147
- radiation losses 147
- , solar 195, 196, 206
- spectra 170
- temperature 196, 200-202
polarizability 44-47, 242, 246, 247
- 249-253, 258, 259, 262, 263
polarization 9, 37, 45, 68, 84, 86, 101
- 136, 237, 241, 244, 247
- , circular 243, 244, 260, 261
- , elliptical 243
- energy 44
- formula 44, 46
- , instrumental 19, 86, 101
- , linear 100, 102, 243, 244, 246, 253
- - 261
- scrambler 86
polarizer 84, 86, 253, 260, 261
population density 196, 197, 203, 204
- - 242
pre-Rydberg states 112, 118, 119
- - 127-130, 132-135, 138-141

quadrupole moment 117, 123, 137
quantum beats 20, 28, 42, 67, 69, 72, 74
- - 75, 248-257
- defect 44, 46, 54
quenching 211-214, 220, 221, 230, 231

refocusing 22-24, 29, 36, 37, 74
renormalization factor 129, 131
repeller ring 25, 26
replenishment ratio 83, 96
Rydberg series or states 43, 111, 115
- - 118, 119

satellites 49, 51, 53, 55, 65, 227, 282
- 289
scattering 5, 15, 37, 50, 53, 71, 75, 270
- 276, 277, 279
selection rules 48, 49, 53, 76, 101, 139
- - 184, 217, 218, 224, 229, 231, 232
- - 268, 290
self-consistent field 111, 141, 155, 156
- - 158-160, 162-165, 170, 171
side-on geometry 34
solar abundances 186-190, 195, 196, 198
- - 201-206
- atmosphere 186-188, 194-196, 201, 204
- - 205, 207
- spectra 147, 185, 186, 188, 193-207
- - 227
- temperature 186, 189, 193-196, 202, 205
spectra, classification of 37-56, 195
- , electron 266, 280-289
- for astrophysics 147, 189-190, 193-207
- , impurities in 9-11, 34, 147
- , molecular 112, 184
- needed 189
- of various elements (optical) 19, 21
- - 34-56, 139, 184, 185, 200-204, 250
- - 251, 253-257
- of various elements (electron) 266
- - 280-289
- , review of 38
- , temperature dependence of 201-206
spectrograph 20, 21, 53
spectrometer (optical) calibration 20, 21
- - 37, 183, 184, 190, 198, 200, 203
- characteristics 8, 18, 34-36, 182-184

spectrometer, Czerny-Turner 19, 233
- , Fabry-Perot 37
- , grazing-incidence 19, 27, 35, 51
- , as monitor 183
- , normal incidence 8, 18
- on satellites 189, 206
- , Paschen-Runge 183
- , Seya-Namioka 19
- slits 8, 18-20, 23, 24, 27, 34, 36, 74
- - 75
- X-ray 53, 231
spectrometer (electron) 269-279
spin-orbit-interaction 53, 55, 169, 173
- - 174, 227, 231, 243, 268, 284, 290
- - 291
spin-orbitals 114-116
- density 117, 123
- symmetry 120
spin-spin 53, 55
- interaction 268, 284, 290, 291
Stark effect 47, 48, 64, 212, 221, 250
- - 251, 253-256, 259, 262, 263
states (also see levels and term)
- , definition of 68
- , identification of 39, 43, 45, 53
- - 162, 279, 281-286, 291
- in various ions 209-233, 282, 285
- - 290-295
- , inner-shell 33, 38, 53, 54, 266, 269
- - 282-286
- , intravalency 116, 118, 119, 121
- - 127-130, 132, 134, 135, 138-141
- , lowest of their symmetry 120
- - 125-127, 131, 133
- , not lowest of their symmetry 120-122
- - 126, 129, 130, 133, 134, 136
- , of high n,ℓ 38, 47
- , range of 279
statistical weight 98, 181, 186
sum rules 84, 101, 135, 169
superposition of configurations 111
- - 157-160, 162-164, 166, 285
target chambers, features of 12-14
term (also see levels or states)
- , definition of 68
- identifications 64
- splittings 116
- , values of 41, 43, 44
time reversal 220
transition(s), classified 39, 40, 47, 53
- - 54, 56, 184
- density 114
- energy 169, 216
- energies, calculated 215
- , forbidden 6, 47, 141
- , hydrogenic 43-48, 212-215, 217-224
- , identification of 40, 41, 43, 44, 46
- - 49, 50, 52, 228, 232
- , induced by rf 9
- , in-shell 111
- integral 149, 154, 157, 164, 165, 170
- - 171, 175
- , intershell 111, 134, 135
- , resonance 40, 49
- , super 112, 119, 120
- , two-electron 49, 53
- , two-photon 212, 218-227
transition probabilities or rates (also see oscillator strengths, lifetimes and A-values) 8, 9, 50, 65, 67-69, 71
- - 84, 150, 179-182, 184-186, 188-190
- - 195, 198, 203, 204, 206, 212, 265, 292
- and line strength 70
- as f(Z) 228
- , calculation of 52, 53, 65, 71, 123
- - 124, 228
- , cascade 80
- , Corliss-Bozman 179, 180
- , compilation of 70, 71, 78, 179
- , corrections to 179, 180, 185, 219
- , estimated 185
- , number measured 70
- , product of 81
- , stimulated 103

transition probabilities (continued)
- , two-photon 218, 219
- , uncertainties in 188

vacuum polarization 209, 210
variational calculations 51, 115, 123
- - 127, 281, 282, 293
- collapse 113, 128-131, 136, 137
vignetting 24, 74

wave functions
- , Hartree-Fock 115
- , hydrogenic 44
wavelengths, calibration of 20, 21, 183
- - 184
- tabulated 184, 185, 195
- , uncertainties in 39, 41, 45, 47, 53

Zeeman effect 28, 37, 64, 84, 257

Titles of Related Interest

DYE LASERS
Topics in Applied Physics, Vol.1
ed. by *F.P. Schäfer*

F.P. Schäfer: Principles of Dye Laser Operation

B.B. Snavely: Continuous-Wave Dye Lasers

C.V. Shank, E.P. Ippen: Mode-Locking of Dye Lasers

K.H. Drexhage: Structure and Properties of Laser Dyes

T.W. Hänsch: Applications of Dye Lasers

HIGH-RESOLUTION LASER SPECTROSCOPY
Topics in Applied Physics, Vol.13
ed. by *K. Shimoda*

K. Shimoda: Introduction

K. Shimoda: Line Broadening and Narrowing Effects

P. Jacquinot: Atomic Beam Spectroscopy

V.S. Letokhov: Saturation Spectroscopy

J.L. Hall, J.A. Magyar: High Resolution Saturated Absorption Studies of Methane and Some Methyl-Halides

V.P. Chebotayev: Three-Level Laser Spectroscopy

S. Haroche: Quantum Beats and Time-Resolved Fluorescence Spectroscopy

N. Bloembergen, M.D. Levenson: Doppler-Free Two-Photon Absorption Spectroscopy

LASER SPECTROSCOPY of Atoms and Molecules
Topics in Applied Physics, Vol.2
ed. by *H. Walther*

H. Walther: Atomic and Molecular Spectroscopy with Lasers

E.D. Hinkley, K.W. Nill, F.A. Blum: Infrared Spectroscopy of Molecules by Means of Lasers

K. Shimoda: Double-Resonance Spectroscopy of Molecules by Means of Lasers

J.M. Cherlow, S.P.S. Porto: Laser Raman Spectroscopy of Gases

B. Decomps, M. Dumont: Linear and Nonlinear Phenomena in Laser Optical Pumping

K.M. Evenson, F.R. Petersen: Laser Frequency Measurements, the Speed of Light, and the Meter

LASER MONITORING OF THE ATMOSPHERE
Topics in Applied Physics, Vol.14
ed. by *E.D. Hinkley*

E.D. Hinkley: Introduction

S.H. Melfi: Remote Sensing for Air Quality Management

V.E. Zuev: Laser-Light Transmission through the Atmosphere

R.T.H. Collis, P.B. Russell: Lidar Measurement of Particles and Gases by Elastic Backscattering and Differential Absorption

H. Inaba: Detection of Atoms and Molecules by Raman Scattering and Resonance Fluorescence

E.D. Hinkley, R.T. Ku, P.L. Kelley: Techniques for Detection of Molecular Pollutants by Absorption of Laser Radiation

R.T. Menzies: Laser Heterodyne Detection Techniques

LASER SPECTROSCOPY, Proceedings of the 2nd International Conference, Mègève, France, June 23-27, 1975
Lecture Notes on Physics, Vol.43
Pp.X+468, 230 figs, 30 tables (5 pages in French) (1975)
ed. by *S. Haroche, J.C. Pebay-Peyroula, T.W. Hänsch, S.E. Harris*